# JMP Start Statistics

# A Guide to Statistics and Data Analysis Using JMP® and JMP IN® Software

## *Third Edition*

John Sall

Lee Creighton

Ann Lehman

SAS Institute Inc.

**THOMSON**

**BROOKS/COLE**

Australia • Canada • Mexico • Singapore • Spain • United Kingdom • United States

**THOMSON**

™

**BROOKS/COLE**

Publisher: Curt Hinrichs
Marketing Team: Joseph Rogove and Jessica Perry
Editorial Assistant: Katherine Brayton
Production Project Manager: Belinda Krohmer
Manuscript Editor: SAS Institute Staff
Permissions Editor: Sommy Ko
Cover Photo: Kazuo Kawai/Photonica
Cover Designer: Bill Reuter
Manufacturing Buyer: Judy Inouye
Compositor: Lee Creighton
Printer: Webcom Limited

Printed in Canada
4 5 6 7 07 06 05

For more information about our products,
contact us at:
**Thomson Learning Academic Resource
Center
1-800-423-0563**
For permission to use material from this
text, contact us by:
**Phone:** 1-800-730-2214
**Fax:** 1-800-730-2215
**Web:** http://www.thomsonrights.com

Brooks/Cole—Thomson Learning
10 Davis Drive
Belmont, CA 94002
Asia
Thomson Learning
5 Shenton Way #01-01
UIC Building
Singapore 068808

Australia/New Zealand
Thomson Learning
102 Dodds Street
Southbank, Victoria 3006
Australia

Canada
Nelson
1120 Birchmount Road
Toronto, Ontario M1K 5G4
Canada

Europe/Middle East/Africa
Thomson Learning
High Holborn House
50/51 Bedford Row
London WC1R 4LR
United Kingdom

Library of Congress Control Number: 2003115076
JMP Start Statistics 3e (book-only): 0-534-99749-X
JMP Start Statistics 3e + JMP-IN 5.1: 0-534-99747-3

# Table of Contents

# 5   What Are Statistics?

# 6   Simulations in JMP

# 9  Comparing Many Means: One-Way Analysis of Variance

# 12 Categorical Models

# 17 Exploratory Modeling

# Preface

JMP® is statistical discovery software. JMP helps you explore data, fit models, discover patterns, and discover points that don't fit patterns. This book is a guide to statistics using JMP.

This book is distributed in two ways:

- It is the documentation for JMP IN software, the student version of JMP. This book covers most of the features of JMP and JMP IN in the context of learning statistics. JMP IN is for student use only.

- It is supplemental documentation for the full, professional version of JMP. The books that come with the full version of JMP are reference manuals; this book focuses on learning statistics.

## The Software

As statistical discovery software, JMP's emphasis is to interactively work with data to make discoveries.

- With graphics, you are more likely to make discoveries. You are also more likely to understand the results.

- With interactivity, you are encouraged to dig deeper and to try out more things that might improve your chances of discovering something important. With interactivity, one analysis leads to a refinement, and one discovery leads to another discovery,.

- With a progressive structure, you build context that maintains a live analysis. You don't have to redo analyses and plots to make changes in them, so details come to attention at the right time.

Software's job is to create a virtual workplace. The software has facilities and platforms where the tools are located and the work is performed. JMP provides the workplace that we think is best for the job of analyzing data. With the right software workplace, researchers will celebrate computers and statistics rather than avoid them.

JMP aims to present a graph with every statistic. You can and should always see the analysis in both ways, statistical text and graphics, without having to ask for it that way. The text and graphs stay together.

JMP is controlled largely through point-and-click mouse manipulation. If you click on a point in a plot, JMP identifies and highlights the point in the plot, and highlights the point in the data table. In fact, JMP highlights the point everywhere else it is represented.

JMP has a progressive organization. You begin with a simple surface at the top, but as you analyze, more and more depth is revealed. The analysis is alive, and as you dig deeper into the data, more and more options are offered according to the context of the analysis.

In JMP, completeness is not measured by the "feature count", but by the range of applications, and the orthogonality of the tools. In JMP, you get more feeling of being in control despite less awareness of the control surface. In JMP, you get a feeling that statistics is an orderly discipline that makes sense, rather than an unorganized collection of methods.

A statistical software package is often the point of entry into the practice of statistics. JMP strives to offer fulfillment rather than frustration, empowerment rather than intimidation.

If you give me a truck, I will find someone to drive it for me. But if you give me a sports car, I will learn to drive it myself. Believe that statistics can be interesting and reachable so that people will want to drive that vehicle.

# *JMP Start Statistics*, Third Edition

Many changes have been made since the second edition of *JMP Start Statistics*. Based on comments and suggestions by teachers, students, and other users, we have expanded and enhanced the book, hopefully to make it more informative and useful.

*JMP Start Statistics* has been updated and revised to feature version 5.1 of JMP IN software. Major enhancements have been made to the product, including new platforms for data mining (recursive partitioning and neural nets), analyses (such as clustering and discriminant analysis) as well as more report options not available in previous versions.

The Time Series chapter has been completely rewritten to focus on ARIMA models.

Building on the success of the classroom exercises from the second edition, we have added more of them. In addition, chapters have been rearranged to streamline their pedagogy, and new sections and chapters have been added where needed.

# SAS Institute

JMP is from SAS Institute, a large private research institution specializing in data analysis software. SAS Institute's principal commercial product is the SAS System, a large software system that performs much of the world's large-scale statistical data processing. JMP is positioned as the small personal analysis tool, involving a much smaller investment than the SAS System.

# This Book

### Software Manual and Statistics Text

This book is a mix of software manual and statistics text. It is designed to be a complete and orderly introduction to analyzing data. It is a teaching text, but is especially useful when used in conjunction with a standard statistical textbook.

### Not Just the Basics

A few of the techniques in this book are not found in most introductory statistics courses, but are accessible in basic form using JMP IN. These techniques include logistic regression, correspondence analysis, principal components with biplots, leverage plots, and density estimation. All these techniques are used in the service of understanding other more basic methods. Where appropriate, we have labeled supplemental material as "Special Topics" so that they are recognized as optional material that is not on the main track.

JMP IN also includes several advanced methods that are not covered in this book, such as nonlinear regression and multivariate analysis of variance. If you are planning to use these features, we recommend that you refer to the help system or the documentation for the professional version of JMP (included on the JMP IN CD).

### Examples Both Real and Simulated

Most examples are real-world applications. A few simulations are included too, so that difference between a true value and its estimate can be discussed, along with the variability in the estimates. Some examples are unusual, calculated to surprise you in the service of emphasizing some important concept. The data for the examples are on the CD, with step-by-step instructions in the text. Example data are also available on the internet at www.jmp.com. JMP IN can also import data from files distributed on diskettes with other textbooks.

## Acknowledgments

The software developers for JMP are John Sall, Katherine Ng, Michael Hecht, Richard Potter, Brian Corcoran, Annie Zangi, Bradley Jones, Craige Hales, Paul Nelson, Chris Gotwalt, and Xan Gregg. Ann Lehman, Lee Creighton, Meredith Blackwelder, and Melanie Drake produce the documentation and help systems.

Thanks also to Kristin Nauta, her sales and marketing staff, and the Statistics R&D group at SAS Institute.

Statistical technical support is provided by Craig DeVault, Duane Hayes, and Kathleen Kiernan. Nicole Jones, Jianfeng Ding, Jim Borek, Kyoko Tidball, and Hui Di provide ongoing quality assurance. Additional testing and technical support is done by Noriki Inoue and Kyoko Takenaka from SAS Japan. Bob Hickey is the release engineer. Previous developers include Eric Wasserman, Charles Soper, and Dave Tilley.

Thank you to the testers for both JMP and JMP IN: Michael Benson, Avignor Cahaner, Howard Yetter, David Ikle, Robert Stine, Andy Mauromoustkos, Al Best, and Chris Olsen. Further acknowledgements for JMP are in the JMP documentation on the installation CD.

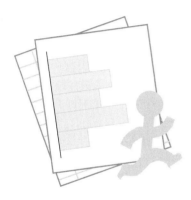

# Preliminaries

## JMP IN Package Contents and Installation

If you purchased JMP IN, the package contains a CD and this book. Installation instructions are contained on the installation CD. The program CD is the same whether you are installing on a PC or on a Macintosh.

Insert the CD and read the readme.txt file, which contains late-breaking news (features that emerged too late to be in this book) or warnings about known problems. It is always a good idea to read this file.

There is also the JMP IN installer on the CD. Begin installation by double-clicking the installer program. Follow the instructions to complete a standard install.

After installation, the Documents folder contains PDF files of the documentation that comes with the professional version of JMP:

- *JMP Introductory Guide*
- *JMP User's Guide*
- *JMP Design of Experiments*
- *JMP Statistics and Graphics Guide*
- *JMP Scripting Language Guide.*

These documents cover all the features in JMP and JMP IN. Printed versions of these books are available from SAS Institute Inc. You can use Adobe Reader to browse these documents and search for topics.

# What You Need to Know

## ...about your computer

Before you begin using JMP IN, you should be familiar with standard operations and terminology such as click, double-click, ⌘-click and option-click on the Macintosh (Control-click and Alt-click under Windows or Linux), shift-click, drag, select, copy, and paste. You should also know how to use menu bars and scroll bars, move and resize windows, and manipulate files. If you are using your computer for the first time, consult the reference guides that came with it for more information.

## ...about statistics

This book is designed to help you learn about statistics. Even though JMP IN has many advanced features, you need no background of formal statistical training to use it. All analysis platforms include graphical displays with options that help you review and interpret the results. Each platform also includes access to help windows that offer general help and some statistical details.

# Learning About JMP

## ...on your own with JMP Help

If you are familiar with Macintosh, Microsoft Windows, or Linux software, you may want to proceed on your own. After you install JMP, you can open any of the JMP sample data files and experiment with analysis tools. Help is available for most menus, options, and reports.

There are several ways to access JMP Help:

- If you are using Microsoft Windows, help in typical Windows format is available under the **Help** menu on the main menu bar.

- On the Macintosh, select the **Contents** command from **Help** menu (OS 9) or **JMP Help** from the help menu (OS X).

- On Linux, select an item from the **Help** menu.

- You can click the **Help** button from launch dialogs whenever you launch an analysis or graph platform.

- After you generate a report, select the help tool ( **?** ) from the **Tools** menu or Toolbar and click the report surface. Context-sensitive help tells about the items in that report window.

## ...hands-on examples

This book, *JMP Start Statistics*, documents JMP by describing its features, and is reinforced with hands-on examples. By following along with these step-by-step examples, you can quickly become familiar with JMP menus, options, and report windows.

Mouse-along steps for example analyses begin with the mouse symbol in the margin, like this paragraph.

## ...reading about JMP

The professional version of JMP version 5 is accompanied by five books—the *JMP Introductory Guide*, the *JMP User's Guide*, *JMP Design of Experiments*, the *Statistics and Graphics Guide*, and the *JMP Scripting Language Guide*. These references cover all the commands and options in JMP and have extensive examples of the **Analyze** and **Graph** menus. These books may be available in printed form from your department, computer lab, or library. They are included as PDF files on the JMP IN CD.

# Chapter Organization

This book contains chapters of documentation supported by guided actions you can take to become familiar with the JMP IN product. It is divided into two parts:

The first five chapters get you started quickly with information about JMP tables, how to use the JMP formula editor, and give an overview of how to obtain results from the **Analyze** and **Graph** menus.

- Chapter 1, "Preliminaries," is this introductory material

- Chapter 2, "JMP Right In," tells you how to start and stop JMP, how to open data tables, and takes you on a short guided tour. You are introduced to the general personality of JMP. You will see how data are handled by JMP. There is an overview of all analysis and graph commands, information about how to navigate a platform of results, and a description of the tools and options available for all analyses. The Help system is covered in detail.

- Chapter 3, "JMP Data Tables" focuses on using the JMP data table. It shows how to create tables, subset, sort, and manipulate them with built-in menu commands, and how to get data and results out of JMP and into a report.

- Chapter 4, "Formula Editor Adventures," covers the formula editor. There is a description of the formula editor components and overview of the extensive functions available for calculating column values.

- Chapter 5, "What Are Statistics?" gives you some things to ponder about the nature and use of statistics. It also attempts to dispel statistical fears and phobias that are prevalent among students and professionals alike.

Chapters 6–21 cover the array of analysis techniques offered by JMP IN. Chapters begin with simple-to-use techniques and work gradually to more complex methods. Emphasis is on learning to think about these techniques and on how to visualize data analysis at work. JMP IN offers a graph for almost every statistic and supporting tables for every graph. Using highly interactive methods, you can learn more quickly and discover what your data have to say.

- Chapter 6, "Simulations in JMP," introduces you to some probability topics by using JMP's scripting language. You learn how to open and execute these scripts.

- Chapter 7, "Univariate Distributions: One Variable, One Sample," covers distributions of continuous and categorical variables and statistics to test univariate distributions.

- Chapter 8, "The Difference Between Two Means," covers $t$-tests of independent groups and tells how to handle paired data. The nonparametric approach to testing related pairs is shown.

- Chapter 9, "Comparing Many Means: One-Way Analysis of Variance," covers one-way analysis of variance, with standard statistics and a variety of graphical techniques.

- Chapter 10, "Fitting Curves Through Points: Regression," shows how to fit a regression model for one factor.

- Chapter 11, "Categorical Distributions," discusses how to think about the variability in single batches of categorical data. It covers estimating and testing probabilities in categorical distributions, shows Monte Carlo methods, and introduces the Pearson and Likelihood ratio chi-square statistics.

- Chapter 12, "Categorical Models," covers fitting categorical responses to a model, starting with the usual tests of independence in a two-way table, and continuing with graphical techniques and logistic regression.

- Chapter 13, "Multiple Regression," describes the parts of a linear model with continuous factors, talks about fitting models with multiple numeric effects, and shows a variety of examples, including the use of stepwise regression to find active effects.

- Chapter 14, "Fitting Linear Models," is an advanced chapter that continues the discussion of Chapter 12, moving on to categorical effects and complex effects, such as interactions and nesting.

- Chapter 15, "Bivariate and Multivariate Relationships," looks at ways to examine two or more response variables using correlations, scatterplot matrices, three-dimensional plots, principal components, and other techniques. Outliers are discussed.

- Chapter 15, "Design of Experiments," looks at JMP IN's built-in commands to generate specified experimental designs. Also, examples of how to analyze common screening and response level designs are covered.

- Chapter 16, "Statistical Quality Control," is a survey of types of control charts available in JMP, and an explanation of when to use them.

- Chapter 17, "Exploratory Modeling," illustrates two common data mining techniques—Neural Nets and Recursive Partitioning.

- Chapter 18, "Discriminant and Cluster Analysis," discusses methods that group data into clumps.

- Chapter 19, "Statistical Quality Control," discusses common types of control charts for both continuous and attribute data.

- Chapter 20, "Time Series," discusses some beginning methods for looking at data with correlations over time.

- Chapter 21, "Machines of Fit" is an essay about statistical fitting that may prove enlightening to those who have a mind for mechanics.

# Typographical Conventions

The following conventions help you relate written material to information you see on your screen:

- Reference to menu names (**File** menu) or menu items (**Save** command), and buttons on dialogs (**OK**), appear in the **Helvetica bold** font.

- When you are asked to choose a command from a submenu, such as **File > Save As**, go to the **File** menu and choose the **Save As** command.

- Likewise, items on popup menus in reports are shown in the **Helvetica bold** font, but you are given a more detailed instruction about where to find the command or option. For example, you might be asked to select the **Show Points** option from the popup menu on the analysis title bar, or select the **Save Predicted** command from the Fitting popup menu on the scatterplot title bar. The popup menus will always be visible as a small red triangle on the platform or on its outline title bars, as circled in the picture below.

- References to variable names data table names and some items in reports show in Helvetica but can appear in illustrations in either a plain or boldface font. These items show on your screen as you have specified in your JMP Preferences.

- Words or phrases that are important, new, or have definitions specific to JMP are in *italics* the first time you see them.

- When there is an action statement, you can follow along with the example by following the instruction. These statements are preceded with a mouse symbol (⊙) in the margin. An example of an action statement is:

  Highlight the Month column by clicking the area above the column name, and then choose **Cols > Column Info**.

- Occasionally, side comments or special paragraphs are included and shaded in gray, or are in a side bar.

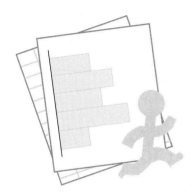

# JMP Right In

**2**

## Hello!

JMP software is so easy to use that after reading this chapter you'll find yourself confident enough to learn everything on your own. We will cover the essentials fast—before you escape this book. This chapter offers you the opportunity to make a small investment in time for a large return later on.

If you are already familiar with JMP IN and want to dive right into statistics, you can skip ahead to Chapters 6–21. You can always return later for more details about using JMP or for more details about statistics.

By the way, JMP is pronounced "jump." We use JMP and JMP IN interchangeably. JMP IN is the student version of JMP.

# First Session

This first section just gets your toes wet. In most of the chapters of this book, you can follow along in a hands-on fashion. Watch for the mouse symbol (🖰) and perform the action it describes. Try it now:

🖰   To start JMP IN, double-click the JMP IN application icon, or select its icon and choose **File** > **Open** from your system's main menu.

When the application is active, you see the JMP menu bar and the JMP Starter window:

**Figure 2.1**   The JMP Main Menu and the JMP Starter

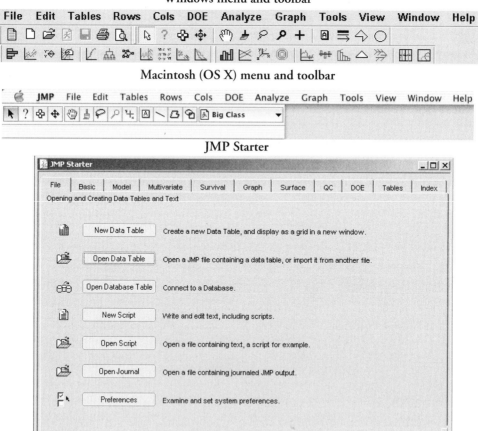

As with other applications, the **File** menu (**JMP** menu on Macintosh) has all the strategic commands, like opening data tables or saving them. To quit JMP, choose the **Exit** (Windows)

or **Quit** (Macintosh) command from this menu. (Note that the **Quit** command is located on the **JMP** menu in Macintosh OS X.)

Start by opening a JMP data table and doing a simple analysis.

## Open a JMP Data Table

When you first start JMP, you are presented with the JMP Starter window, a window that allows quick access to JMP's most frequently used features. Rather than start with a blank file or import data from text files, open a JMP data table from the collection of sample data tables that comes with JMP IN.

🖰   Choose the **Open** command in the **File** menu (choose **File > Open**).

🖰   When the Open File dialog appears, as shown in **Figure 2.2** or **Figure 2.3**, select Big Class.jmp from the list of sample data files. You may have to navigate through your directories to find the directory with the sample data.

🖰   Select Big Class and click **Open** (Windows and Macintosh) or **Finish** (Linux) on the dialog.

**Figure 2.2**   Open File Dialog (Windows)

**Figure 2.3**  Open File Dialog (Macintosh)

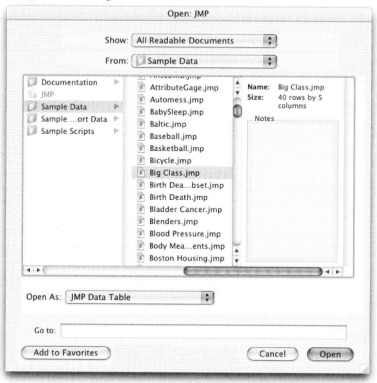

**Figure 2.4**   Open File Dialog (Linux)

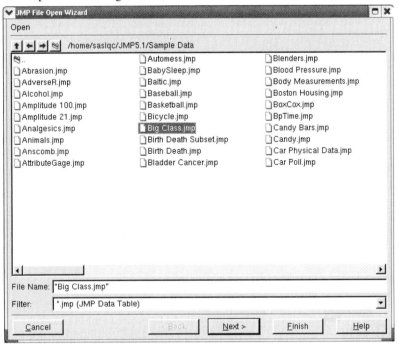

You should now see the presentation of the table with columns called name, age, sex, height, and weight (shown in **Figure 2.5**).

In Chapter 3, "Data Tables, Reports, and Scripts" on page 25, you learn the facilities of the data table, but for now let's try an analysis.

**Figure 2.5**  Partial Listing of the Big Class Data Table

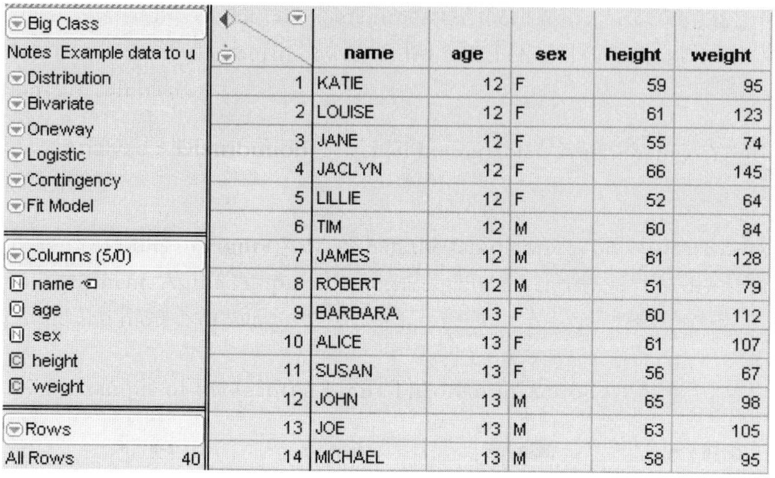

| | | name | age | sex | height | weight |
|---|---|---|---|---|---|---|
| ▽ Big Class | ◆ ▽ | | | | | |
| Notes Example data to u | | | | | | |
| ▽ Distribution | 1 | KATIE | 12 | F | 59 | 95 |
| ▽ Bivariate | 2 | LOUISE | 12 | F | 61 | 123 |
| ▽ Oneway | 3 | JANE | 12 | F | 55 | 74 |
| ▽ Logistic | 4 | JACLYN | 12 | F | 66 | 145 |
| ▽ Contingency | 5 | LILLIE | 12 | F | 52 | 64 |
| ▽ Fit Model | 6 | TIM | 12 | M | 60 | 84 |
| | 7 | JAMES | 12 | M | 61 | 128 |
| ▽ Columns (5/0) | 8 | ROBERT | 12 | M | 51 | 79 |
| ▯ name ◁ | 9 | BARBARA | 13 | F | 60 | 112 |
| ▢ age | 10 | ALICE | 13 | F | 61 | 107 |
| ▯ sex | 11 | SUSAN | 13 | F | 56 | 67 |
| ▢ height | 12 | JOHN | 13 | M | 65 | 98 |
| ▢ weight | 13 | JOE | 13 | M | 63 | 105 |
| ▽ Rows | 14 | MICHAEL | 13 | M | 58 | 95 |
| All Rows | 40 | | | | | |

## Launch an Analysis Platform

What is the distribution of the weight and age columns in the table?

🖱 Click on the **Analyze** menu and choose the **Distribution** command.

This is called *launching* the Distribution platform. The launch dialog (**Figure 2.6**) now appears, prompting you to choose the variables you want to analyze.

🖱 Click on weight to highlight it in the variable list on the left of the dialog.

🖱 Click **Y, Columns** to move it into the list of variables on the right of the dialog, which are the variables to be analyzed.

🖱 Similarly, select the age variable and add it to the analysis variable list.

The term *variable* is often used to designate a column in the data table. Picking variables to fill roles is sometimes called *role assignment*.

You should now see the completed launch dialog shown in **Figure 2.6**.

🖱 Click **OK**, which closes the launch dialog and performs the Distribution analysis.

**Figure 2.6**  Distribution launch dialog

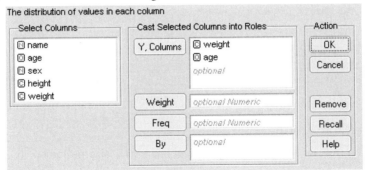

The resulting window shows the distribution of the two variables, weight and age as in
**Figure 2.7**.

**Figure 2.7**  Histograms from the Distribution platform

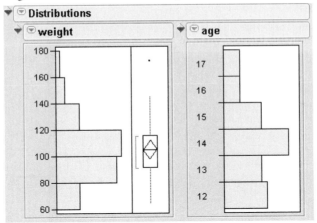

# Interact with the Surface of the Report

All JMP reports start with a basic analysis, which are then worked with interactively. This
allows you to dig into a more detailed analysis, or customize the presentation. It's a live object,
not a dead report.

## Row Highlighting

🖰   Click on one of the histogram bars, for example, the age bar for 12-year-olds.

The bar is highlighted, along with portions of the bars in the other histogram and certain
rows in the data table corresponding to that histogram bar. This is the dynamic linking of the

rows in the plots and data tables. Later you will see other ways of selecting and working with attributes of the rows of a table.

**Figure 2.8**  Highlighted bars and data table rows

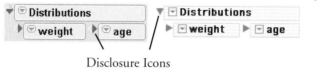

On the right of the weight histogram is a box plot with a single point near the top.

🖰  Move the mouse over that point to see the label, LAWRENCE, appear in a popup box.

🖰  Click on the point in the plot.

The point highlights and the corresponding row is highlighted in the data table.

## Disclosure Icons

Each report title is part of the analysis presentation outline. Click on the diamond on the side of each report title to alternately open and close the contents of that outline level.

**Figure 2.9**  Disclosure icons for Windows (left) and Macintosh OS X (Right)

Disclosure Icons

## Contextual Popup Menus

There is a small red triangle on the title bar at the top of the analysis window that accesses popup menu commands for the analysis. This popup menu has commands specific to the platform, but applies to all the analyses in the window. For example, you can change the orientation of the graphs in the Distribution platform by checking or unchecking **Display Options** > **Horizontal Layout** (**Figure 2.10**).

🖐 Click on one of the menus next to weight or age and select **Display Options** > **Horizontal Layout**.

**Figure 2.10** Display Options menu

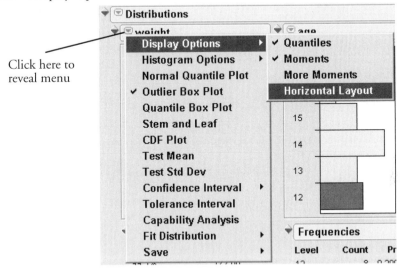

Click here to reveal menu

In this same popup menu, you find options for performing further analyses or saving parts of the analysis in several forms. Whenever you see a red triangle, there are more options available. The options are specific to the context of the outline level where they are located. These options are explained in later sections of this book.

If you want to resize the graph windows in an analysis, move your mouse over the side or corner of the graph. The cursor changes to a double arrow, which lets you to drag the borders of the graph to the position you want.

## Special Tools

When you need to do something special, pick a tool in the tools menu (or in a tool palette) and click or drag inside the analysis.

The grabber( 🖐 ) is for grabbing objects.

🖐 Select this tool, then click and drag in a histogram.

The brush ( 🖌 )is for highlighting all the data in an rectangular area.

🖐 Try getting this tool and dragging in the histogram. Option-drag or Alt-drag to change the size of the rectangle.

The lasso ( 🔘 )is for roping points in order to select them. We'll use this later in scatterplots.

The crosshairs ( + )are for sighting along lines in a graph.

The magnifier ( 🔍 )is for zooming in to certain areas in a plot. This tool is for scatterplots. Hold down the ⌘ or Alt key and click to restore the original scaling.

The question mark ( ? )is for getting help on the analysis platform surface.

   🖱   Get the question mark tool and click on different areas in the Distribution platform.

The selection tool ( ✛ )is for picking out an area to copy so that you can paste its picture into another application. Hold down the Option or Alt key to select in a different way. See the chapter "Data Tables, Reports, and Scripts" on page 25 for details.

In JMP, the surface of an analysis platform bristles with interactivity. Launching an analysis is just the starting point. You then explore, evaluate, follow clues, dig deeper, get more details, fuss with the presentation, and so on.

# Modeling Type

Notice in the previous example that there are different kinds of graphs and reports for **weight** and **age**. This is because the variables can be assigned different *modeling types*. The **weight** column has a *continuous* modeling type, so JMP treated the numbers as values from a continuous scale. The **age** column has an *ordinal* modeling type, so JMP treated its values as labels of discrete categories.

Here is a brief description of the three modeling types:

- *Continuous* are numeric values used in directly in an analysis.
- *Ordinal* values are category labels, but their order is meaningful.
- *Nominal* values are treated as unordered, categorical names of levels.

The ordinal and nominal modeling types are treated the same in most analyses, and are often referred to collectively as *categorical*.

You change the modeling type using the Columns panel at the left of the data grid. Notice the 🔲 beside the column heading for **age**. This icon is a popup menu.

   🖱   Click on the 🔲 to see the menu for choosing the modeling type for a column.

**Figure 2.11**   Modeling type popup menu on the Columns panel

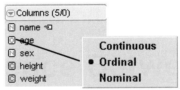

Why does JMP make a fuss over modeling type? For one thing, it's a convenience feature. You are telling JMP ahead of time how you want the column treated so that you don't have to say it again every time you do an analysis. It also helps reduce the number of commands you need to learn. Instead of two distribution platforms, one for continuous variables and a different one for categorical variables, a single command performs the anticipated analysis based on the modeling type you assigned.

You can change the modeling type whenever you want the variable treated differently. For example, if you wanted to find the mean of **age** instead of categorical frequency counts, simply change the modeling type from ordinal to continuous and repeat the analysis.

The following sections demonstrate how the modeling type affects the kind of analysis from several platforms.

## Analyze and Graph

The **Analyze** and **Graph** menus, shown here with their equivalent toolbars, launch interactive platforms to analyze data.

**Figure 2.12**   Analyze and Graph menus

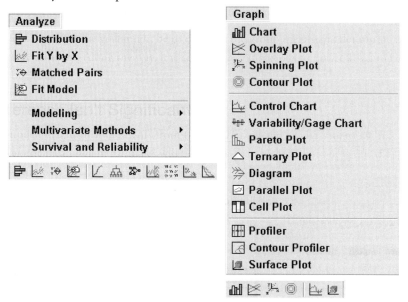

The **Analyze** menu is for statistics and data analysis. The **Graph** menu is for specialized plots. That distinction, however, doesn't prevent analysis platforms from being full of graphs, nor the graph platforms from computing statistics. Each platform provides a context for sets of related statistical methods and graphs. It won't take long to learn this short list of platforms. The next sections briefly describe the **Analyze** and **Graph** commands.

## The Analyze Menu

**Distribution** is for univariate statistics, which describe the distribution of values for each variable, one at a time, using histograms, box plots, and other statistics.

**Fit Y by X** is for bivariate analysis. A bivariate analysis describes the distribution of a Y variable as it depends on the value of the X variable. The continuous or categorical modeling type of the Y and X variables leads to the four analyses: scatterplot with regression curve fitting, one-way analysis of variance, contingency table analysis, and logistic regression.

**Matched Pairs** compares means between two response columns using a paired *t*-test. Often the two columns represent measurements on the same subject before and after some treatment.

**Fit Model** launches a general fitting platform for linear models such as multiple regression, analysis of variance models, and others.

### Modeling

**Nonlinear** fits models that are nonlinear in their parameters, using iterative methods.

**Partition** recursively partitions values, similar to CART™ and CHAID™.

**Neural Net** implements a standard type of neural network.

**Time Series** lets you explore, analyze and forecast univariate time series taken over equally spaced time periods. The analysis begins with a plot of the points in the time series with autocorrelations and partial autocorrelations, and can fit ARIMA, seasonal ARIMA, and smoothing models.

### Multivariate Methods

**Multivariate** describes relationships among variables, focusing on the correlation structure: correlations and other measures of association, scatterplot matrices, multivariate outliers, and principal components.

**Cluster** allows for *k*-means and hierarchical clustering. Normal mixtures and Self-Organizing Maps (SOMs) are found in this platform.

**Discriminant** fits discriminant analysis models.

**PLS** implements partial least-squares analyses.

**Item Analysis** analyzes questionnaire or test data using Item Response Theory.

### Survival and Reliability

**Survival /Reliability** models the time until an event, allowing censored data. This kind of analysis is used in both reliability engineering and survival analysis.

**Fit Parametric Survival** opens the fit model dialog to model parametric (regression) survival curves.

**Fit Proportional Hazards** opens the Fit Model dialog to fit the Cox proportional hazards model.

**Recurrence Analysis** analyzes repairable systems.

## The Graph Menu

**Chart** gives many forms of charts such as bar, pie, line, and needle charts.

**Overlay Plot** overlays several numeric *y*-variables, with options to connect points, show a step plot, needle plot, and others. It is possible to have two *y*-axes in these plots.

**Spinning Plot** shows a three-dimensional rotating scatterplot with options to see principal components and biplots.

**Contour Plot** constructs a contour plot for one or more response variables, y, for the values of two x variables. **Contour Plot** assumes the x values lie in a rectangular coordinate system, but the observed points do not have to form a grid.

**Control Chart** monitors a process through time to watch for it going out-of-control.

**Variability/Gage Chart** is used for analyzing measurement systems. Data can be continuous measurements or attributes.

**Pareto Plot** creates a bar chart (Pareto chart) that displays the severity (frequency) of problems in a quality-related process or operation. Pareto plots compare quality-related measures or counts in a process or operation. The defining characteristic of Pareto plots is that the bars are in descending order of values, which visually emphasizes the most important measures or frequencies.

**Ternary Plot** constructs a plot using triangular coordinates. The ternary platform uses the same options as the contour platform for building and filling contours. In addition it a specialized crosshair tool that lets you read the triangular axis values.

**Diagram** is used to construct *Ishikawa charts*, also called *fishbone charts*, or *cause-and-effect diagrams*. These charts are useful when organizing the sources (causes) of a problem (effect), perhaps for brainstorming, or as a preliminary analysis to identify variables in preparation for further experimentation.

**Parallel Plot** shows connected-line plots of several variables at once.

**Cell Plot** produces a "heat map" of column, assigning colors based on a gradient (for continuous variables) or according to a level (of categorical variables)

**Profiler** is available for tables with columns whose values are computed from model prediction formulas. Usually, profiler plots appear in standard least squares reports, where they are a menu option. However, if you save the prediction equation from the analysis, you can access the prediction profile independent of a report from the **Graph** menu and look at the model using the response column with the saved prediction formula.

**Contour Profiler** works the same as the **Profiler** command. It is usually accessed from the Fit Model platform when a model has multiple response. However, if you then save the prediction formulas for the responses, you can access the Contour Profiler at a later time from the **Graph** menu and specify the columns with the prediction equations as the response columns

**Surface Plot** draws a three-dimensional, rotatable surface.

# Navigating Platforms and Building Context

The first few times JMP is used, most people have navigational questions: How do I get a particular graph? For example, how do I produce a histogram? How do I get a *t*-test?

The strategy for approaching JMP analyses is to build an analysis context. Once you build that context, then the graphs and statistics become easily available–often they happen automatically, without having to ask for them specifically.

There are three keys for establishing the context:

- The *Modeling Type* identifies a variable as either continuous or categorical.
- The *X or Y Role* identifies whether the variable is a response (Y) or a factor (X).
- The *analysis platform* is the general approach and character of the analysis.

Once you settle on a context, commands appear in logical places.

# Contexts for a Histogram

Suppose you want to display a histogram. In other packages, you might find a histogram command in a graph menu, but in JMP you need to think of the context. You want a histogram so that you can see the distribution of values. So, launch the Distribution platform in the **Analyze** menu. Once you are there, then there are many graphs and reports available for focusing on the distribution of values.

Occasionally, you may want the histogram as a business graph. Then, instead of using the Distribution platform, use the Chart platform in the **Graph** menu.

# Contexts for the *t*-test

Suppose you want a *t*-test. Other packages might have a *t*-test command. JMP has many *t*-test commands because there are many contexts in which this statistic is used. So first, you have to build the context of your situation.

If you want the *t*-test to test a single variable's mean against a hypothesized value, then you are focusing on the distribution, so you launch the Distribution platform. On the title bar of the distribution report is a popup menu with the command **Test Mean**. This command gives you a *t*-test, as well as the option to see a nonparametric test.

If you want the *t*-test to compare the means of two independent groups, then you have two variables to set the context—perhaps a continuous Y response and a categorical X factor. Since the analysis deals with two variables, use the Fit Y By X platform. You want to find the mean Y response for each group identified by the X factor. If you launch the Fit Y by X platform, you'll see the side-by-side comparison of the two distributions, and you can use the **t test** or **Means/Anova/Pooled t** command from the popup menu on the analysis title bar.

If you want to compare the means of two continuous responses that form matched pairs, there are several ways to build the appropriate context. You can make a third data column to form the difference of the responses, and use the Distribution platform to do a *t*-test that the mean of the differences is zero. Or you can use the **Matched Pairs** command to launch the Matched Pairs platform for the two variables. In Chapter 6, "The Difference Between Two Means," you will learn more ways to do a *t*-test.

## Contexts for a Scatterplot

Suppose you want a scatterplot of two variables. Think of it more generally as a bivariate analysis, which suggests using the Fit Y by X platform. With two continuous variables, the Fit Y By X platform produces a scatterplot. You can also fit regression lines inside this scatterplot from the same report.

You might also consider the **Overlay Plot** command in the **Graph** menu. As a graph platform, it does not compute regressions, but will overlay multiple Y's in the same graph and connect the points.

If you have a whole series of scatterplots for many variables in mind, you are looking for many bivariate associations, part of the Multivariate platform. A matrix of scatterplots appears automatically for the Multivariate platform.

## Contexts for Nonparametric Statistics

There is no separate platform for nonparametric statistics. However, there are many standard nonparametric statistics in JMP, positioned by context. When you test a mean in the Distribution platform, you have the option to do a (nonparametric) Wilcoxon signed-rank test. When you do a *t*-test or one-way ANOVA in the Fit Y by X platform, you also have three optional nonparametric tests. The Wilcoxon rank sum test given there is equivalent to the Mann-Whitney *U*-test. If you want a nonparametric measure of association, like Kendall's τ or Spearman's correlation, look in the Multivariate platform.

# The Personality of JMP

Here are some reasons why JMP is different from other packages:

*Graphs are in the service of Statistics (and vice versa).* The goal of JMP is to provide a graph for every statistic, presented with the statistic. The graphs shouldn't appear in separate windows, but rather should work together. In the Analysis platforms, the graphs tend to follow the statistical context. In the Graph platforms, the statistics tend to follow the graphical context.

*JMP encourages good data analysis.* In the example presented in this chapter, you didn't have to ask for a histogram because it appeared when you launched the Distribution platform. The Distribution platform was designed that way, because in good data analysis you always examine a graph of a distribution before you start doing statistical tests on it. This encourages responsible data analysis.

*JMP allows you to make discoveries.* JMP was developed with the charter to be "Statistical Discovery Software." After all, you want to find out what you didn't know, as well as try to prove what you already know. Graphs attract your attention to an outlier or other unusual feature of the data that might prove valuable to discovery. Imagine Marie Curie using a computer for her pitchblende experiment. If software had given her only the end results, rather than showing her the data and the graphs, she might not have noticed the discrepancy that lead to the discovery of radium.

*JMP bristles with interactivity.* In some products, you have to specify exactly what you want ahead of time because often that is your last chance before you do the analysis. JMP is interactive, so everything is open to change and customization at any point in the analysis. It is easier to remove a histogram when you don't want it than decide ahead of time that you want one.

*You can see your data from multiple perspectives.* Did you know that a $t$-test for two groups is a special case of an $F$-test for several groups? With JMP, you tend to get general methods that are good for many situations, rather than specialty methods for special cases. You also tend to get several ways to test the same thing. For two groups, there is a $t$-test and equivalent $F$-test. When you are ready for more, there are three nonparametric tests to use in the same situation. You also can test for and adjust for different variances across the groups. And there are two graphs to show you the separation of the means. Even after you perform statistical tests, there are multiple ways of looking at the results, in terms of the $p$-value, the confidence intervals, least significant differences, the sample size, and least significant number. With this much statistical breadth, it is good that commands appear as you qualify the context, rather than you having to select multiple commands from a single menu bar. JMP unfolds the details progressively, as they become relevant.

# 3

# Data Tables, Reports, and Scripts

## Overview

JMP data are organized as rows and columns of a table referred to as the *data table*. The columns have names and the rows are numbered. An open data table is kept in memory and you communicate with it through an active window. The window displays a data grid and panels with information about the data table. You can open as many data tables in a JMP session as memory allows.

Commands in the **File**, **Edit**, **Tables**, **Rows**, and **Cols** menus give you a broad range of data handling operations and file management tasks, such as data entry, data validation, text editing, and extensive table manipulation.

In particular, the **Tables** menu has commands that perform a wide variety of data management tasks on JMP data tables. These commands let you sort, subset, stack or split table columns, join two tables side by side, concatenate multiple tables end to end, and transpose tables. You can also create summary tables for group processing and summary statistics.

The purpose of this chapter is to tell you about JMP data tables and give a variety of hands-on examples to help you get comfortable handling table operations.

# The Ins and Outs of a JMP Data Table

A JMP data table appears as a data grid. Using this view, you can do a variety of table management tasks such as editing cells, creating, rearranging or deleting rows and columns, taking a subset, sorting, or combining tables. **Figure 3.1** identifies active areas of a spreadsheet.

There are a few basic things to keep in mind:

- Column names can have as many as 31 characters and can use any keyboard character, including spaces. The size and font for names and values is a preference setting you control.

- You can drag column boundaries and enlarge the column to view long values.

- There is no limit to the number of rows or columns in a data table. However, the table must fit in memory.

**Figure 3.1**   Active areas of a JMP spreadsheet

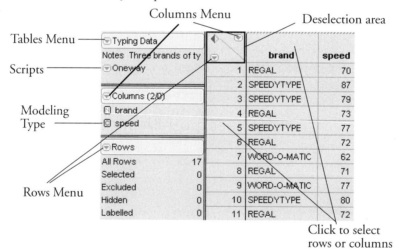

## Selecting and Deselecting Rows and Columns

Many actions, including commands from the **Rows** and **Cols** menus, operate only on selected rows and columns. The selection of rows and columns is done by highlighting them. To highlight a row, click the space that contains the row number. To highlight a column, click the background area above the column name. These areas are shown in **Figure 3.1**.

To extend a selection of rows or columns, drag (in the selection area) across the range of rows or columns you want, or Shift-click the first and last row or column of the range. Ctrl-click (⌘-click on the Macintosh) to make a discontiguous selection. To select rows and columns at the same time, drag across table cells in the spreadsheet.

To deselect a row or column, Ctrl-click (⌘-click on the Macintosh) on the row or column selection area. Click the triangular rows or columns area in the upper-left corner of the spreadsheet to deselect all rows or columns at once.

## Mousing Around a Spreadsheet: Cursor Forms

To navigate in the spreadsheet, you need to understand how the cursor works in each part of the spreadsheet.

🖰   To experiment with the different cursor forms, open the Typing.jmp sample table, move the mouse around on the surface are illustrated in **Figure 3.1**, and see how the cursor changes to the forms listed next.

### Arrow cursor ( ⌖ )

When a data table is the active window, the cursor displays as a standard arrow when it is anywhere in the table panels to the left of the data grid, except when it is on a red triangle popup menu icon or a diamond-shaped disclosure icon. It is also a standard arrow when it is in the upper-left corner of the data grid.

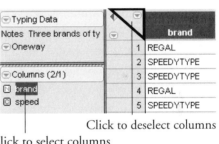

Click to deselect columns

Click to select columns
Double-click to edit column name

### I-beam cursor ( ⌶ )

The cursor is an I-beam when it is over text in the data grid, highlighted column names in the data grid, or column panels. To edit text in the data grid, position the I-beam next to characters and click to highlight the cell. Once the cell is highlighted, simply begin typing to edit the cell's contents. Alternatively, double-click and drag the I-beam to select a portion of the text for replacement. By default, the entire character string is selected. Use the keyboard to make changes after positioning the I-beam. To edit a column name, first click the column selection area to highlight the column. As with cell entries, simply begin typing once the column is highlighted.

### Fat Plus cursor (⊕)

The cursor becomes a large thick plus when you move it into a column or row selection area. Use the fat-plus cursor to select a single row or column. Shift-click a beginning and ending row (or a beginning and ending column) to select an entire range. Ctrl-click (⌘-click on the Macintosh) to select multiple rows or columns that are not contiguous.

The fat plus tool is also used to select areas of reports to copy and past to other locations See "Copy, Paste, and Drag Data" on page 41 for details on this use of the fat plus tool.

### Double Arrow cursor ( ↔ )

The cursor changes to a filled double arrow when on a column boundary. Drag this cursor left or right to change the width of a spreadsheet column.

### List Check and Range Check cursors ( ↧ ⊥ )

The cursor changes when it moves over values in columns that have data validation in effect (automatic checking for specific values). It becomes a small, downward-pointing arrow on a column with *list checking* and a large double I–beam on a column with *range checking*. When you click, the value highlights and the cursor becomes the standard I–beam; you enter or edit data as usual. However, you can only enter data values from a list or range of values you pre-specify.

### Popup Pointer cursor ( ☝ )

The cursor changes to a finger pointer over any popup menu icon or diamond-shaped disclosure icon. Click the disclosure icon to open or close a window panel or report outline; click to select a popup menu item icon.

☝   After you finish exploring, choose **File** > **Close** to close the Typing.jmp table.

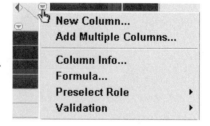

# Creating a New JMP Table

Hopefully, most data will reach you already in electronic form. However, if you have to key in data, JMP provides a spreadsheet with familiar data entry features. A short example shows you how to start from scratch.

Suppose data values are blood pressure readings collected over six months and recorded in a notebook page as shown in **Figure 3.2.**

**Figure 3.2**   Notebook of Raw Study Data Used to Define Rows and Columns

Blood Pressure Study

| Month | Control | Placebo | 300mg | 450mg |
|-------|---------|---------|-------|-------|
| March | 165 | 163 | 166 | 168 |
| April | 162 | 159 | 165 | 163 |
| May | 164 | 158 | 161 | 153 |
| June | 162 | 161 | 158 | 151 |
| July | 166 | 158 | 160 | 148 |
| August | 163 | 158 | 157 | 150 |

## Define Rows and Columns

JMP data tables have rows and columns, which represent *observations* and *variables* respectively, in statistical terms. The raw data in **Figure 3.2** are arranged as five columns (month and four treatment groups) and six rows (months March through August). The first line in the notebook names each column of values that can be used as column names in a JMP table. To enter this data into JMP, you first need a blank data table.

- Choose **File > New > Data Table** to see a new empty data table. A new untitled table appears with one column and no rows. On the Macintosh, **File > New** creates a new empty data table.

### The Add Columns Command
- Choose **Cols > Add Multiple Columns** and respond to the Add Columns dialog by requesting 5 new columns.

The default column names are Column 1, Column 2, and so on, but you can change them by typing in the editable column fields.

To edit a column name, first click the column selection area to highlight the column. Then, begin typing the name of the column. The column name starts changing, with the insertion point showing as a blinking vertical bar. You can also drag the I-beam to select a portion of the text for replacement. Use the Tab key to move from one column to the next.

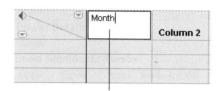

Highlight the column, then begin typing.

🖑    Type the names from the data journal (Month, Control, Placebo, 300 mg, and 450 mg) into the columns headers of the new table.

## Set Column Characteristics

Columns can have different characteristics. By default, their modeling type is continuous, so they expect numeric data. However, in this example, the Month column will hold a non-numeric character variable.

🖑    Double-click the column name area for Month to activate the column info dialog.

🖑    In the Column Info dialog, use the Data Type popup menu to change Month to a character variable, then click OK.

**Figure 3.3**  Column Info dialog

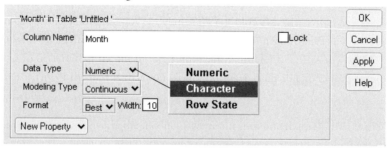

You can also use the Column Info dialog to change the other column characteristics and to access the JMP calculator for computing column values.

## Add Rows

Adding new rows is easy.

🖑    Choose **Rows** > **Add Rows** and ask for six new rows.

Alternatively, if you double-click anywhere in the body of the table, the table automatically fills with new rows through the position of the cursor.

The last step is to give the table a name and save it.

🖰   Choose **File** > **Save As** to name the table BP Study.jmp. You may also navigate to another folder if you want to save this data somewhere else.

The data table is now ready to hold data values. **Figure 3.4** summarizes the table evolution so far.

**Figure 3.4**   JMP Data Table with New Rows, Columns, and Names

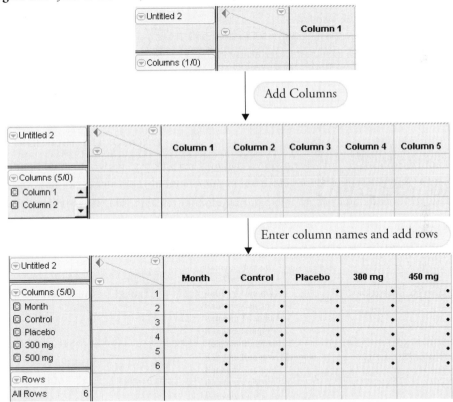

## Enter Data

Entering data into the data table requires typing values into the appropriate table cells. To enter data into the data table, do the following:

🖰   Move the cursor into a data cell and double-click to begin editing the cell.

A blinking vertical bar appears in the cell.

🖰    Key in the data from the notebook (**Figure 3.2**).

If you make a mistake, drag the I–beam across the incorrect entry to highlight it and type the correction over it. Your result should look like the table in **Figure 3.5**. The Tab and Return keys are useful keyboard tools for data entry:

- Tab moves the cursor one cell to the right. Shift-Tab moves the cursor one cell to the left. Moving the cursor with the Tab key automatically wraps it to the beginning of the next (or previous) row. Tabbing past the last table cell creates a new row.

- Enter (or Return) either moves the cursor down one cell or one cell to the right, based on the setting in JMP's Preferences.

**Figure 3.5**    Finished Blood Pressure Study Table

| | | Month | Control | Placebo | 300 mg | 450 mg |
|---|---|---|---|---|---|---|
| 1 | March | | 165 | 163 | 166 | 168 |
| 2 | April | | 162 | 159 | 155 | 163 |
| 3 | May | | 164 | 158 | 161 | 153 |
| 4 | June | | 162 | 161 | 158 | 151 |
| 5 | July | | 166 | 158 | 160 | 148 |
| 6 | August | | 163 | 158 | 157 | 150 |

Columns (5/0): Month, Control, Placebo, 300 mg, 450 mg
Rows — All Rows 6

## The New Column Command

In the first part of this example, you used the **Add Cols** command from the **Cols** menu to create a group of new columns in a data table. Often you need to add a single new column with specific characteristics.

Continuing with the current example, suppose you learn that the blood pressure readings were taken at one lab, called "Accurate Readings Inc.," during March and April but at another location called "Most Reliable Measurements Ltd." for the remaining months of the study. You want to include this information in the data table.

🖰    Begin by choosing **Cols > New Column**, which displays a New Column dialog like the one shown in **Figure 3.6**.

The New Column dialog lets you set the new column's characteristics.

🖰    Type a new name, Location, in the Column Name area.

🖱   Because the actual names of the location are characters, select **Character** from the Data Type popup menu as shown in **Figure 3.6**.

Notice that the Modeling Type then automatically changes to **Nominal**.

The default field length is 16 characters, but suppose you need to enter 31 characters for one of the locations.

🖱   Type 31 into the Field Width box.

When you click **OK**, the new column appears in the table, where you can enter data as previously described.

**Figure 3.6**   The New Column Dialog

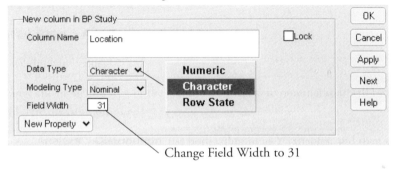

Change Field Width to 31

# Plot the Data

There are many ways to check the data for errors. One way is to generate a plot and see that there are no obvious anomalous values. Let's experiment with the **Chart** command in the **Graph** menu.

We will plot the months along the horizontal (x) axis and the columns of blood pressure statistics for each treatment group up the vertical (y) axis.

🖱   Choose **Graph** > **Chart**.

🖱   Select Month as the X variable.

🖱   Highlight all four continuous columns and select **Data** from the **Statistics** drop-down list.

When you click **OK** you first see a bar chart.

**Figure 3.7**  Initial bar chart

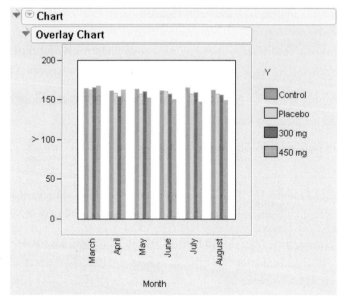

Now, use some options.

⊕ Click the popup menu icon on the title bar of the chart to see a variety of options.

⊕ Make sure the **Overlay** option is checked, and then select **Y Options** > **Line Chart** to see the chart shown in **Figure 3.8** (the plot doesn't appear to have much to say yet).

**Figure 3.8**   Line Chart for Blood Pressure Values over Month

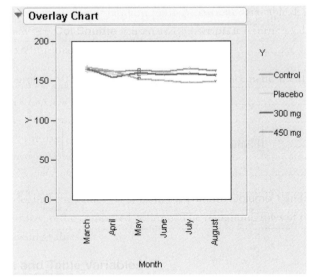

By default, *y*-axis scaling begins at zero. To present easy-to-read information, the *y*-axis needs to be rescaled:

- Double-click anywhere in the *y*-axis area to bring up the Rescale dialog box.

Based on what you can see in **Figure 3.8**, the plotted values range from about 145 to 175.

- Type these values into the Axis Rescale dialog as the minimum and maximum.

- Change the increment for the tick marks from 50 to 5, which divides the range into six intervals.

- Click **OK**.

Use the Annotate tool ( A ) to annotate the chart with captions as shown in **Figure 3.9**.

- Select the Annotate tool and click in the chart where you want to insert the caption.

- Type "Comparison of Treatment Groups" and click outside the caption.

- Resize the caption by clicking and dragging on the border. Move the caption by clicking and dragging in its interior.

**Figure 3.9**   Line Chart with Modified *y*-Axis

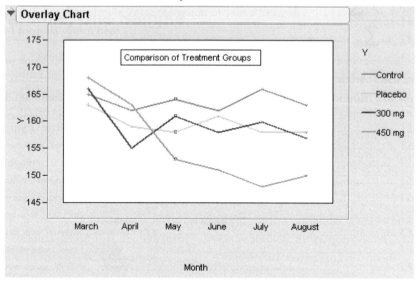

## Importing Data

The **File** > **Open** command displays a specialized dialog that lets you locate the file you want to open and tell JMP the file format of the incoming file. The **Open** command then reads the file into a JMP data table.

JMP directly reads JMP data tables, JMP journal files, JMP script files, SAS transport files, text files with any column delimiter, Excel files, and flat-file database files. To open database files, you must have an appropriate ODBC driver installed on your system. In addition, the Windows version of JMP can read and write SAS data sets.

If you indicate what kind of file to expect with an appropriate **Files of Type** (Windows), **Show** (Macintosh), or **Filter** (Linux) selection, JMP gives helpful information when possible. The example in **Figure 3.10** shows an Open Data File dialog when the **Files of Type** drop-down list is changed from the default (**Data Files**) to **JMP data tables**. The dialog shows the table notes (if there are any).

If the incoming file is not a JMP data table, and you choose to look at all files, JMP looks at the type of file given by the 3-character extension appended to its file name and opens it accordingly. To examine all files, choose * (Linux), *.* (Windows), or All Readable Files (Macintosh). This works as long as the file has the structure indicated by its name.

**Figure 3.10**   The File Open dialog to read a JMP table

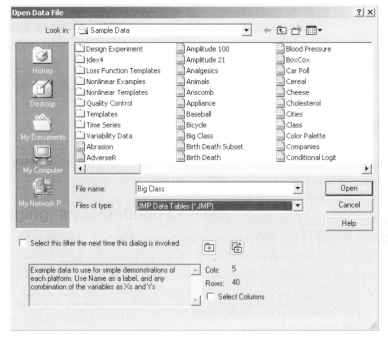

## Importing Text Files

To import a text file, select one of the following:

- On Windows, select **Text Import** or **Text Import Preview** in the **Files of Type** list.

- On the Macintosh, select **Text** from the **Show** list, then **Text Data** or **Text Data with Preview** from the **Open As** list (**Figure 3.11**).

- On Linux, select **.txt (Fixed Width)** or **.txt (Delimited)** from the **Filter** menu. Note that you are immediately presented with the dialog in **Figure 3.12**.

**Figure 3.11**   Macintosh Text Import

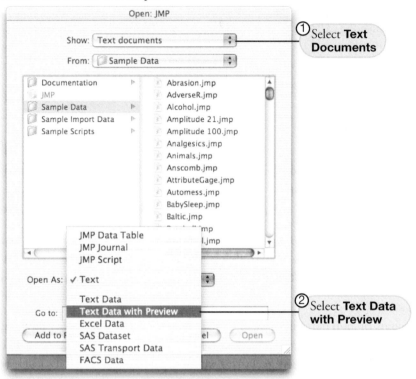

JMP attempts to discern the arrangement of text data. This is adequate for a rectangular text files with no missing fields, a consistent field delimiter, and an end-of-line delimiter.

**Note:** If double-quotes are encountered when importing text data, JMP changes the delimiter rules to look for an end double-quote. Other text delimiters, including spaces embedded within the quotes, are ignored and treated as part of the text string.

If you want to see a preview of an incoming text file, choose **Text Import Preview**. Otherwise, JMP imports the data based on your text import preferences (**File > Preferences** on Windows, **Edit > Preferences** or **JMP > Preferences** on Macintosh). Linux works slightly differently. If you want to open a text file immediately, select the **Finish** button on the Open dialog. To see a preview, select the **Next** button.

With previewed files, the panel shown here appears when you click **Open**. It asks whether the data have delimited fields or fixed-width fields.

If you select **Fixed Width**, a dialog opens to allow you to specify the columns for each variable.

If you click **Delimited**, the data arrangement is described as shown in **Figure 3.12**. This dialog begins

with settings from your Preferences file. It also shows the column names, data types, and the first two rows of data. In **Figure 3.12**, preferences are set that indicate the incoming table contains column headers to be used as the column names; the column names are name, age, sex, and height. If no column names are indicated, the Name fields are called Column 1, Column 2, and so on.

You can also identify one or more end-of-field delimiters, end-of-line delimiters, choose the option to **Strip enclosing quotes**, and see how many rows and columns will be read.

**Figure 3.12**  Import Delimited Field Text File

## Importing Microsoft Excel Files

JMP has the ability to directly import Microsoft Excel worksheets and workbooks.

On Windows, you must have the Excel ODBC driver installed. Macintosh systems read Excel files without an ODBC driver. Single Excel worksheets are imported using one of the following.

• On Windows, choose **Excel Files(*.xls)** from the **Files of Type** drop down list.

• On the Macintosh, select **Excel documents** from the **Show** drop-down list.

JMP can also import Excel workbooks that contain several data tables inside them.

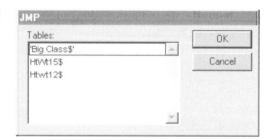

• On Windows, after selecting **Excel Files(*.xls)** from the **Open** dialog and double-clicking on the desired workbook, JMP will present the dialog shown in the figure shown to the right. This dialog shows all the worksheets in the selected workbook, included in the sample data as Students.xls. Select the one to open and click **OK** to import the data into JMP.

• On the Macintosh, all tables are opened as separate JMP data tables.

## Using ODBC

On Windows or Linux, you can open files for any format that has a corresponding ODBC driver on your system.

Use the standard **File > Open** command to access flat-file databases like Microsoft Access, Microsoft FoxPro, and dBase. Installed ODBC drivers appear at the end of the Files of Type list.

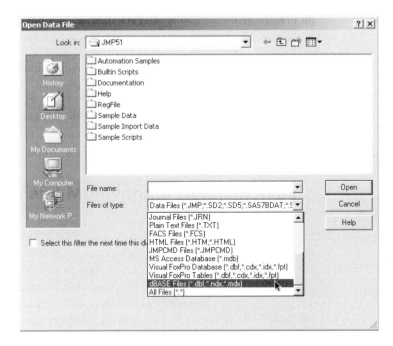

Use the **Open Database** command to import data from relational databases, like dBII and Oracle. Details of using the **Open Database** command are in the *JMP User's Guide*.

## Copy, Paste, and Drag Data

You can use the standard copy and paste operations to move data and graphical displays within JMP and from JMP to other applications. The following commands in the **Edit** menu let you move data around:

### Copy

The **Copy** command in the **Edit** menu copies the values of selected data cells from the active data table to the clipboard. If you do not select rows, **Copy** copies entire rows. Likewise, you can copy values from whole columns if no rows are selected. If you select both rows and columns, **Copy** copies the subset of cells defined by their intersection. Data you cut or copy to the clipboard can be pasted into JMP tables or into other applications.

⊕   If you want to copy part of an analysis window, use the fat-plus area selection tool in the **Tools** menu. Click on the area you want to copy to select and highlight it. Shift-click to extend the selection. If nothing is selected, the **Copy** command copies the entire window to the clipboard. If you right-click (Control-click on the Macintosh) when this tool is active, a popup menu gives you commands to **Copy**, **Paste**, or **Select All**.

### Paste

The **Paste** command copies data from the clipboard into a JMP data table. **Paste** can be used with the **Copy** command to duplicate rows, columns, or any subset of cells defined by selected rows and columns.

To transfer data from another application into a JMP data table, first copy the data to the clipboard from within the other application. Then use the **Paste** command to copy the values to a JMP data table. Rows and columns are automatically created as needed. If you choose **Paste** while holding down the Shift (Windows and Linux) or Option (Macintosh) key, the first line of information on the clipboard is used as column names in the new JMP data table.

To duplicate an entire row or column:

1. Select and **Copy** the row or column to be duplicated.

2. Select an existing row or column to receive the values.

3. Select **Paste**.

To duplicate a subset of values defined by selecting both rows and columns, follow the previous steps, but select the same arrangement of rows and columns to receive the copied values as originally contained them. If you paste data with fewer rows into a destination with more rows, the source values recycle until all receiving rows are filled.

### Drag

You can also move or duplicate rows and columns by dragging. Hold the mouse down in the selection area of one or more selected rows or columns and drag them to the position in the data table where you want them. Use Ctrl-drag (Option-drag on the Macintosh) to duplicate rows and columns instead of moving them.

# Moving Data Out of JMP

Two questions that come up early with JMP (or with any application) are "Can I get my data back out of JMP?" and "How do I get results out of JMP?"

The **Save As** command writes the active data table to a file after prompting you for a name and file type. It can save the data table as a JMP file, convert it to a SAS Transport file, Excel file, text file, or save it in any database format available on your system.

The **Save As** command has the same options for saving or exporting data as the **Open** command has for importing data.

## Windows

JMP can save data in any of the following formats.

- **JMP Data Tables** save the table in JMP format. This is the default **Save As** option.

- **SAS Transport Files** converts a JMP data table to SAS transport file format and saves it in a SAS transport library. The **Append To** option appends the data table to an existing SAS transport library. If you don't use **Append To**, a new SAS transport library is created using the name and location you give. If you do not specify a new file name, the SAS transport library replaces the existing JMP data table.

- **Excel Files** saves data tables in Microsoft Excel .xls format. The resulting file is directly readable by most versions of Excel, including Excel 97, 98, and 2000.

- **Text Export Files** converts data from a JMP file to standard text format, with rows and columns. The **Options** button in the **Save As** dialog displays choices to describe specific text arrangements:

    **Export Column Names to Text File** has **Yes** and **No** radio buttons to request that JMP column names be written as the first record of the text file, or that no labels or header information be saved with the data.

    **End of Field** and **End of Line** designate the characters to identify the end of each field and end of line in the saved text file. These options are described previously in the section "Importing Data" on page 36.

- JMP can also save data in formats readable by databases with ODBC drivers installed on your system, such as **Microsoft FoxPro**, **dBase**, and **Microsoft Access**.

**Figure 3.13**  Save As Dialog and Text Formatting Options (Windows)

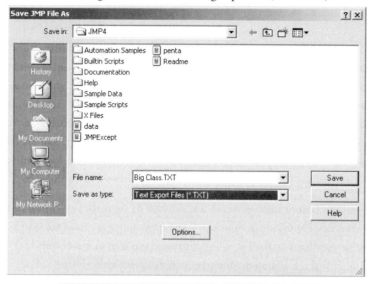

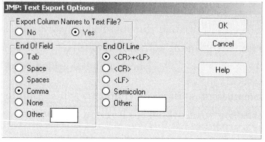

## Macintosh

JMP can save data as a JMP data table, SAS Transport file, or an Excel file. For example, to save a text file, select **File > Save As**, then select **Text Data** from the **Format** drop-down list (see **Figure 3.14**). The file is saved using the settings set in Preferences.

## Linux

JMP can save data as a JMP data table, text file, or SAS Transport file. For example, to save a text file, select **File > Save As**, then select **.txt (Delimited File)** from the **Save As Type** menu.

**Figure 3.14**  Save As Text (Macintosh)

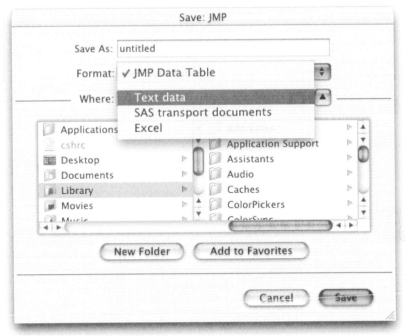

# Working with Graphs and Reports

You can use standard copy and paste operations to move graphical displays and statistical reports from JMP to other applications. Although the **Edit** menu includes both **Cut** and **Copy** commands, they both perform the same tasks in report windows. **Cut** copies all or selected parts of the active report window into the clipboard, including the images.

## Copy and Paste

When you copy from a report window, the image is stored in the clipboard. If you want to copy part of a report window, use the fat-plus tool (see left) from the **Tools** menu or toolbar. Click on the area you want to copy, shift-click to extend the selected area, use the **Copy** command to copy the selected area to the clipboard, then use the **Paste** command to paste the results into a JMP journal or another application.

## Drag Report Elements

Any element in a JMP report window that can be selected can be dragged. When you drag report elements within the report frame, they are copied to the destination area where you drop them. As you drag an element, a visual cue shows where the element is to be dropped, and the edge of the report frame shows if you attempt to drag outside it.

You can copy and paste any report element to other applications, and drag and drop JMP reports and graphs to any other application that supports drag-and-drop operations.

The format used when pasting depends on the application you paste into. If the application has a **Paste Special** command, you can select a paste format such as: rich text, which includes pictures (RTF), unformatted text (TXT), picture (PICT or WMF), bitmap (BMP), and enhanced picture (EMF).

To delete a copied report element, select it and press the Delete key on the keyboard.

# Context Menu Commands

Right-click (Control-click on the Macintosh) on a report window to see the context popup menu shown in the following examples. The context menu changes depending on where you click. If you are not over a display element with its own context menu, the menu for the whole platform is shown.

## Context Commands for Report Tables

By default, the tables in results have no formatting to separate rows and columns. Some (or, in many cases, all) available columns for the report are showing. Context menu items for report tables let you tailor the appearance and content of the tables as follows.

- **Table Style** lets you enhance the appearance of a table by drawing borders or other visual styles to the table rows and columns. The example shown to the right has beveled column separators.

- **Columns** lets you choose which columns you want to show in the analysis table. Analysis tables often have many columns, some of which may be initially hidden. The leftmost table in **Figure 3.15** is a Parameter Estimates table showing only the estimate name, the estimate itself, and the probability associated with the estimate. The standard error and chi-square values are hidden.

- **Sort by Column** lets you sort the rows of a report table. This command displays a list of visible columns in a report and you choose one or more columns as sort variables. The middle table in **Figure 3.15** is the Parameter Estimates table on the left sorted by **Prob>ChiSq**.

- **Make into Data Table** lets you create a JMP data table from any analysis table. The rightmost data table in **Figure 3.15** is the JMP data table result of sorted Parameter Estimates table with hidden columns.

**Figure 3.15**   Results of Context Commands for Analysis Tables

| Parameter Estimates | | |
| --- | --- | --- |
| Term | Estimate | Prob>ChiSq |
| Intercept[13] | 3.32387173 | 0.0268 |
| Intercept[14] | 4.29921448 | 0.0051 |
| Intercept[15] | 5.76150124 | 0.0004 |
| Intercept[16] | 6.96581366 | <.0001 |
| Intercept[17] | 7.93933219 | <.0001 |
| weight | -0.0483808 | 0.0011 |

| Parameter Estimates | | |
| --- | --- | --- |
| Term | Estimate | Prob>ChiSq |
| Intercept[13] | 3.32387173 | 0.0268 |
| Intercept[14] | 4.29921448 | 0.0051 |
| weight | -0.0483808 | 0.0011 |
| Intercept[15] | 5.76150124 | 0.0004 |
| Intercept[16] | 6.96581366 | <.0001 |
| Intercept[17] | 7.93933219 | <.0001 |

| | Term | Estimate | Prob>ChiSq |
| --- | --- | --- | --- |
| 1 | Intercept[13] | 3.32387173 | 0.02675324 |
| 2 | Intercept[14] | 4.29921448 | 0.00512396 |
| 3 | weight | -0.0483808 | 0.00111523 |
| 4 | Intercept[15] | 5.76150124 | 0.00044907 |
| 5 | Intercept[16] | 6.96581366 | 0.00006952 |
| 6 | Intercept[17] | 7.93933219 | 0.00001994 |

- **Make into Matrix** lets you store a report table as a matrix. When selected, the dialog shown here appears, allowing you to designate the name of the matrix and where it should be stored.

# Juggling Data Tables

Each of the following examples uses commands from the **Tables**, **Rows**, or **Cols** menus.

## Data Management

Suppose you have the following situation. Person A began a data entry task and entered state names in order of ascending auto theft rates. Then Person B took over the data entry, but mistakenly entered the auto theft rates in alphabetical order by state.

🖱 To see the result, open the Automess.jmp sample table.

Could this ever really happen? Never underestimate the diabolical convolution of data that can appear in an electronic table, and hence a circulated report. Always check your data with common sense.

| | State | Auto theft |
| --- | --- | --- |
| 1 | SOUTH DAKOTA | 348 |
| 2 | NORTH DAKOTA | 565 |
| 3 | WYOMING | 863 |
| 4 | WEST VIRGINIA | 289 |
| 5 | IDAHO | 1016 |
| 6 | IOWA | 428 |

To put the data into its correct order, you need make a copy of the Automess table, sort it in ascending order, and join the sorted result with the original table:

🖱 With Automess.jmp active, choose **Tables > Subset** and press **OK**.

This automatically creates a duplicate table when no rows or columns are selected in the original table.

The table name is Subset of Automess.JMP, but you don't need to give this table an official name because it is only temporary.

🖰    With this subset table active, choose **Tables** > **Sort**.

🖰    When the Sort dialog appears, choose Auto theft as the sort variable and click **Sort**.

There is now an untitled table that is sorted by auto theft rates in ascending order.

🖰    Close Subset of Automess.JMP, as it is no longer needed.

Now you want to join the incorrectly sorted Automess.JMP table with the correctly sorted Untitled table. You do this as follows:

🖰    Choose **Tables** > **Join**.

🖰    When the Join dialog appears, note which table is listed next to the word Join (either Automess or Untitled), click the other table in the list of tables.

🖰    Because you don't want all the columns from both tables in the final result, click the **Select Columns** button.

The variables from both tables appear in list boxes.

🖰    Select State from the Automess table, Auto theft from the Untitled table, and click **Add**.

🖰    Click **Done** to close the Select Columns dialog.

🖰    When the Join dialog again appears, click the **Join** button.

Check the new joined data table: the first row is South Dakota with a theft rate of 110, and the last row is the District of Columbia with a rate of 1336. If you want to keep this table, use **Save As** and specify a name and disk location for it.

## Give New Shape to a Table: Stack Columns

A typical situation occurs when response data are recorded in two columns and you need them to be stacked into a single column. For example, suppose you collect three months of data and enter it in three columns. If you then want to look at quarterly figures, you need to change the data arrangement so that the three columns stack into a single column. You can do this with the **Stack** command in the **Cols** menu.

Here is an example of stacking columns.

🖱  Open the Chezsplt.jmp sample data to see the table on the left in **Figure 3.16**.

This sample data (McCullagh and Nelder, 1983) has columns for four kinds of cheese, labeled A, B, C, and D. In a taste test, judges ranked the cheeses on an ordinal scale from 1 to 9 (1-awful, 9-wonderful). The Response column shows these ratings. The counts for each cheese for each ranking of taste are the body of the table. Its form looks like a two-way table, but to analyze this contingency table JMP needs to see the cheese categories in a single column. To rearrange the data:

🖱  Choose **Tables** > **Stack**.

🖱  In the dialog that appears, select the cheeses (A, B, C, and D) from the **Columns** list and add them to the **Stack** list. Leave everything else as is.

🖱  Click **OK** to see the table on the right in **Figure 3.16**.

The _ID_ column shows the cheeses, and the _Stacked_ column is now the count variable for the response categories.

🖱  Right-click (Control-click on the Mac) in the _Stacked_ column, and select **Preselect Role** > **Freq** to change the _Stacked_ column to represent frequency.

This causes the values _Stacked_ in the Untitled table to be interpreted by analyses as the number of times that row's response value occurred.

**Figure 3.16**  Stack Columns Example

| Chezsplt | | | | | | | | | Untitled 2 | | | | | |
|---|---|---|---|---|---|---|---|---|---|---|---|---|---|---|
| Notes  This sampl | | | **Response** | **A** | **B** | **C** | **D** | | Source | | | **Response** | **_ID_** | **_Stack_** |
| | | 1 | 1 | 0 | 6 | 1 | 0 | | | | 1 | 1 | A | 0 |
| | | 2 | 2 | 0 | 9 | 1 | 0 | | | | 2 | 1 | B | 6 |
| Columns (5/0) | | 3 | 3 | 1 | 12 | 6 | 0 | | Columns (3 | | 3 | 1 | C | 1 |
| Response | | 4 | 4 | 7 | 11 | 8 | 1 | | Response | | 4 | 1 | D | 0 |
| A | | 5 | 5 | 8 | 7 | 23 | 3 | | _ID_ | | 5 | 2 | A | 0 |
| B | | 6 | 6 | 8 | 6 | 7 | 7 | | _Stack_ | | 6 | 2 | B | 9 |
| C | | 7 | 7 | 19 | 1 | 5 | 14 | | | | 7 | 2 | C | 1 |
| D | | 8 | 8 | 8 | 0 | 1 | 16 | | | | 8 | 2 | D | 0 |
| | | 9 | 9 | 1 | 0 | 0 | 11 | | | | 9 | 3 | A | 1 |
| | | | | | | | | | | | 10 | 3 | B | 12 |
| | | | | | | | | | | | 11 | 3 | C | 6 |

To see how response relates to type of cheese:

🖱  Choose **Analyze** > **Fit Y by X**.

In the Fit Y by X Launch Dialog select Response as **Y, Response** and _ID_ as **X, Factor**.

Since its role was pre-selected, _Stacked_ is already designated as **Freq**.

When you click **OK,** the contingency table platform appears with a Mosaic plot, Crosstabs table, Tests table, and popup menu options. You can use the Question Mark tool in the Tools menu and click on the platform surface to find more information about the platform components. A simplified version of the Crosstabs table is shown in **Figure 3.17**. The Cheese data is used again later for further analysis.

**Figure 3.17** Contingency Table for the Cheese Data

**Contingency Table**

| | | Response | | | | | | | | | |
|---|---|---|---|---|---|---|---|---|---|---|---|
| | Count | 1 | 2 | 3 | 4 | 5 | 6 | 7 | 8 | 9 | |
| _ID_ | A | 0 | 0 | 1 | 7 | 8 | 8 | 19 | 8 | 1 | 52 |
| | B | 6 | 9 | 12 | 11 | 7 | 6 | 1 | 0 | 0 | 52 |
| | C | 1 | 1 | 6 | 8 | 23 | 7 | 5 | 1 | 0 | 52 |
| | D | 0 | 0 | 0 | 1 | 3 | 7 | 14 | 16 | 11 | 52 |
| | | 7 | 10 | 19 | 27 | 41 | 28 | 39 | 25 | 12 | 208 |

Extra Credit: For practice, see if you can use the **Split** command on the stacked data table to reproduce a copy of the Chezsplt table.

# The Summary Command

One of the most powerful and useful commands in the **Tables** menu is the **Summary** command.

**Summary** creates a JMP window that contains a summary table. This table summarizes columns from the active data table, called its *source table*. It has a single row for each level of a grouping variable you specify. A grouping variable divides a data table into groups according to each of its values. For example, a gender variable can be used to group a table into males and females.

When there are several grouping variables (for example, gender and age), the summary table has a row for each combination of levels of all variables. Each row in the summary table identifies its corresponding subset of rows in the source table. The columns of the summary table are summary statistics you request.

## Create a Table of Summary Statistics

The example data used to illustrate the **Summary** command is the JMP table called Companies.jmp (see **Figure 3.18**).

   🖱 Open the Companies.jmp sample table.

It is a collection of financial information for 32 companies (Forbes 1990). The first column (Type) identifies the type of company with values "Computer" or "Pharmaceut." The second column (Size Co) categorizes each company by size with values "small," "medium," and "big." These two columns are typical examples of grouping information.

**Figure 3.18**   JMP Table to Summarize

| | Type | Size Co | Sales ($M) | Profits ($M) | # Employ | profit/ emp | Assets | %profit/ sales |
|---|---|---|---|---|---|---|---|---|
| 1 | Computer | small | 855.1 | 31.0 | 7523 | 4120.70 | 615.2 | 3.63 |
| 2 | Pharmaceut | big | 5453.5 | 859.8 | 40929 | 21007.11 | 4851.6 | 15.77 |
| 3 | Computer | small | 2153.7 | 153.0 | 8200 | 18658.54 | 2233.7 | 7.10 |
| 4 | Pharmaceut | big | 6747.0 | 1102.2 | 50816 | 21690.02 | 5681.5 | 16.34 |
| 5 | Computer | small | 5284.0 | 454.0 | 12068 | 37620.15 | 2743.9 | 8.59 |
| 6 | Pharmaceut | big | 9422.0 | 747.0 | 54100 | 13807.76 | 8497.0 | 7.93 |
| 7 | Computer | small | 2876.1 | 333.3 | 9500 | 35084.21 | 2090.4 | 11.59 |
| 8 | Computer | small | 709.3 | 41.4 | 5000 | 8280.00 | 468.1 | 5.84 |
| 9 | Computer | small | 2952.1 | -680.4 | 18000 | -37800 | 1860.7 | -23.05 |
| 10 | Computer | small | 784.7 | 89.0 | 4708 | 18903.99 | 955.8 | 11.34 |
| 11 | Computer | small | 1324.3 | -119.7 | 13740 | -8711.79 | 1040.2 | -9.04 |
| 12 | Pharmaceut | medium | 4175.6 | 939.5 | 28200 | 33315.60 | 5848.0 | 22.50 |
| 13 | Computer | big | 11899.0 | 829.0 | 95000 | 8726.32 | 10075.0 | 6.97 |
| 14 | Computer | small | 873.6 | 79.5 | 8200 | 9695.12 | 808.0 | 9.10 |

(Companies table panel: Notes Selected Dat; Columns (8/0): Type, Size Co +, Sales ($M), Profits ($M), # Employ, profit/ emp +, Assets, %profit/ sales +; Rows: All Rows 32)

   🖱 Choose **Tables > Summary**.

   🖱 When the Summary dialog appears (**Figure 3.19**), select the variable Type in the Columns list of the dialog and click **Group** to see it in the grouping variables list.

You can select as many grouping variables as you want.

   🖱 Click **OK** to see the summary table.

**Figure 3.19** Summary Dialog and Summary Table

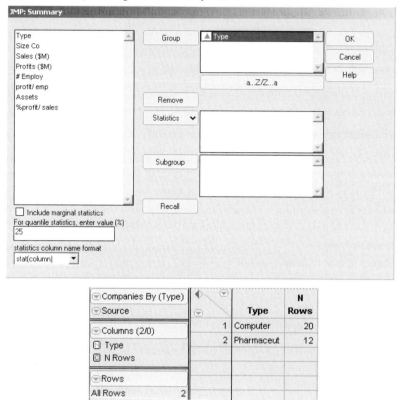

The new summary table appears in an active window. This table is linked to its source table. When you highlight rows in the summary table, the corresponding rows are also highlighted in its source table.

Initially, a summary displays frequency counts (N Rows) for each level of the grouping variables. This example shows 20 computer companies and 12 pharmaceutical companies. However, you can add columns of descriptive statistics to the table. The **Statistics** popup menu in the Summary dialog lists standard univariate descriptive statistics.

To add summary statistics to an existing summary table, follow these steps:

- Use the **Add Statistics Column** command accessed by the popup icon in the upper left-hand corner of the summary table.

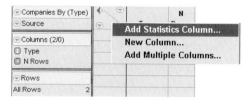

This command displays the Summary dialog again.

- 🖰  Select any numeric column from the source table columns list.

- 🖰  Select the statistic you want from the **Stats** popup menu on the dialog.

- 🖰  If necessary, repeat to add more statistics to the summary table.

- 🖰  Click **OK** to add the columns of statistics to the summary table.

The table in **Figure 3.20** shows the sum of Profits($M) in the summary table grouped by Type.

**Figure 3.20**   Expanded Summary Table

| | Type | N Rows | Sum(Profits ($M)) |
|---|---|---|---|
| 1 | Computer | 20 | 4817.3 |
| 2 | Pharmaceut | 12 | 8280.9 |

Another way to add summary statistics to a summary table is with the **Subgroup** button in the Summary dialog. This method creates a new column in the summary table for each level of the variable you specify with **Subgroup**. The subgroup variable is usually nested within all the grouping variables.

# Working With Scripts

JMP contains a full-fledged scripting language, used for automating repetitive tasks and scripting instructional simulations. Several scripts are featured throughout this book to demonstrate statistical concepts.

Scripts are stored in two formats.

- attached to a data table
- as a stand-alone scripting file.

Scripts that are attached to a data table are displayed in the Tables panel, as shown to the right. This sample is from the Big Class.jmp sample data table, showing six scripts that have been saved with it.

To run an attached script,

- 🖰  Click the button beside the script's name

🖰   Select **Run Script** from the menu that appears.

Stand-alone scripts are stored as simple text files. They may be opened and run independently of a data table.

## Opening and Running Scripts on Windows

To open and run a stand-alone script on Windows,

🖰   Select **File > Open**.

🖰   Change the **Files of Type** drop-down list to **JSL Scripts (\*.JSL)**, **\*.JSL (JSL Script)** or **JMP Files (\*.JMP, \*.JSL, \*.JRN).** The one you pick depends on whether you're running Windows, Linux, or Macintosh.

🖰   Double-click the name of the script to open.

The script opens in a script editor window. To execute the script,

🖰   Select **Edit > Run Script** or press the shortcut key **Ctrl-R**.

## Opening and Running Scripts on the Macintosh

To open and run a stand-alone script on the Macintosh,

🖰   Select **File > Open**

🖰   Make sure the **Show** list displays **All Readable Documents**.

🖰   Double-click on the script to open.

The script opens in a script editor window. To execute the script,

🖰   Select **Edit > Run Script** or press the shortcut key **⌘-R**.

As an example, use the Julia Sets.jsl script stored in the Sample Scripts folder.

🖰   Open and run the Julia Sets.jsl script.

The resulting window (shown in **Figure 3.21**) illustrates several key features of a typical instructional script.

**Figure 3.21**   Julia set script window

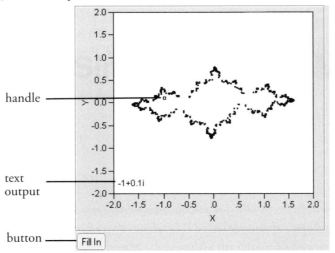

- A *handle* is a draggable script element that updates the display as it is dragged. In this case, the handle represents the seed value for the Julia set, whose shape changes based on the value of the handle's coordinates.

- *Text output* is sometimes drawn directly on the graphics screen rather than displayed as reports below the window. This script shows the seed value (an imaginary number) in the lower left corner of the window.

- *Buttons* reveal options, set conditions, or trigger actions in the script. In this example, the **Fill In** button draws more detail of the Julia set.

To practice with these elements,

  Click and drag the handle to different places in the window and observe how the Julia set changes.

  When the Julia set has an interesting shape, press the **Fill In** button to reveal its details.

# Formula Editor Adventures

# Overview

What is the Formula Editor?

Each column has access to the Formula Editor. The JMP Formula Editor is a powerful tool for building formulas that calculate values for each cell in a column. The Formula Editor window operates like a pocket calculator with buttons, displays, and an extensive list of easy-to-use features for building formulas.

JMP formulas can use information from other columns in the data table, built-in functions, and constants. Formulas can be simple expressions of numeric, character, or row state constants or can contain complex evaluations based on conditional clauses.

When you create a formula, that formula becomes an integral part of the data table. The formula is stored as part of the column when you save the data table, and it is retrieved when you reopen the data table. You can examine or change a column's formula at any time by opening it into a Formula Editor window.

A column whose values are computed using a formula is both *linked* and *locked*. It is linked to (or dependent on) all other columns that are part of its formula. Its values are automatically recomputed whenever you edit the values in these columns. It is also locked so that its data values cannot be edited, which would invalidate its formula.

This chapter describes Formula Editor features and gives a variety of examples. See the online *JMP User's Guide* for a complete list of Formula Editor functions.

# The Formula Editor Window

The JMP Formula Editor is a window for creating or modifying a formula. You can open the Formula Editor window for a column three ways:

- Select **Formula** from the **Cols** menu for one or more selected columns.

- Select **Formula** from the **New Property** popup menu in a New Column dialog and click the **Edit Formula** button that appears.

- Right-click (Windows and Linux) or Control-click (Macintosh) in the heading of a column and select **Formula** from the context menu that appears. This opens the Formula Editor window without first opening the Column Info dialog.

The Formula Editor window is divided into two areas: the *control panel* consisting of the column list and function browser, and the *formula display*, an area for editing formulas. **Figure 4.1** shows the parts of the Formula Editor. The Formula Editor control panel is composed of buttons (**OK**, **Apply**, **Help**), selection lists for variables and functions, and a keypad. The formula display is an editing area you use to construct and modify formulas.

**Figure 4.1**   The Formula Editor Window

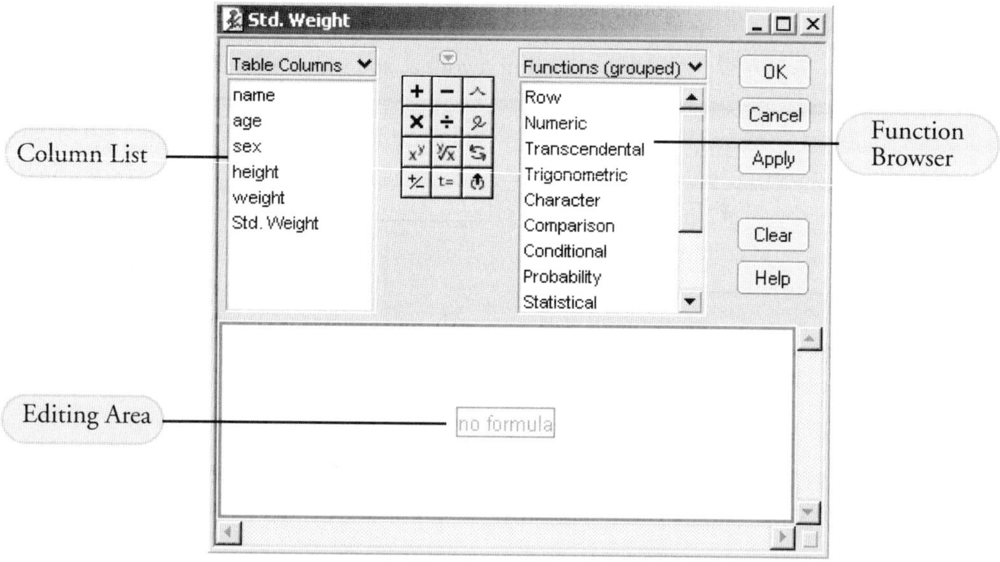

The sections that follow show you a simple example, define Formula Editor terminology, and give the details you need to use the control panel and the formula display.

# A Quick Example

The following example gives you a quick look at the basic features of the Formula Editor. Suppose you want to compute a standardized value. That is, for a numeric variable $x$ you want to compute

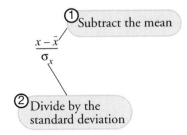

for each row in a data table.

🖲  For this example, open the Students.jmp data table.

It has a column called **weight**, and you want a new column that uses the above formula to generate standardized weight values.

🖲  Begin by choosing **Cols > New Column**, which displays a New Column dialog like the one shown in **Figure 4.2**.

The New Column dialog lets you set the new column's characteristics.

🖲  Type the new name, Std. Weight, in the **Column Name** area.

The other default column characteristics define a numeric continuous variable and are correct for this example.

**Figure 4.2**  The New Column Dialog

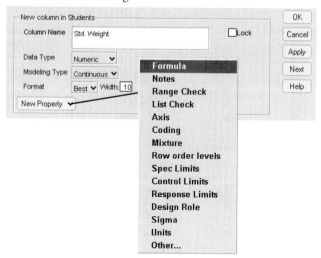

🖰    Select **Formula** from the **New Property** popup menu.

This opens the Formula Editor window shown in **Figure 4.1**.

Next, enter the formula that standardizes the weight values by following the steps in **Figure 4.3**.

**Figure 4.3**   Entering a formula

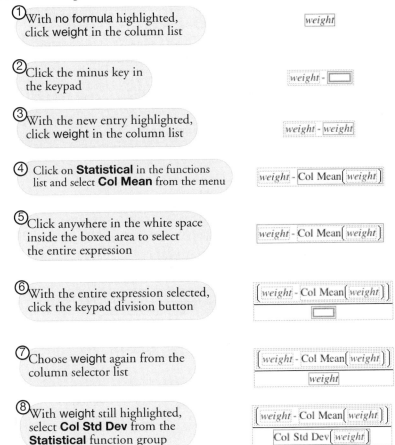

① With no formula highlighted, click **weight** in the column list

② Click the minus key in the keypad

③ With the new entry highlighted, click **weight** in the column list

④ Click on **Statistical** in the functions list and select **Col Mean** from the menu

⑤ Click anywhere in the white space inside the boxed area to select the entire expression

⑥ With the entire expression selected, click the keypad division button

⑦ Choose **weight** again from the column selector list

⑧ With weight still highlighted, select **Col Std Dev** from the **Statistical** function group

You have now entered your first formula.

☞   Close the Formula Editor window by clicking the **OK** button.

The new column fills with values. If you change any of the weight values, the calculated Std. Weight values automatically recompute.

If you make a mistake entering a formula, choose **Undo** from the **Edit** menu. **Undo** reverses the effect of the last command. There are other editing commands to help you modify formulas, including **Cut**, **Copy**, and **Paste**. The Delete key removes selected expressions. If you need to rearrange terms or expressions, you can select and drag to move formula pieces.

This example may be all you need to proceed. However, the rest of the chapter covers details about the Formula Editor, and gives a variety of other examples. Complete documentation of the Formula Editor is found in the *JMP User's Guide*.

# Formula Editor: Pieces and Parts

This section begins with Formula Editor terminology used when discussing examples. The Formula Editor has distinct areas, so we also describe its geography. This section gives a brief description of all the function categories. Later sections give examples of specific functions.

## Terminology

The following list is a glossary of terminology used in discussions about the Formula Editor.

- A *function* is a mathematical or logical operation that performs a specific action on one or more arguments. Functions include most items in the function browser and all keypad operators.

- An *expression* is a formula (or any part of it) that can be highlighted as a single unit, including terms, empty terms, and functions grouped with their arguments.

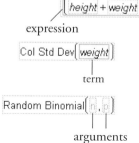

- A *term* is an indivisible part of an expression. Constants and variables are terms.

- An *argument* is a constant, a column, or expression (including mathematical operands) that is operated on by the function.

- An *empty term* is a placeholder for an expression, represented by a small empty box.

- A *missing value* in a table cell shows as a missing value mark (a large dot) for numeric data, or a null character string for character data.

**Figure 4.4**   Missing Values

missing string

missing numeric value

## The Formula Editor Control Panel

The top part of the Formula Editor is called the *control panel*. It is composed of buttons and selection lists as illustrated in **Figure 4.5**.

**Figure 4.5**   The Formula Editor Control Panel

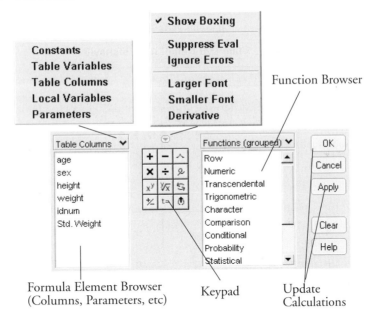

Function Browser

Formula Element Browser (Columns, Parameters, etc)

Keypad

Update Calculations

Some of the Formula Editor features such as those in the keypad are like those on a hand-held calculator. Other features are unique to the JMP Formula Editor. Here is a brief description of the Formula Editor control panel areas shown in **Figure 4.5**.

The *Formula Element Browser* displays selection lists of table columns, constants, variables or parameters. By default, the list of table columns is visible. You can change the kind of elements listed by choosing from the popup menu at the top of the formula element browser.

To choose a formula element, select an expression in the formula editing area, then click the select an element in the function list (see **Figure 4.5**).

The *keypad* is a set of buttons used to build formulas. Some of the buttons, such as the arithmetic operators, are familiar. Others have special functions, described in the next section.

The *function browser* groups collections of functions and features in lists organized by topic. To enter a function in a formula, select an expression and click any item in one of the function browser topics. You can also see a list of ungrouped functions in alphabetical order by selecting **Functions (All)** from the popup menu above the function list.

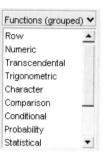

The **Help** button displays a help window with information to help you use the Formula Editor.

The spreadsheet columns automatically fill with calculated values whenever you change a formula and then close the Formula Editor window or make it inactive. Use the **Apply** button to calculate a column's values if you want the Formula Editor window to remain open.

The popup menu above the keypad (shown in **Figure 4.5**) has these commands.

- **Show Boxing** outlines specific terms within the formula. Boxing is important when you want to select and modify a specific portion of a formula, or need to determine the order of evaluation that takes place.

- **Suppress Eval** suppresses formula evaluation unless you specifically click **Apply** on the Formula Editor. This is a useful formula development mode for building complex formulas. You can turn off evaluation and build sections of a formula, and evaluate only to test it. In particular, you can close the Formula Editor and reopen it at a later time to continue building a formula without the formula attempting to evaluate.

- **Ignore Errors** suppresses error messages while a formula is under development. This is useful in situations where you want to see an evaluation for some rows and don't want to see an error message for every row where the formula evaluation finds problems. If you don't select **Ignore Errors**, an error message dialog appears when there is an error and asks if you want to ignore further errors. This has the same effect as the **Ignore Errors** menu selection.

- **Larger Font** increases the font size of the displayed formula.

- **Smaller Font** decreases the font size of the displayed formula.

- **Derivative** takes the first derivative of the entire formula. To use this command, first select a variable for the derivative to be taken with respect to. Then, select the **Derivative** command from the menu. This procedure is illustrated in **Figure 4.6**.

**Figure 4.6**   Derivatives in the Formula Editor

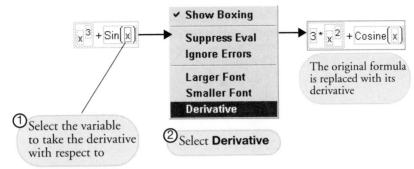

# The Keypad Functions

The keypad is composed of common operators (referred to as *functions* in all Formula Editor documentation). Enter a keypad function by selecting an expression in the formula display and clicking on the appropriate keypad buttons.

### Arithmetic keys

The four arithmetic functions work as they normally do on a pocket calculator.

### Insert and Delete keys

The **Insert** button inserts a new empty formula clause or function argument. To insert a clause into a formula, first select the existing clause or argument you want the new element to follow. When you click the **Insert** button, the new clause appears and is selected. You can also insert a new clause or argument by typing a comma. The **Delete** button empties the selected box, or if it is already empty removes it as an argument.

### Raise to a Power

The general exponential function raises a given value to a specified power. It has an exponent of two by default. Select the exponent and double click on it to change its value.

### Root

The root function calculates the specified root of the radicand. It has an implied index of 2, which is not displayed. To change the index to another value, highlight the argument of the root (the part under the radical) and hit the Insert key. A box appears where you can enter the desired index value.

### Switch Terms

The switch terms keypad function looks at the operator that is central to the selected expression and switches the expressions on either side of that operator.

### Unary Sign Function

The unary sign function inverts the sign of its argument. Apply the function to a selected variable expressions or use it to enter negative constants.

### Local Variable Assignment Key

This keypad function creates a local variable and assigns it the value of the selected expression. Its value can be as simple as an empty term, or as complicated as a complex formula.

### Peel Expression

To use this function, begin by selecting any expression. When you click the **Delete** button, the selected expression is deleted leaving a selected empty term in its place. This process repeats each time the key is clicked. In this way, you can delete a formula term by term, in the precedence order of the formula, beginning with the first term you select. See the section "Tips on Editing a Formula" on page 86 for a demonstration of peeling expressions.

# The Formula Display Area

The formula display is the area where you build and view a formula. To compose a formula, select expressions in the formula display and apply functions and terms from the formula control panel.

Functions always operate upon selected expressions, terms always replace selected expressions, and arguments are always grouped with functions. To find which expressions serve as a function's arguments, select that function in the formula. When the **Show Boxing** option is in effect, the boxed groupings also show how order of precedence rules apply and show which arguments will be deleted if you delete a function.

# Function Browser Definitions

The function browser groups the Formula Editor functions by topic. To enter a function, highlight an expression and click any item in the function browser topics. Examples of some commonly-used functions are included later in this chapter.

The function categories are briefly described in the following list. They are presented in the order you find them in the function browser.

- **Row** lists miscellaneous functions such as **Lag**, **Dif**, **Subscript**, **Row** (the current row number), and **NRow** (the total number of rows).

- **Numeric** lists commonly-used functions such as **Round**, **Floor**, **Ceiling**, **Modulo** and **Absolute Value**.

- **Transcendental** supports logarithmic functions for any base, functions for combinatorical calculations, the beta function, and several gamma functions.

- **Trigonometric** lists the standard trigonometric and hyperbolic functions such as sine, cosine, tangent, inverse functions, and their hyperbolic equivalents.

- **Character** lists functions that operate on character arguments for trimming, finding the length of a string, and changing numbers to characters or characters to numbers.

- **Comparison** are the standard logical comparisons such as less than, less than or equal to, not equal to, and so forth.

- **Conditional** are the logical functions **Not**, **And**, and **Or**. They also include programming-like functions such a **If/then/else**, **Match**, and **Choose**.

- **Statistical** lists functions that calculate standard statistical quantities such as the mean or standard deviation, both down columns or across rows.

- **Probability** lists functions that compute probabilities and quantiles for the Beta, chi-square, *F,* gamma, Normal, Student's *t,* and a variety of other distributions.

- **Random** is a collection of functions that generate random numbers from a variety of distributions.

- **Date Time** are functions that require arguments with the date data type, which is interpreted as the number of seconds since January 1, 1904. You assign Date as the Data Type in the New Column or Column Info dialog. Date functions return values such as day, week, or month of the year, compute dates, and can find date intervals.

- **Row State** lists functions that assign or detect special row characteristics called row states. Row states include color, marker, label, hidden (in plots), excluded (from analyses), and selected or not selected.

- **Assignment** functions work in place. That is, the result returned by the operation (on the right of the operator) is stored in the argument on the left of the operator and replaces its current value. They are named constants that you create and can use in any formula.

## Row Function Examples

To do the next examples, create an empty data table and insert some rows and columns.

🖰 Choose **File > New > Data Table**. (Choose **File > New** on the Macintosh)

🖰 When the new table appears, choose **Rows** > **Add Rows** and ask for ten rows.

🖰 Choose **Cols > Add Multiple Columns** and ask for 9 new columns.

The first category in the function browser is called **Row**. When you click **Row** in the function browser, you see the list of functions shown to the right. These functions are very commonly used.

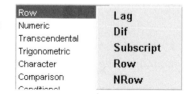

### Lag(column,1)

The Lag function returns the value of the first argument in the row defined by the current row less the second argument. The default lag is one, which you can change to any number. The value returned for any lag that identifies a row number less than one is missing. Note that $\text{Lag}(X,n)$ gives the same result as the subscripted notation, $X_{\text{Row}()-n}$. But Lag is more general, supporting entire expressions as well as simple column names.

### Row()

is the current row number when an expression is evaluated for that row. You can incorporate this function in any expression, including those used as column name subscripts (discussed in the next section). The default subscript of a column name is the current row number unless otherwise specified.

### Dif(column,n)

Returns the difference between the value of the column in the current row and the value $n$ rows previous.

### NRow ()

is the total number of rows in the data table.

### Subscript

enables you to use a column's value from a row other than the current row. Highlight a column name in the formula display and click **Subscript** to display a placeholder for the subscript. The placeholder can be changed to any numeric expression. Subscripts that evaluate to nonexistent row numbers produce missing values. A column name without a subscript refers to the current row. To remove a subscript from a column, select the subscript and delete it. Then delete the empty box that remains. The formula

$$\text{Count}_{\text{Row}()} - \text{Count}_{\text{Row}()-1}$$

uses subscript of $Row()-1$ to calculate the difference of two successive rows in a column named Count. The following formula calculates values for a a column called Fib, which, after the formula is evaluated, contains the terms of the Fibonacci series (each value is the sum of the two preceding values in the calculated column).

$$\text{If}\begin{cases} Row()<=2 \Rightarrow 1 \\ \text{else} \qquad \Rightarrow Fib_{Row()-1} + Fib_{Row()-2} \end{cases}$$

The first two rows have the value 1.

Each other row is the sum of the previous two rows.

It shows the use of subscripts to do recursive calculations. A recursive formula includes the name of the calculated column, subscripted such that it references previously evaluated rows.

## Using a Subscript

Use a subscript to refer to a specific row of the subscripted column. A simple use for subscripts is to create a lag variable. A *lag variable* is a column whose rows have values that are previous values of another column. Follow these steps to create a lag variable:

- Give the first column in the empty data table the name RowID.

- With the column selected, choose **Cols** > **Column Info** and select Formula from the list of column properties, as shown previously in **Figure 4.2**.

- Click **Row** in the Function Browser, select **Row** from its list of functions, and click **OK** to close the Formula Editor window.

- Name the second column TotRows, open the Formula Editor, select **NRow** from the list of **Row** functions, and click **OK**.

- Name the third column Lag.

Build the lag formula with these steps for the Lag column.

- Open the formula editor.

- Click Row ID in the list of columns.

- Select **Subscript** from the list of **Row** functions.

- Click the minus sign on the Formula Editor keypad, or key in a minus from the keyboard.

🖑  Click **Row** from the **Rows** functions for the empty term on the left of the minus sign in the subscript and type "1" in the empty term on the right of the minus sign.

You should now see the lag formula $RowID_{Row()-1}$.

🖑  Click **OK** on the Formula Editor to see the data table results shown in **Figure 4.7**.

Do not be alarmed with the invalid row number error message. Simply click Continue to continue with the formula evaluation. The Lag function refers to the row previous to the current row; at row 1, this results in referring to the non-existent row 0, which produces a missing value in the data table.

Note that the values in RowID and Lag are functions of individual rows, but the constant value in the TotRows column is a function of the data table.

**Figure 4.7**   Formula Example

| | RowID | TotRows | Lag |
|---|---|---|---|
| 1 | 1 | 10 | • |
| 2 | 2 | 10 | 1 |
| 3 | 3 | 10 | 2 |
| 4 | 4 | 10 | 3 |
| 5 | 5 | 10 | 4 |
| 6 | 6 | 10 | 5 |
| 7 | 7 | 10 | 6 |
| 8 | 8 | 10 | 7 |
| 9 | 9 | 10 | 8 |
| 10 | 10 | 10 | 9 |

# Conditional Expressions and Comparison Operators

This function category has many familiar programming functions. This section shows examples of conditionals used with comparison operators.

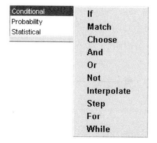

The most basic and general conditional function is the If function. Its arguments are called if, then, and else *clause*s. When you highlight an expression and click **If**, the Formula Editor creates a new conditional expression like the one shown to the right. It has one If argument (a conditional expression denoted *expr*) and one then argument. A conditional expression is usually a comparison, like $a < b$. However, any expression that evaluates as a numeric value can be used as a conditional expression. Expressions that evaluate as zero or missing are false. All other numeric expressions are true.

If you need more than one then statement, click the insert icon on the keypad (or type a comma, its keyboard equivalent) to add a new argument. Use the delete icon on the keypad (or hit the Delete key on your keyboard) to remove unwanted arguments.

## Using the If function

To create your own Fibonacci sequence:

- Name a blank data table column Fib.

- Right-click (Control-click on the Macintosh) and select **Formula** from the resulting menu.

Enter the Fibonacci formula using the steps shown in **Figure 4.8**.

**Figure 4.8**   Entering the Fibonacci formula

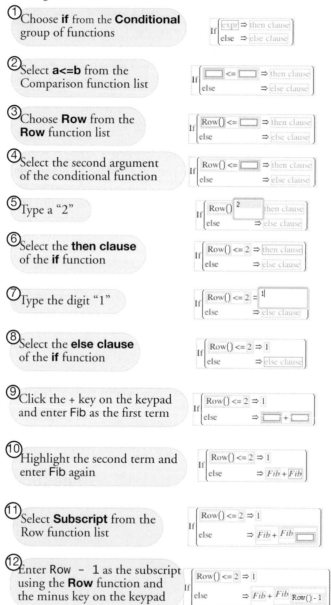

① Choose **if** from the **Conditional** group of functions

② Select **a<=b** from the Comparison function list

③ Choose **Row** from the **Row** function list

④ Select the second argument of the conditional function

⑤ Type a "2"

⑥ Select the **then clause** of the **if** function

⑦ Type the digit "1"

⑧ Select the **else clause** of the **if** function

⑨ Click the + key on the keypad and enter Fib as the first term

⑩ Highlight the second term and enter Fib again

⑪ Select **Subscript** from the Row function list

⑫ Enter Row - 1 as the subscript using the **Row** function and the minus key on the keypad

When you close the Formula Editor by clicking **OK**, the formula you entered generates the values shown in **Figure 4.9**.

**Figure 4.9**   Results of the Formula Example

| | RowID | TotRows | Lag | Fib | Group |
|---|---|---|---|---|---|
| 1 | 1 | 10 | • | 1 | 1 |
| 2 | 2 | 10 | 1 | 1 | 0 |
| 3 | 3 | 10 | 2 | 2 | 1 |
| 4 | 4 | 10 | 3 | 3 | 0 |
| 5 | 5 | 10 | 4 | 5 | 1 |
| 6 | 6 | 10 | 5 | 8 | 0 |
| 7 | 7 | 10 | 6 | 13 | 1 |
| 8 | 8 | 10 | 7 | 21 | 0 |
| 9 | 9 | 10 | 8 | 34 | 1 |
| 10 | 10 | 10 | 9 | 55 | 0 |

The Fibonacci sequence has many interesting and easy to understand properties that you can find discussed in number theory textbooks.

🖰   For practice, create the values in the Group column shown in **Figure 4.9**. Use **Modulo** from the **Numeric** functions with RowID as its argument, as shown to the right.

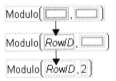

## Using the Match Function

A common use for conditional functions is to re-code variables.

Often, a numeric coding variable represents a descriptive character value. The following example uses the Match function for re-coding (the If function could also be used for re-coding).

When you select **Match** from the **Conditional** list, the Formula Editor shows a single Match condition with an empty expression and an empty then term. You add and

delete clauses in a Match conditional the same way as in the **If** conditional described previously: select a then clause and click the add or delete button. The Match conditional compares an expression to a list of clauses and returns the value of the result expression for the first matching argument encountered. With Match, you provide the matching expression only once and then give a match for each argument.

As an example, open the Hot dogs.jmp data table. Suppose you want to create a value to re-code the protein/fat ratio into categories "Very Lean", "Lean", "Average", and so on (see the formula below).

🖰   Create a new character column and use the following formula.

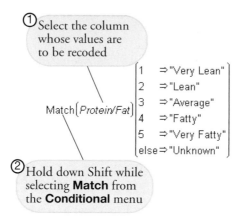

① Select the column whose values are to be recoded

$$\text{Match}\left(\textit{Protein/Fat}\right)\begin{bmatrix} 1 & \Rightarrow\text{"Very Lean"} \\ 2 & \Rightarrow\text{"Lean"} \\ 3 & \Rightarrow\text{"Average"} \\ 4 & \Rightarrow\text{"Fatty"} \\ 5 & \Rightarrow\text{"Very Fatty"} \\ \text{else} & \Rightarrow\text{"Unknown"} \end{bmatrix}$$

② Hold down Shift while selecting **Match** from the **Conditional** menu

**Note:** `Match` evaluates faster and uses less memory than an equivalent `If`.

## Summarize Down Columns or Across Rows

The Formula Editor evaluates statistical functions differently from other functions. Most functions evaluate data for the current row only. However, all **Statistical** functions require a set of values upon which to operate. Some **Statistical** functions compute statistics for the set of values in a column, and other functions compute statistics for the set of arguments you provide.

**Figure 4.10**   Statistical Functions

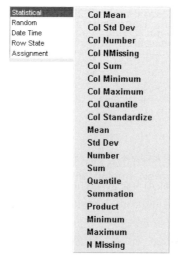

The functions with names prefaced by "Col" (Col Mean, Col Sum, and so on) always evaluate for all of the rows in the column. Thus, used alone as a column formula, these functions produce a constant value for each row. The "Col" functions accept only a single argument, which can be a column name, or an expression involving any combination of column names, constants, or other functions.

The other statistical functions (Mean, Std Dev, and so on) accept multiple arguments that can be variables, constants, and expressions.

The Sum and Product functions always evaluate for an explicit range of values that you specify.

### The Quantile Function

The Col Quantile function computes a quantile for a column of $n$ nonmissing values. The Col Quantile function's quantile argument (call it $p$) represents the quantile percentage divided by 100.

The $p$th quantile value is calculated using the formula $I = p(N + 1)$ where $p$ is the quantile and $N$ is the total number of nonmissing values. If $I$ is an integer, then the quantile value is $y_p = y_i$. If $I$ is not an integer, then the value is interpolated by assigning the integer part of the result to $i$, and the fractional part to $f$, and by applying the formula

$$q_p = (1 - f)y_i + (f)y_{i+1}$$

The following are example quantile formulas for a column named age.

Col Quantile (age, 1) finds the maximum age.

Col Quantile (age, 0.75) calculates the upper quartile age.

Col Quantile (age, 0.5) calculates the median age.

Col Quantile (age, 0.25) calculates the lower quartile age.

Col Quantile (age, 0.0) calculates the minimum age.

### Using the Summation Function

$$\sum_{i=1}^{NRow()} body$$

The **Summation** ($\Sigma$) function uses the summation notation shown to the left. To calculate a sum, select **Summation** from the **Statistical** function list and choose a variable or create an expression as its argument. The **Summation** function repeatedly evaluates the expression for the index you apply to the body of the function from the lower summation limit specified to the upper limit, and then adds the nonmissing results together to determine the final result. You can replace the index $i$, the index constant 1, and the upper limit, NRow(), with any expressions appropriate for your formula. Use the **Subscript** function in the **Row** function category to create a subscript for the body of the summation.

For example, the summation shown to the right computes the total of all revenue values for row 1 through the current row number, filling the calculated column with the cumulative totals of the revenue column.

$$\sum_{i=1}^{NRow()} Revenue_i$$

Let's see how to compute a moving average using the summation function.

- ⌐ Open the XYZ Stock Averages.JMP sample table.

- ⌐ Create a new column called Moving Avg and select **Formula** from the New Properties list in the New Column.

- ⌐ Use the dialog to change the format from **Best** to **Fixed Dec** to help make give the numeric representation in the data table two decimal places.

When you specify Formula as a new column property and click **Edit Formula**, the new column appears in the table with missing values, and its Formula Editor window opens. You should see a table like the one shown in **Figure 4.11**.

**Figure 4.11**   Example Table for Building a Moving Average

| | Date | DJI High | DJI Close | DJI Low | XYZ | Moving Avg |
|---|---|---|---|---|---|---|
| 1 | 04/15/1991 | 2957.18 | 2933.17 | 2896.29 | 62.250 | • |
| 2 | 04/16/1991 | 2995.79 | 2986.88 | 2912.13 | 64.250 | • |
| 3 | 04/17/1991 | 3030.45 | 3004.46 | 2963.12 | 63.250 | • |
| 4 | 04/18/1991 | 3027.72 | 2999.26 | 2976.24 | 61.000 | • |
| 5 | 04/19/1991 | 3000.25 | 2965.59 | 2943.56 | 59.625 | • |

A *moving average* is the average of a fixed number of consecutive values in a column, updated for each row. The following example shows you how to compute a 10-day moving average for the XYZ stock. This means that for each row the Formula Editor computes of the sum of the current XYZ value with the 9 preceding values, then divides that sum by 10.

- ⌐ Because you only want to compute the moving average starting with the 10th row, begin by selecting the conditional If function.

- ⌐ With the If expression highlighted, select **a>b** from the **Comparison** function category, which will be used to determine the row number.

🖱 For the left side of the comparison, select **Row** from the **Row** functions.

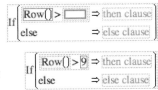

🖱 Highlight the right side of the comparison and type in the number 9. The If expression should now appear as Row()>9.

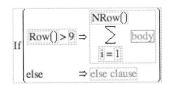

🖱 Now highlight the then clause and begin the formula to compute the ten-day moving average by selecting the **Summation** function from the **Statistical** function category. Highlight the body of the summation and click XYZ in the column selector list.

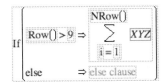

Now tailor the summation indices to sum just the 10 values you want:

🖱 Highlight the summation body, XYZ.

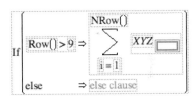

🖱 Select **Subscript** from the **Row** function category.

An empty subscript now appears with the summation body.

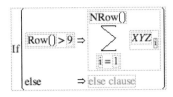

🖱 To assign the subscript either key the letter "i", or drag the "i" from the lower limit of the summation into the empty subscript.

🖱 Highlight the 1 in the lower summation limit and press the Delete key to change it to an empty term.

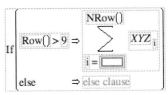

🖰 Enter the expression Row()–9 inside the
parentheses, using the **Row** selection in the
**Row** function category.

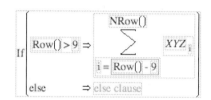

🖰 Click the upper index to highlight it and
select **Row** from the **Row** functions.

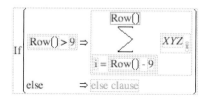

To finish the moving average formula, you want to
divide the sum by 10 but not start the averaging
process until you actually have 10 values to work
with.

🖰 Click in the summation to highlight the whole
summation expression.

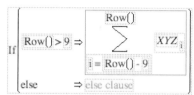

🖰 Click the divide operator on the control
panel, and then enter the constant 10 into
the highlighted denominator that appears.

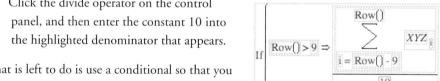

All that is left to do is use a conditional so that you
don't compute anything for the first 9 values in the
table:

🖰 When you click **Apply** or close the Formula Editor, the Moving Avg column fills with
values.

Now generate a plot to see the result of your efforts:

🖰 Choose **Graph > Overlay Plot**, and select Date as X and both XYZ and Moving Avg as
Y.

🖰 When you click **OK**, select **Y Options > Connect Points** from the platform drop-
down menu.

You then see the plot in **Figure 4.12**, which compares the XYZ stock price with its ten-day
moving average.

**Figure 4.12**   Plot of Stock Prices and Their Moving Average

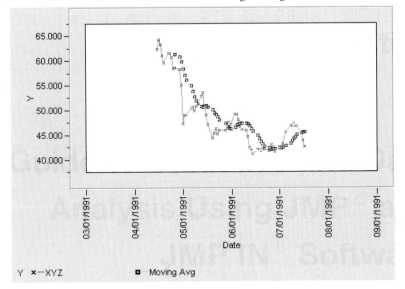

## Random Number Functions

Random number functions generate real numbers by essentially "rolling the dice" within the constraints of the specified distribution. You can use the random number functions with a default 'seed', that provides a pseudo-random starting point for the random series, or use the `Random Reset` function and give a specific starting seed.

Random numbers in JMP are calculated with the Mersenne-Twister technique, far more efficient than in older versions of JMP. However, if you want to use the older functions, they

are provided as `RandomSeededNormal` and `RandomSeededUniform`. Their seed is set using the `Random Seed` function.

Each time you click **Apply** in the Formula Editor window, random number functions produce a new set of numbers. This section shows examples of two commonly used random functions, `Uniform` and `Normal`.

### The Uniform Distribution

The `Uniform` function generates random numbers uniformly between 0 and 1. This means that any number between 0 and 1 is as likely to occur as any other. The result is an approximately even distribution. You can use the `Uniform` function to generate any set of numbers by modifying the function with the appropriate constants.

You can see simulated distributions using the `Uniform` function and the Distribution platform.

 🖰  Choose **File > New**, to create a new data table.

 🖰  Select (highlight) Column 1 and choose **Cols > Column Info**.

 🖰  When the Column Info dialog appears, select **Formula** from the New Properties popup menu and click **Edit Formula** button.

 🖰  When the Formula Editor window opens, select **Random Uniform**, from the **Random** function list in the function browser, and then close the Formula Editor.

 🖰  Choose **Cols > New Column** to create a second column.

Follow the same steps as before except modify the Uniform function to generate the integers from 1 to 10 as follows.

 🖰  Click **Random** in the function browser and select **Random Uniform** from its list.

 🖰  Click the multiply sign on the Formula Editor key pad and enter 10 as the multiplier

 🖰  Select the entire formula and click the addition sign on the Formula Editor keypad

 🖰  Enter 1 in the empty argument term of the plus operator.

> **Note:** JMP has a `Random Integer(n)` function that selects integers from a uniform distribution from 1 to **n**. It could be used here for the same effect. We're using the `Random Uniform` function to illustrate how to manipulate a random number by multiplying and adding constants. You can see an example of the `Random Integer` function in "Rolling Dice" on page 104.

The next steps are the key to generating a uniform distribution of integers (as opposed to real numbers as in Column 1):

🖱   Click to select the entire formula.

🖱   Select the **Floor** function from the **Numeric** function list.

The final formula is

$$Floor(Random\ Uniform()*10+1)$$

🖱   Close the Formula Editor.

You now have a table template for creating two uniform distributions. All you have to do is add as many columns as you want.

🖱   Choose **Rows > Add Rows** and add 500 rows.

The table will fill with values.

🖱   Change the modeling type of the column of integers to nominal so it will be treated as a discrete distribution.

🖱   Choose **Analyze > Distribution**, use both columns as Y variables, and then click **OK**.

You will see two histograms similar to those shown in **Figure 4.13**. The histogram on the left represents simple uniform random numbers and the histogram on the right shows random integers from 1 to 10.

**Figure 4.13**   Example of Two Uniform Distribution Simulations

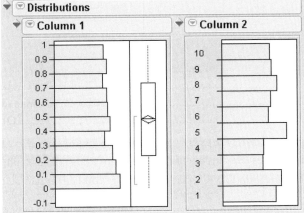

## The Normal Distribution

**Random Normal** generates random numbers that approximate a Normal distribution with a mean of 0 and variance of 1. The Normal distribution is bell-shaped and symmetrical. You can modify the **Random Normal** function with arguments that specify a Normal distribution with a different mean and standard deviation.

As an exercise, follow the same instructions described previously for the Uniform random number function.

🖰 Create a table with a column for a standard Normal distribution using the `Random Normal()` function.

The Random Normal function takes two optional arguments. The first specifies the mean of the distribution; the second, specifies the standard deviation of the distribution.

🖰 Create a second column for a random Normal distribution with mean 30 and standard deviation 5.

The modified Normal formula is

$$\text{Random Normal(30, 5)}$$

**Figure 4.14** shows the Distribution platform for these Normal simulations.

**Figure 4.14**    Illustration of Normal Distributions

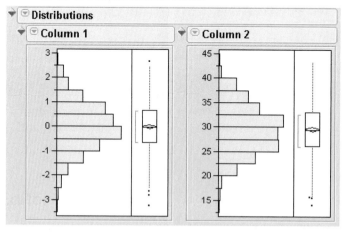

### The Col Shuffle Command

**Col Shuffle** selects a row number at random from the current data table. Each row number is selected only once. When **Col Shuffle** is used as a subscript, it returns a value selected at random from the column that serves as its argument.

For example, to identify a 50% random sample without replacement, use the following formula:

$$\text{If}\left(\text{Row}() < \frac{\text{NRow}()}{2} \Rightarrow Column\ 1_{\text{Col Shuffle}()}\right)$$

This formula chooses half the values ($n/2$) from the column Column 1 and assigns them to the first half of the rows in the computed column. The remaining rows of the computed column fill with missing values.

### Local Variables and Table Variables

*Local variables* lets you define temporary numeric variables to use in expressions. Local variables exist only for the column in which they are defined.

To create a new local variable, use the $\boxed{\text{t=}}$ button on the formula editor keypad. This button adds a temporary local variable to the formula editing area, which appears as a command ending in a semicolon. Alternatively, you can select **Local Variables** from the Formula Elements popup menu, select **New Local** and complete the dialog that appears.

By default, local variables have the names t1, t2, and so on, and have missing values. Local variables appear in a formula as bold italic terms.

Optionally, you can create a local variable, change a local variable name and assign a starting value in the Local Variable dialog. To use the Local Variable dialog, select **Local Variables** from the Formula Elements popup menu and complete the dialog, as illustrated here.

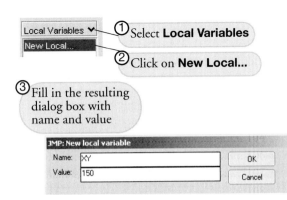

As an example, suppose you have variables $x$ and $y$ and you want to compute the slope in a simple linear regression of $y$ on $x$ using the standard formula shown here.

$$\frac{\sum(x-\bar{x})(y-\bar{y})}{\sum(x-\bar{x})^2}$$

One way to do this is to create two local variables, called XY and Xsqrd, as described in the numerator and denominator in the equation above. Then assign them to the numerator and the denominator calculations of the slope formula. The slope computation is simplified to XY divided by Xsqrd.

$$XY= \sum_{i=1}^{NRow()} \left[\left(X_i\text{-Col Mean}(X)\right)*\left(Y_i\text{-Col Mean}(Y)\right)\right];$$

$$Xsqrd= \sum_{i=1}^{NRow()} \left(X_i\text{-Col Mean}(X)\right)^2;$$

$$\frac{XY}{Xsqrd};$$

The **Local Variables** command in the Formula Editor popup menu lists all the local variables you create.

*Table variables* are available to the entire table. Table variable names are displayed in the Tables panel at the left of the data grid. The Formula Editor can refer to a table variable in a formula.

Many of the sample data files have a table variable called Notes. The **Table Variables** command in the Formula Elements popup menu lists all the Table variables that exist for a table. You can create additional Table variables with the **New Table Variable** command in the Tables panel of the data table, or edit the values of existing Table Variables.

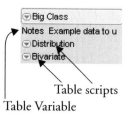

Table scripts

Table Variable

# Tips on Building Formulas

## Examining Expression Values

Once JMP has evaluated a formula, you can select an expression to see its value. This is true for both parameters and expressions that evaluate to a constant value. To do this, select the expression you want to know about and right-click(PC) or Control-click(Mac) on it. This displays a popup menu as shown here. When you select **Evaluate**, the current value of the selected expression shows until you move the cursor.

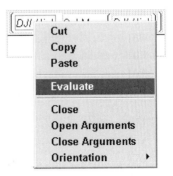

## Cutting, Dragging, and Pasting Formulas

You can cut or copy entire formulas or an expression, and paste it into another formula display. Or, you can drag any selected part of a formula to another location within the same formula. When you place the arrow cursor inside an expression and click, the expression is highlighted. When the cursor is over a selected area, it changes to a hand cursor. As you drag across the formula, destination expressions are highlighted, and then you can drag the selected expression to the new location you want. When you drag, the selected expression is copied to the new location where it replaces the existing expression when the mouse button is released.

When you copy (or drag) an expression from one data table to another, it expects to find the same column names. If a formula column name does not appear in the table, an error alerts you when the formula attempts to evaluate.

## Selecting Expressions

You can use the keyboard arrow keys to select expressions for editing or to view the grouping of terms within a formula when parentheses are not present or the boxing option is not in effect. You can click on any single term in an expression to select it for editing.

Once an operand is selected, the left and right arrow keys move the selection across other associative operands within the expression. The left arrow highlights the next formula element to the left of the currently highlighted term, or extends the selection to include an additional term that is part of a group.

**Tip:** Turn the boxing option on to see how the elements and terms are grouped in a formula you create. It is often easier to leave the boxing option on while creating a formula.

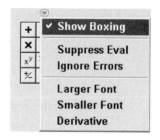

The up arrow extends the current selection by adding the next operand and operator of the formula term to the selection. The down arrow reduces the current selection by removing an operand and operator from the selection.

## Tips on Editing a Formula

If you need to change a formula, highlight the column and select **Formula** from the **Cols** menu. Alternatively, right-click at the top of the column or on the column name in the Columns panel and select **Formula** from the context menu that appears.

Deleting a function also deletes its arguments. Deleting a required argument or empty term from a function sometimes deletes the function as well. You can save complicated expressions that serve as arguments and paste them where needed. Use the **Copy** command to copy the arguments to the clipboard.

Another editing technique that is sometimes useful is to peel a function from its argument as shown in **Figure 4.15**. To peel a function from a single argument, first click to select the function. Then, choose the **peel** button from the keypad, as shown in **Figure 4.15**.

After you complete formula changes, the new values fill the column automatically when you click **Apply** or close the Formula Editor window.

Once you have created a formula, you can change values in columns that are referenced by your formula. JMP automatically recalculates all affected values in the formula's column.

**Figure 4.15**   Peeling Expressions or Arguments

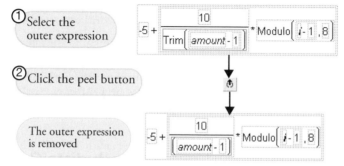

# Exercises

1.  The file Pendulum.jmp contains the results of an experiment in a physics class comparing the length of a pendulum to its period (the time it takes the pendulum to make a complete swing). Calculations were made for very short pendulums (2 cm) up to very long (20 m) pendulums. We will use the calculator to determine a model to predict the period of a pendulum from its length.

    (a) Produce a scatterplot of the data by selecting **Analyze > Fit Y By X** and selecting Period as the Y variable and Length as the X Variable

    (b) Create a new column named Transformed Period that contains a formula to take the square root of the Period column. Produce a scatterplot of Transformed Period vs. Length. Is this the graph of a linear function?

    (c) Try other transformations (natural log of Period, reciprocal of Period, or square of Period, for example) until the scatterplot looks linear.

    (d) Find the line of best fit for the linear transformed data by selecting **Fit Line** from the popup menu beside the title of the scatterplot.

    (e) This line is not the fit of the original data, but of the transformed data. Substitute a term representing the transformation you did to linearize the data (for example, if the square root transformation made the data linear, substitute $\sqrt{\text{Period}}$ into the regression equation) and solve the equation for Period.

    A Physics textbook reveals the relationship between period and length to be

    $$\text{Period} = \frac{2\pi}{\sqrt{g}}\sqrt{\text{Length}} \text{ where } g\text{=}9.8\frac{m}{s^2}.$$

    (f) Create a new column to use this formula to calculate the theoretical values. then, construct another column to calculate the difference between the observed values of the students and the theoretical values.

    (g) Examine a histogram of these differences. Does it appear that there was a trend in the observations of the students?

2.  Is there a correlation among the mean, minimum, maximum, and standard deviation of a set of data? To investigate, create a new data table in JMP with these characteristics:

    (a) Create ten columns of data named Data 1 through Data 10, each with the formula Random Uniform(). Add 500 rows to this data table.

(b)  Create four columns to hold the four summary statistics of interest, one each for the mean, minimum, maximum, and standard deviation. Create a formula in each column to calculate the appropriate statistic of the ten data rows.

(c)  Select the Multivariate platform in the **Analyze** menu and include the four summary statistics as the Y's in the resulting dialog box. Pressing **OK** should produce 16 scatterplots. Which statistics seem to show a correlation?

(d)  As an extension, select two of the statistics that seem to show a correlation. Produce a single scatterplot of these two statistics using the Fit Y By X platform in the **Analyze** menu. From the red arrow drop-down menu beside the title of the plot, select **Nonpar Density**. Then choose **Save Density Grid** from the popup menu beside the density legend.

(e)  Finally, select **Graph > Spinning Plot** and include the first three columns of the saved density grid. You should now see a 3D scatterplot of the correlation, with a peak where the data points overlapped each other. Use the hand tool to move the spinning plot around.

3.  Make a data table consisting of a single column that holds the Fibonacci sequence (whose formula is shown on page 69). Label this column **Fib**.

(a)  Add a new column called **Ratio**, and give it the following formula to take the ratio of adjacent rows. **Note:** This will produce an error alert when evaluated for the first row. Click **OK** on the alert dialog and all will be well.

$$\frac{Fib}{Fib_{Row()-1}}$$

What value does this ratio converge to?

(b)  A more genearalized Fibonacci sequence uses values aside from 1 as the first two elements. A Lucas sequence uses the same recursive rule as the Fibonacci, but uses different starting values. Create a column to hold a Lucas sequence beginning with 2 and 5 called **Lucas** with the following formula.

$$\text{If} \begin{bmatrix} Row()==1 \Rightarrow 2 \\ Row()==2 \Rightarrow 5 \\ else \quad \Rightarrow Lucas_{Row()-1} + Lucas_{Row()-2} \end{bmatrix}$$

(c)   Create a column to calculate the ratio of two successive terms of the Lucas sequence in part (b). Is it the same as the number in part (a)?

(d)   Create a Lucas sequence starting with the values 1, 2. Calculate the ratio of successive terms and compare it to the answer in part (c).

(e)   There are innumerable other properties of the Fibonacci sequence. For example, add a column that contains a formula that finds the average of the first $n$ terms of the Fibonacci sequence, and comment on the result. HINT: Create a column holding the formula

$$\frac{\sum_{i=1}^{\text{Row}()} Fib}{\text{Row}()}$$

# What Are Statistics?

## Overview

Statistics are numbers, but the practice of statistics is the craft of measuring imperfect knowledge. That's one definition, and there are many more.

This chapter is a collection of short essays to get you started on the many ways of statistical thinking and to get you used to the terminology of the field.

# Ponderings

## The Business of Statistics

The discipline of statistics provides the framework of balance sheets and income statements for scientific knowledge. Statistics is an accounting discipline, but instead of accounting for money, it is accounting for scientific credibility. It is the craft of weighing and balancing observational evidence. Scientific conclusions are based on experimental data in the presence of uncertainty, and statistics is the mechanism to judge the merit of those conclusions—the statistical tests are like credibility audits. Of course, you can juggle the books and make poor science sometimes look better than it is. However, there are important phenomena that you just can't uncover without statistics.

A special joy in statistics is when it can be used as a discovery tool to find out new phenomena. There are many views of your data—the more perspectives you have on your data, the more likely you are to find out something new. Statistics as a discovery tool is the auditing process that unveils phenomena that are not anticipated by a scientific model and are unseen with a straightforward analysis. These anomalies lead to better scientific models.

Statistics fits models, weights evidence, helps identify patterns in data and then helps find data points that don't fit the patterns. Statistics is induction from experience; it is there to keep score on the evidence that supports scientific models.

Statistics is the science of uncertainty, credibility accounting, measurement science, truth-craft, the stain you apply to your data to reveal the hidden structure, the sleuthing tool of a scientific detective.

Statistics is a necessary bureaucracy of science.

## The Yin and Yang of Statistics

There are two sides to statistics.

First, there is the Yang of statistics, a shining sun. The Yang is always illuminating, condensing, evaporating uncertainty, and leaving behind the precipitate of knowledge. It pushes phenomena into forms. The Yang is out to prove things in the presence of uncertainty and ultimately compel the data to confess its form, conquering ignorance headlong. The Yang demolishes hypotheses by ridiculing their improbability. The Yang mechanically cranks through the ore of knowledge and distills it to the answer.

On the other side, we find the contrapositive Yin, the moon, reflecting the light. The Yin catches the shadow of the world, feeling the shape of truth under the umbra. The Yin is forever looking and listening for clues from the creator, nurturing seeds of pattern and anomaly into maturing discoveries. The Yin whispers its secrets to our left hemisphere. It unlocks doors for us in the middle of the night, planting dream seeds, making connections. The Yin draws out the puzzle pieces to tantalize our curiosity. It teases our awareness and tickles our sense of mystery until the climax of revelation—Eureka!

The Yin and Yang are forever interacting, catalyzed by Random, the agent of uncertainty. As we see the world reflected in the pool of experience, the waters are stirred, and in the agitated surface we can't see exactly how things are. Emerging from this, we find that the world is knowable only by degree, that we have knowledge in measure, not in absolute.

## The Faces of Statistics

Everyone has a different view of statistics.

| Match the definition on this side.... | with someone likely to have said it on this side |
|---|---|
| 1. The literature of numerical facts. | a. Engineer |
| 2. An applied branch of mathematics. | b. Original meaning |
| 3. The science of evidence in the face of uncertainty. | c. Social scientist |
| 4. A digestive process that condenses a mass of raw data into a few high-nutrient pellets of knowledge. | d. Philosopher |
| 5. A cooking discipline with data as the ingredients and methods as the recipes. | e. Economist |
| 6. The calculus of empiricism. | f. Computer scientist |
| 7. The lubricant for models of the world. | g. Mathematician |
| 8. A calibration aid. | h. Physicist |
| 9. Adjustment for imperfect measurement. | i. Baseball fan |
| 10. An application of information theory. | j. Lawyer |
| 11. Involves a measurable space, a sigma algebra, and Lebesgue integration. | k. Joe College |
| 12. The nation's state. | l. Politician |
| 13. The proof of the pudding. | m. Businessman |
| 14. The craft of separating signal from noise. | n. Statistician |
| 15. A way to predict the future. | |

An interesting way to think of statistics is as a toy for grown-ups; remember that toys are proxies that children use to model the world. Children use toys to learn behaviors and develop explanations and strategies, as aids for internalizing the external. This is the case with statistical models. You model the world with a mathematical equation, and then see how the model stacks up to the observed world.

Statistics lives in the interface of the real world data and mathematical models, between induction and deduction, empiricism and idealism, thought and experience. It seeks to balance real data and a mathematical model. The model addresses the data and stretches to fit. The model changes and the change of fit is measured. When the model doesn't fit, the data suspends from the model and leaves clues. You see patterns in the data that don't fit—this leads to a better model, and points that don't fit into patterns can lead to important discoveries.

# Don't Panic

Some university students have a panic reaction to the subject of statistics. Yet most science, engineering, business, and social science majors usually have to take at least one statistics course. What are some of the sources of our phobias about statistics?

### Abstract Mathematics

Though statistics can become very mathematical to those so inclined, applied statistics can be used effectively with only a very basic level of mathematics. You can talk about statistical properties and procedures without plunging into abstract mathematical depths. In this book, we are interested in looking at applied statistics.

### Lingo

Statisticians often don't bother to translate terms like 'heteroschedasticity' into 'varying variances' or 'multicollinearity' into 'closely related variables.' Or, for that matter, further translate 'varying variances' into 'difference in the spread of values between samples,' and 'loosely related variables' into 'variables that give similar information.' We will tame some of the common statistical terms in the discussions that follow.

### Awkward Phrasing

There is a lot of subtlety in statistical statements that can sound awkward, but the phrasing is very precise and means exactly what it says. Sometimes statistical statements include multiple negatives. For example, "The statistical test failed to reject the null hypothesis of no effect at the specified alpha level." That is a quadruple negative statement—count the negatives: 'fail,' 'reject,' 'null,' and 'no effect.' You can reduce the negatives by saying "the statistical results are not significant" as long as you are careful not to confuse that with the statement "there is no effect." Failing to prove something does not prove the opposite!

### A Bad Reputation

The tendency to assume the proof of an effect because you cannot statistically prove the absence of the effect is the origin of the saying, "Statistics can prove anything." This is what happens when you twist a term like 'nonsignificant' into 'no effect.' This idea is common in a courtroom; you can't twist the phrase "there is not enough evidence to prove beyond reasonable doubt that the accused committed the crime" with "the accused is innocent." What nonsignificant really means is that there is not enough data to show a significant effect—it does not mean that you are certain there is no effect at all.

### Uncertainty

Although we are comfortable with uncertainty in ordinary daily life, we are not used to embracing it in our knowledge of the world. We think of knowledge in terms of hard facts and solid logic, though much of our most valuable real knowledge is less than solid. We can say when we know something for sure (yesterday it rained), and we can say when we don't know (don't know whether it will rain tomorrow). But when we describe knowing something with incomplete certainty, it sounds apologetic or uncommitted. For example, it sounds like a form of equivocation to say that there is a 0.9 chance that it will rain tomorrow. Yet much of what we think we know is really just that kind of uncertainty.

# Preparations

A few fundamental concepts will prepare you for absorbing details in upcoming chapters.

# Three Levels of Uncertainty

Statistics is about uncertainty, but there are several different levels of uncertainty that you have to keep in separate accounts:

### Random Events

Even if you know everything possible about the world, you still have events you can't predict. You can see an obvious example of this in any gambling casino. You can be an expert at playing blackjack, but the randomness of the card deck renders the outcome of any game indeterminate. We make models with random error terms to account for uncertainty due to randomness. Some of the error term may be due to ignoring details; some may be measurement error; but much of it is attributed to inherent randomness.

### Unknown Parameters

Not only are you uncertain how an event is going to turn out, you often don't even know what the numbers (parameters) are in the model that generates the events. You have to estimate the parameters and test if hypothesized values of them are plausible, given the data. This is the chief responsibility of the field of statistics.

### Unknown Models

Sometimes you not only don't know how an event is going to turn out, and you don't know what the numbers are in the model, but you don't even know if the form of the model is right.

Statistics is very limited in its help for certifying that the model is correct. Most statistical conclusions assume that the model is correct. The correctness of the model is the responsibility of the subject-matter science. Statistics might give you clues if the model is not carrying the data very well. Statistical analyses can give diagnostic plots to help you see patterns that could lead to new insights, to better models.

## Probability and Randomness

In the old days, statistics texts all began with chapters on probability. Today, many popular statistics books discuss probability in later chapters. We skip the balls-in-urns entirely, though probability is the essence of our subject.

Randomness makes the world interesting and probability is needed as the measuring stick. Probability is the aspect of uncertainty that allows the information content of events to be measured. If the world were deterministic, then the information value of an event would be zero because it would already be known to occur; the probability of the event occurring would be 1. The sun rising tomorrow is a nearly deterministic event and doesn't make the front page of the newspaper when it happens. The event that happens but has been attributed to having probability near zero would be big news. For example, the event of extraterrestrial intelligent life forms landing on earth would make the headlines.

Statistical language uses the term probability on several levels:

- When we make observations or collect measurements, our responses are said to have a *probability distribution*. For whatever reason, we assume that something in the world adds randomness to our observed responses, which makes for all the fun in analyzing data that has uncertainty in it.

- We calculate statistics using probability distributions, seeking the safe position of maximum likelihood, which is the position of least improbability.

- The significance of an event is reported in terms of probability. We demolish statistical null hypotheses by making their consequences look incredibly improbable.

## Assumptions

Statisticians are naturally conservative professionals. Like the boilerplate of official financial audits, statisticians' opinions are full of provisos such as "assuming that the model is correct, and assuming that the error is Normally distributed, and assuming that the observations are independent and identically distributed, and assuming that there is no measurement error, and assuming...." Even then the conclusions are hypothetical, with phrases like "if you say the hypothesis is false, then the probability of being wrong is less than 0.05."

Statisticians are just being precise, though they sound like they are combining the skills of equivocation like a politician, techno-babble like a technocrat, and trick-prediction like the Oracle at Delphi.

### Ceteris Paribus

A crucial assumption is the *ceteris paribus* clause, which is Latin for other things being equal. This means we assume that the response we observed was really only affected by the model and random error; all other factors that might affect the response were maintained at the same controlled value across all observations or experimental units. This is, of course, often not the case, especially in observational studies, and the researcher must try to make whatever adjustments, appeals, or apologies to atone for this. When statistical evidence is admitted in court cases, there are endless ways to challenge it, based on all the assumptions that may have been violated.

### Is the Model Correct?

The most important assumption is that your model is right. There are no easy tests for this assumption. Statistics almost always measure one model against a submodel, and these have no validity if neither model is appropriate in the first place.

### Is the Sample Valid?

The other supremely important issue is that the data relate to your model; that is, that you have collected your data in a way that is fair to the questions that you will be asking it. If your sample is ill-chosen, or if you have skewed your data by rejecting data in a process that relates to its applicability to the questions, then your judgments will be flawed. If you have not taken careful consideration of the direction of causation, you may be in trouble. If taking a response affects the value of another response, then they are not independent of each other, which can affect the study conclusions.

In brief, are your samples fairly taken and are your experimental units independent?

## Data Mining?

One issue that most researchers are guilty of to a certain extent is stringing together a whole series of conclusions and assuming that the joint conclusion has the same confidence as the individual ones. An example of this is data mining, in which hundreds of models are tried until one is found with the hoped-for results. Just think about the fact that if you collect purely random data, you will find a given test significant at the 0.05 level about 5% of the time. So you could just repeat the experiment until you get what you want, discarding the rest. That's obviously bad science, but something similar often happens in published studies. This multiple-testing problem remains largely unaddressed by statistical literature and software except for some special cases such as means comparisons, a few general methods that may be inefficient (Bonferroni's adjustment), and expensive, brute-force approaches (resampling methods).

Another problem with this issue is that the same kind of bias is present across unrelated researchers because nonsignificant results are often not published. Suppose that 20 unrelated researchers do the same experiment, and by random chance one researcher got a 0.05- level significant result. That's the result that gets published.

In light of all the assumptions and pitfalls, it is appropriate that statisticians are cautious in the way they phrase results. Our trust in our results has limits.

# Statistical Terms

Statisticians are often unaware that they use certain words in a completely different way than other professionals. In the following list, some definitions will be the same as you are used to, and some will be the opposite:

### Model

The statistical model is the mathematical equation that predicts the response variable as a function of other variables, together with some distributional statements about the random terms that allow it to not fit exactly. Sometimes this model is taken very casually in order to look at trends and tease out phenomena, and sometimes the model is taken very seriously.

### Parameters

To a statistician, parameters are the unknown coefficients in a model, to be estimated and to test hypotheses about. They are the indices to distributions; the mean and standard deviation are the location and scale parameters in the Normal distribution family.

Unfortunately, engineers use the same word (parameters) to describe the factors themselves.

Statisticians usually name their parameters after Greek letters, like mu($\mu$), sigma($\sigma$), beta($\beta$), and theta($\theta$). You can tell where statisticians went to school by which Greek and Roman letters they use in various situations. For example, in multivariate models, the L-Beta-M fraternity is distinguished from C-Eta-M.

### Hypotheses

In science, the hypothesis is the bright idea that you want to confirm. In statistics, this is turned upside down because it uses logic analogous to a proof-by-contradiction. The so-called *null hypothesis* is usually the statement that you want to demolish. The usual null hypothesis is that some factor has no effect on the response. You are of course trying to support the opposite, which is called the *alternative hypothesis*. You support the alternative hypothesis by statistically rejecting the null hypothesis.

### Two-Sided versus One-Sided, Two-Tailed versus One-Tailed

Most often, the null hypothesis can be stated as some parameter in a model being zero. The alternative is that it is not zero, which is called a two-sided alternative. In some cases, you may be willing to state the hypothesis with a one-sided alternative, for example that the parameter is greater than zero. The one-sided test has greater power at the cost of less generality. These terms have only this narrow technical meaning, and it has nothing to do with common English phrases like presenting a one-sided argument (prejudiced, biased in the everyday sense) or being two-faced (hypocrisy or equivocation). You can also substitute the word "tailed" for "sided." The idea is to get a big statistic that is way out in the tails of the distribution where it is highly improbable. You measure how improbable by calculating the area of one of the tails, or the other, or both.

### Statistical Significance

Statistical significance is a precise statistical term that has no relation to whether an effect is of practical significance in the real world. Statistical significance usually means that the data gives you the evidence to believe that some parameter is not the null value. If you have a ton of data, you can get a statistically significant test when the values of the estimates are practically zero. If you have very little data, you may get an estimate of an effect that would indicate enormous practical significance, but it is supported by so little data that it is not statistically significant. A nonsignificant result is one that might be the result of random variation rather than a real effect.

### Significance Level, *p*-value, α-level

To reject a null hypothesis, you want small *p*-values. The *p-value* is the probability of being wrong if you declare an effect to be non-null; that is, the probability of rejecting a 'true' null hypothesis. The *p*-value is sometimes labeled the *significance probability*, or sometimes labeled more precisely in terms of the distribution that is doing the measuring. The *p*-value labeled "Prob>|t|" is read as "the probability of getting a greater t (in absolute value)." The α-level is your standard of the *p*-value that you claim, so that *p*-values below this reject the null hypothesis (that is, they show that there is an effect).

### Power, β level

*Power* is how likely you are to detect an effect if it is there. The more data you have, the more statistical power. The greater the real effect, the more power. The less random variation in your world, the more power. The more sensitive your statistical method, the more power. It you had a method that always declared an effect significant, regardless of the data, it would have a perfect power of 1 (but it would have an $\alpha$-level of 1, too, the probability of declaring significance when there was no effect). The goal in experimental design is usually to get the most power you can afford, given a certain $\alpha$-level. It is not a mistake to connect the statistical term power with the common sense of power as persuasive ability. It has nothing to do with work or energy, though.

## Confidence Intervals

A confidence interval is an interval around a parameter estimate that has a given probability of containing the true value of a parameter. Most often the probability is 95%. Confidence intervals are now considered one of the best ways to report results. It is expressed as a percentage of $1 - \alpha$, so an 0.05 alpha level for a two-tailed $t$-quantile can be used for a 95% confidence interval. (For linear estimates, it is constructed by multiplying the standard error by a $t$-statistic and adding and subtracting that to the estimate. If the model involves nonlinearities, then the linear estimates are just approximations, and there are better confidence intervals called *profile likelihood confidence intervals*. If you want to form confidence regions involving several parameters, it is not valid to just combine confidence limits on individual parameters.)

## Biased, Unbiased

An *unbiased estimator* is one where the expected value of an estimator is the parameter being estimated. It is considered a desirable trait, but not an overwhelming one. There are cases when statisticians recommend biased estimators. The maximum likelihood estimator of the variance has a small but nonzero bias, for example.

## Sample Mean versus True Mean

The *sample mean* is the one you calculate from your data—the sum divided by the number. It is a statistic, that is, a function of your data. The *true mean* is the expected value of the probability distribution that generated your data. You usually don't know the true mean, and that is why you collect data, so you can estimate the true mean with the sample mean.

## Variance and Standard Deviation, Standard Error

*Variance* is the expected squared deviation of a random variable from its expected value. It is estimated by the sample variance. *Standard deviation* is the square root of the variance, and we prefer to report it because it is in the same units as the original random variable (or response values). The sample standard deviation is the square root of the sample variance. The term standard error describes an estimate of the standard deviation of another (unbiased) estimate.

## Degrees of Freedom

The specific name for a value indexing some popular distributions of test statistics. It is called degrees of freedom because it relates to differences in numbers of parameters that are or could be in the model. The more parameters a model has, the more freedom it has to fit the data better. The DF (degrees of freedom) for a test statistic is usually the difference in the number of parameters between two models.

# Simulations in JMP

## Overview

Although JMP is primarily used for analysis of existing data, it is also an excellent simulation platform. Using data tables and scripts, you can use JMP's random number generator to produce many realizations of a random process, then analyze the results. Known as Monte Carlo simulations, these simulations often give insight into otherwise difficult problems.

This chapter gives examples of simple simulations. After reading it, you can experiment with them to create your own simulations.

# Rolling Dice

A simple example of a Monte Carlo simulation from elementary probability is rolling a six-sided die and recording the results over a long period of time. Of course, it is impractical to physically roll a die repeatedly, so JMP is used to simulate the rolling of the die.

The assumption that each face has an equal probability means that we should simulate the rolls using a function that draws from a uniform distribution. We could use the **Random Uniform()** function, which pulls random real numbers from the (0,1) interval. However, JMP has a special version of this function for cases where we want random integers (in this case, random integers from 1 to 6).

 Open the **Dice Rolls.jmp** data table.

The table consists of a column to hold the random integers named **Dice Roll**. Each row of the data table represents a single roll of the die. A second column is used to keep a running average of all the rolls up to that point.

**Figure 6.1**   Dice Rolls.jmp data table

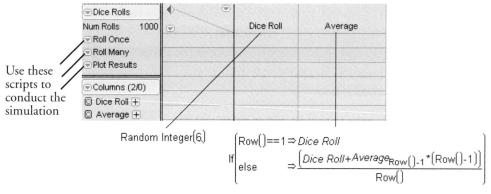

The law of large numbers states that as we increase the number of observations, the average should approach the true theoretical average of the process. In this case, we expect the average to approach (1+2+3+4+5+6)/6, or 3.5.

 Click on the red triangle beside the **Roll Once** script in the side panel of the data table.

This adds a single roll to the data table. Note that this is equivalent to adding rows through the **Rows > Add Rows** command. It is included as a script simply to lessen the number of mouse clicks needed to perform the function.

 Repeat this process several times to add ten rows to the data table.

 After ten rows have been added, run the **Plot Results** script in the side panel of the data table.

This produces a control chart of the results. Note that the results fluctuate fairly widely at this point.

**Figure 6.2**   Plot of Results After Ten Rolls

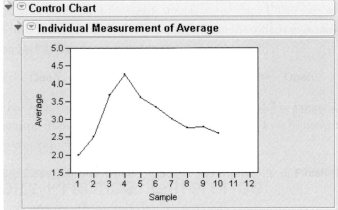

 Run the **Roll Many** script in the side panel of the data table.

This adds many rolls at once. In fact, it adds the number of rows specified in the table variable **Num Rolls** each time it is clicked. To add more or fewer rolls at one time, adjust the value of the **Num Rolls** variable.

Also note that the control chart has automatically updated itself. The chart reflects the new observations just added.

 Continue adding points until there are 2000 points in the data table.

**Figure 6.3** Observed mean approaches theoretical mean

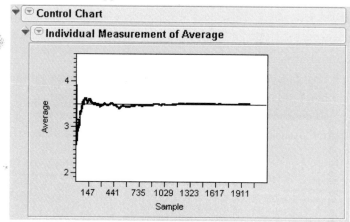

The control chart shows that the mean is leveling off, just as the law of large numbers predicted, at the value 3.5. In fact, you can add a horizontal line to the plot to emphasize this point.

🖰 Double-click the *y*-axis to open the axis specification dialog.

🖰 Enter values into the dialog box as shown in **Figure 6.4**.

**Figure 6.4** Adding a reference line to a plot

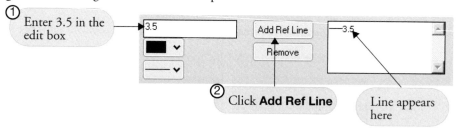

Although this is not a complicated example, it shows how easy it is to produce a simulation based on random events. In addition, this data table could be used as a basis for other simulations.

## Rolling several dice

If you need to roll more than one die at a time, simply copy and paste the formula from the existing Die Roll column into other columns.

# Flipping Coins, Sampling Candy, or Drawing Marbles

The techniques for rolling dice can easily be extended to other situations. Instead of displaying an actual number, use JMP to re-code the random number into something else.

For example, suppose you want to simulate coin flips. There are two outcomes that (in a fair coin) occur with equal probability. One way to simulate this is to draw random numbers from a uniform distribution, where all numbers between 0 and 1 occur with equal probability. If the selected number is below 0.5, declare that the coin landed heads up. Otherwise, declare that the coin landed tails up.

   Create a new data table with two columns.

   In the first column, enter the following formula.

$$\text{If} \begin{bmatrix} \text{Random Uniform()} < 0.5 \Rightarrow \text{"H"} \\ \text{else} \qquad\qquad\quad \Rightarrow \text{"T"} \end{bmatrix}$$

   Add rows to the data table to see the column fill with coin flips.

Extending this to sampling candies of different colors is easy. Suppose you have a bag of multi-colored candies with the following distribution.

| Color | Percentage |
|---|---|
| Blue | 10% |
| Brown | 10% |
| Green | 10% |
| Orange | 10% |
| Red | 20% |
| Yellow | 20% |
| Purple | 20% |

Also, suppose you had a column named t that held random numbers from a uniform distribution. Then an appropriate JMP formula could be

$$\text{If} \begin{cases} t<0.1 \Rightarrow \text{"Blue"} \\ t<0.2 \Rightarrow \text{"Brown"} \\ t<0.3 \Rightarrow \text{"Green"} \\ t<0.4 \Rightarrow \text{"Orange"} \\ t<0.6 \Rightarrow \text{"Red"} \\ t<0.8 \Rightarrow \text{"Yellow"} \\ \text{else} \quad \Rightarrow \text{"Purple"} \end{cases}$$

Note that JMP assigns the value associated with the first condition that is true. So, if t=0.18, "Brown" is assigned and no further formula evaluation is done.

Or, you could use a slightly more complicated formula using a local variable to combine the random number and candy selection into one row.

$$t=\text{Random Uniform}();$$
$$\text{If} \begin{cases} t<0.1 \Rightarrow \text{"Blue"} \\ t<0.2 \Rightarrow \text{"Brown"} \\ t<0.3 \Rightarrow \text{"Green"} \\ t<0.4 \Rightarrow \text{"Orange"} \\ t<0.6 \Rightarrow \text{"Red"} \\ t<0.8 \Rightarrow \text{"Yellow"} \\ \text{else} \quad \Rightarrow \text{"Purple"} \end{cases};$$

# Probability of Making a Triangle

Suppose you randomly pick two points along a line segment. Then, break the line segment at these two points, forming three line segments. What is the probability that a triangle can be formed from these three segments? (Isaac, 1995)

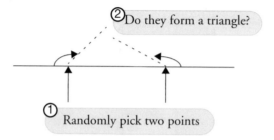

This situation is simulated in the Triangle Probability.jsl script. Upon running the script, a data table is created to hold the simulation results. The initial window is shown in **Figure 6.5**.

For each of the two selected points, a dotted circle indicates the possible positions of the "broken" line segment that they determine.

**Figure 6.5**   Initial triangle probability window

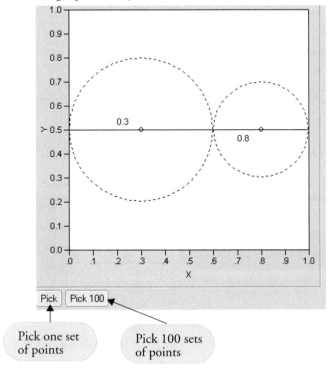

To use this simulation,

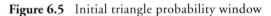   Click the **Pick** button to pick a single pair of points.

Two points are selected and their information is added to a data table. The results after five simulations are shown in **Figure 6.6**.

**Figure 6.6**   Triangle simulation after 5 iterations

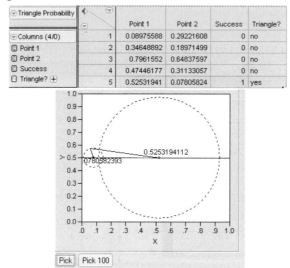

To get an idea of the theoretical probability, you need many rows in the data table.

🖰   Click the Pick 100 button a couple of times to generate a large number of samples.

When finished,

🖰   Select **Analyze > Distribution** and select Triangle? as the **Y, Columns** variable.

🖰   Click **OK** to see the distribution report in **Figure 6.7**.

**Figure 6.7**   Triangle Probability Distribution Report

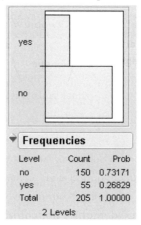

It appears (in this case) that about one quarter of the samples result in triangles. To investigate whether there is a relationship between the two selected points and their formation of a triangle,

 Select **Rows > Color or Mark by Column.**

When the selection dialog appears,

 Select the Triangle? column, making sure the **Set Color By Value** checkbox is checked.

This puts a different color on each row depending on whether it formed a triangle or not.

 Select **Analyze > Fit Y By X**, assigning Point1 to Y and Point2 to X.

This reveals a scatterplot that clearly shows the pattern.

**Figure 6.8**  Scatterplot of Point1 by Point2

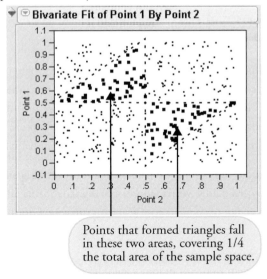

Points that formed triangles fall in these two areas, covering 1/4 the total area of the sample space.

The entire sample space is in a unit square, and the points that formed triangles occupy one fourth of that area. This means that there is a 25% probability that two randomly selected points form a triangle.

Analytically, this makes sense. If the two randomly selected points are $x$ and $y$, letting $x$ represent the smaller of the two, then we know $0 < x < y < 1$, and the three segments have length $x$, $y - x$, and $1 - y$ (see **Figure 6.9**).

**Figure 6.9**  Illustration of points

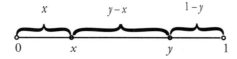

To make a triangle, the sum of the lengths of any two segments has to be larger than the third, so we can put the following conditions on the three points.

$$x + (y - x) > 1 - y$$
$$(y - x) + (1 - y) > x$$
$$(1 - y) + x > y - x$$

Elementary algebra simplifies these inequalities to

$$x < 0.5$$
$$y > 0.5$$
$$y - x < 0.5$$

which explain the upper triangle in **Figure 6.8**. Repeating the same argument with $y$ as the smaller of the two variables explains the lower triangle.

# Confidence Intervals

Introductory students of statistics are often confused by the exact meaning of confidence intervals. For example, they frequently think that a 95% confidence interval contains 95% of the data. They do not realize that the confidence measurement is on the test methodology itself.

To demonstrate the concept, use the Confidence.jsl script. Its output is shown in **Figure 6.10**.

**Figure 6.10**  Confidence Interval Script

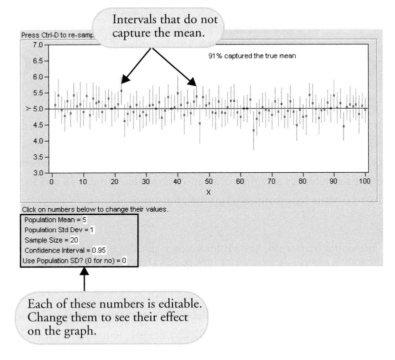

The script draws 100 samples of sample size 20 from a Normal distribution. For each sample, the mean is computed with a 95% confidence interval. Each interval is graphed, in gray if the interval captures the overall mean and in red if it doesn't.

Press Ctrl+D (⌘+D on the Macintosh) to generate another series of 100 samples. Each time, note the number of times the interval captures the theoretical mean. The ones that don't capture the mean are due only to chance, since we are randomly drawing the samples. For a 95% confidence interval, we expect that around 5 will not capture the mean, so seeing a few is not remarkable.

This script can also be used to illustrate the effect of changing the confidence interval.

🖱   Change the confidence interval to 0.5.

This shrinks the size of the confidence intervals on the graph.

The **Use Population SD** interval allows you to use the population standard deviation in the computation of the confidence intervals (rather than the one from the sample). When enabled, all the confidence intervals are of the same size.

# Univariate Distributions: One Variable, One Sample

## Overview

This chapter introduces statistics in the simplest possible setting—the distribution of values in one variable. The Distribution command in the **Analyze** menu launches JMP's *Distribution platform*, which is used to describe the distribution of a single column of values from a table, using graphs and summary statistics.

This chapter also introduces the concept of the distribution of a statistic, and how confidence intervals and hypothesis tests can be obtained.

# Looking at Distributions

Let's take a look at some actual data and start noticing aspects of its distribution.

🖰  Begin by opening the data table called **Birth Death.jmp**, which contains the 1976 birth and death rates of 74 nations (**Figure 7.1**).

🖰  Choose **Analyze > Distribution**.

🖰  Select **birth**, **death**, and **Region** columns as the Y variables and click **OK**.

**Figure 7.1**    Partial Listing of the **Birth Death.jmp** data table

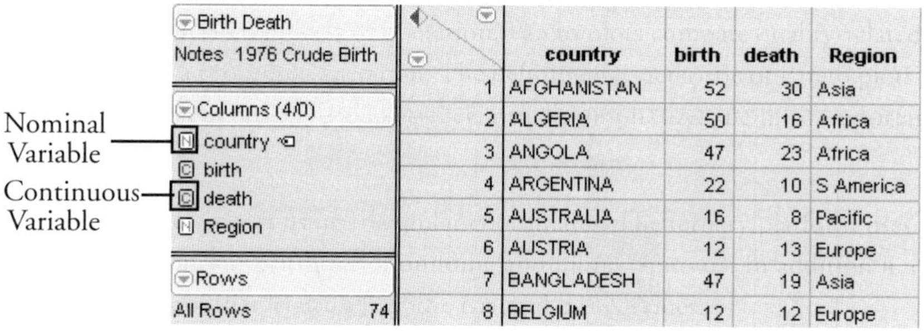

When you see the report (**Figure 7.2**), be adventuresome: scroll around and click in various places on the surface of the report. Notice that a histogram or statistical table can be opened or closed by clicking the disclosure button on its title bar.

🖰  Open and close tables, and click on bars until you have the configuration shown in **Figure 7.2**.

**Figure 7.2**   Histograms, Quantiles, Moments, and Frequencies

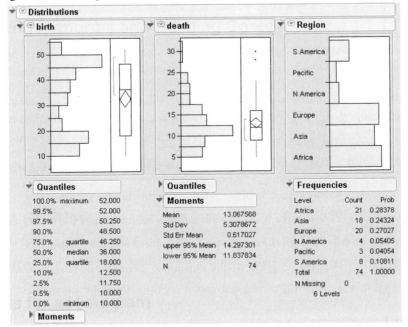

Note that there are two kinds of analyses:

- The analyses for **birth** and **death** are for continuous distributions. These are the kind of reports you get when the column in the data table has the continuous modeling type. The **C** next to the column name in the columns panel indicates that this variable is continuous.

- The analysis for **Region** is for a categorical distribution. These are the kind of graphs and reports you get when the column in the data table has the modeling type of nominal or ordinal, **N** or **O** next to the column name in the Columns panel.

You can change the modeling type of any variable in the Columns panel to control which kind of report you get.

For continuous distributions, the graphs give a general idea of the shape of the distribution. The **death** data cluster together with most values near the center. Distributions such as this one with one peak are called *unimodal*. However, the **birth** data have a different distribution. There are many countries with high birth rates, many with low birth rates, but few with middle rates. Therefore, the **birth** data has two peaks, and is referred to as *bimodal*.

The text reports for **birth** and **death** show a number of measurements concerning the distributions. There are two broad families of measures:

- *Quantiles* are the points at which various percentages of the total sample are above or below.

- *Moments* combine the individual data points to form descriptions of the entire data set. These combinations are usually simple arithmetic operations that involve sums of values raised to a power. Two common moments are the *mean* and *standard deviation*.

The report for the categorical distribution focuses on frequency counts. This chapter concentrates on continuous distributions and postpones the discussion of categorical distributions until Chapter 9, "Categorical Distributions."

Before going into the details of the analysis, let's review the distinctions between the properties of a distribution and the estimates that can be obtained from a distribution.

## Probability Distributions

A *probability distribution* is the mathematical description of how a random process distributes its values. Continuous distributions are described by a *density function*. In statistics, we are often interested in the probability of a random value falling between two values described by this density function (for example, "What's the probability that I will gain between 100 and 300 points if I take the SAT a second time?"). The probability that a random value falls in a particular interval is represented by the area under the density curve in this interval, as illustrated in **Figure 7.3**.

**Figure 7.3**  Continuous Distribution

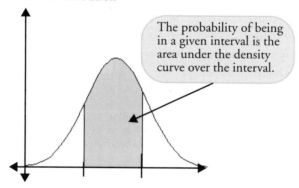

The probability of being in a given interval is the area under the density curve over the interval.

The density function describes all possible values of the random variable, so the area under the whole density curve must be 1. In fact, this is a defining characteristic of all density functions.

In order for a function to be a density function, it must be non-negative and the area underneath the curve must be 1.

These mathematical probability distributions are useful because they can model distributions of values in the real world. This book won't show the formulas for any distributions, but you should learn their names and their uses.

## True Distribution Function or Real-World Sample Distribution

Sometimes it is hard to keep straight when you are referring to the real data sample and when you are referring to its abstract mathematical distribution.

This distinction of the *property* from its *estimate* is crucial in avoiding misunderstanding. Consider the following problem:

How is it that statisticians talk about the variability of a mean, that is, the variability of a single number? When you talk about variability in a sample of values, you can see the variability because you have many different values. However, by computing a mean, the entire list of numbers has been condensed to a single number. How can you talk about the variability of the mean itself?

To get the idea of variance, you have to separate the abstract quality from its estimate. When you do statistics, you are assuming that the data come from a process that has a random element to it. Even if you have a single response value (like a mean), there is variability associated with it, a magnitude whose values are possibly unknown.

For instance, suppose you are interested in finding the average height of males in the United States. You decide to compute the mean of a sample of 100 people. If you replicate this experiment several times on different samples, do you expect to get the same mean for every sample you pick? Of course not. There is variability in the sample means. It is this variability that statistics tries to capture—even if you don't replicate the experiment. Statistics can estimate the variability in the mean, even if it only has a single experiment to examine. The variability in the mean is called the *standard error* of the mean.

If you take a collection of values from a random process, sum them, and divide by the number of them, you have calculated a mean. You can then calculate the variance associated with this single number. There is a simple algebraic relationship between the variability of the responses (the standard deviation of the original data) and the variability of the sum of the responses divided by *n* (the standard error of the mean). Complete details follow in the section "Standard Error of the Mean" on page 134.

**Table 7.1.**    Properties of Distribution Functions and Samples

| Concept | Abstract mathematical form; probability distribution | Numbers from the real world; data; sample |
|---|---|---|
| Average | Expected value or true mean, the point that balances each side of the density | Sample mean, the sum of values divided by the number of values |
| Median | Median, the mid-value of the density area, where 50% of the density is on either side | Sample median, the middle value where 50% of the data are on either side |
| Quantile | The value where some percent of the density is below it | Sample quantile, the value for which some percent of the data are below it. For example, the 90th percentile represents a point where 90 percent of the variables are below it |
| Spread | Variance, the expected squared deviation from the expected value | Sample variance, the sum of squared deviations from the sample mean divided by $n - 1$ |
| General Properties | Any function of the distribution: parameter, property | Any function of the data: estimate, statistic |

The statistic from the real world estimates the parametric value from the distribution.

# The Normal Distribution

The most notable continuous probability distribution is the *Normal distribution*, also known as the *Gaussian distribution*, or the *bell curve*, like the one shown in **Figure 7.4**.

**Figure 7.4**   Standard Normal Density Curve

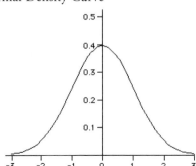

It is an amazing distribution.

Mathematically, its greatest distinction is that it is the most random distribution for a given variance. (It is "most random" in a very precise sense, having maximum expected unexpectedness or entropy.) Its values are as if they had been realized by adding up billions of little random events.

It is also amazing because so much of real world data are Normally distributed. The Normal distribution is so basic that it is the benchmark used as a comparison with the shape of other distributions. Statisticians describe sample distributions by saying how they differ from the Normal. Many of the methods in JMP serve mainly to highlight how a distribution of values differs from a Normal distribution. However, the usefulness of the Normal doesn't end there. The Normal distribution is also the standard used to derive the distribution of estimates and test statistics.

The famous *Central Limit Theorem* says that under various fairly general conditions, the sum of a large number of independent and identically distributed random variables is approximately Normally distributed. Because most statistics can be written as these sums, they are Normally distributed if you have enough data. Many other useful distributions can be derived as simple functions of Normal random distributions.

Later in this chapter, you will meet the distribution of the mean and learn how to test hypotheses about it. The next sections introduce the four most useful distributions of test statistics: the Normal, Student's *t*, chi-square, and *F* distributions.

# Describing Distributions of Values

The following sections take you on a tour of the graphs and statistics in the JMP Distribution platform. These statistics try to show the properties of the distribution of a sample, especially these four focus areas:

- *Location* refers to the center of the distribution.

- *Spread* describes how concentrated or "spread out" the distribution is.

- *Shape* refers to symmetry, whether the distribution is unimodal, and especially how it compares to a Normal distribution.

- *Extremes* are outlying values far away from the rest of the distribution.

## Generating Random Data

Before getting into more real data, let's make some random data with familiar distributions, and then see what an analysis reveals. This is an important exercise because there is no other way to get experience on the distinction between the true distribution of a random process and the distribution of the values you get in a sample.

In Plato's mode of thinking, the "true" world is some ideal form, and what you perceive as real data is only a shadow that gives hints at what the true world is like. Most of the time the true state is unknown, so an experience where the true state is known will be valuable.

In the following example, the true world will be a distribution, and you use JMP's random number generator to obtain realizations of the random process to make a sample of values. Then you will see that the sample mean of those values is not exactly the same as the true mean used to generate the original distribution. This distinction is fundamental to what statistics is all about.

To create your own random data,

- Open Randdist.Jmp.

This data table has four columns, but no rows. The columns contain formulas used to generate random data having the distributions Uniform, Normal, Exponential, and Dbl Exponential (double exponential).

- **Choose Rows > Add Rows** and enter 1000 to see a table like the one in **Figure 7.5**.

Adding rows generates the random data using column formulas. Note that your random results will be a little different from those shown in **Figure 7.5** because the random number generator produces a slightly different set of numbers each time a table is created.

**Figure 7.5**  Partial Listing of the Randdist Data Table

| | Uniform | Normal | Exponential | Dbl Expon |
|---|---|---|---|---|
| 1 | 0.040324 | -0.57437 | 0.085626 | 1.4802 |
| 2 | 0.538689 | 1.859345 | 4.102436 | 0.165275 |
| 3 | 0.129768 | 0.041063 | 0.353676 | -0.35035 |
| 4 | 0.568301 | -0.16652 | 1.873607 | 3.914928 |
| 5 | 0.76171 | 0.393797 | 0.197711 | 0.98341 |
| 6 | 0.571125 | -0.6395 | 1.131872 | 0.376832 |
| 7 | 0.476427 | 0.183686 | 2.03773 | -0.25875 |
| 8 | 0.057852 | -0.21408 | 1.999124 | 0.302307 |

Randdist
Columns (4/0)
◻ Uniform ⊞
◻ Normal ⊞
◻ Exponential ⊞
◻ Dbl Expon ⊞
Rows
All Rows    1000

🖰  To look at the distributions of the columns in the Randdist.jmp table, choose **Analyze** > **Distribution**, and select the four columns as Y variables.

🖰  Click **OK**.

The resulting analysis automatically shows a number of graphs and statistical reports. Further graphs and reports can be obtained by using the popup menus located on the title bar of each analysis. The following sections examine these graphs and the text reports available in the Distribution platform.

## Histograms

A *histogram* defines a set of intervals and shows how many values in a sample fall into each interval. It shows the shape of the density of a batch of values.

Try out the following histogram features:

🖰  Click in a histogram bar.

As the bar is highlighted, the corresponding portions of bars in other histograms are also highlighted, as are the corresponding data table rows. When you do this with real data, you can see *conditional distributions*—the distributions in other variables corresponding to a subset of the selected variable's distribution.

🖰  Double-click on a histogram bar to produce a new JMP table that is a subset corresponding to that bar.

🖑    On the original Distribution plot, choose the **Normal** option from the **Fit Distribution** command on the menu at the left of the report titles.

This superimposes over the histogram the Normal density corresponding to the mean and standard deviation in your sample. **Figure 7.6** shows the histograms from the analysis with Normal curves superimposed on them.

**Figure 7.6**    Histograms of Various Continuous Distributions

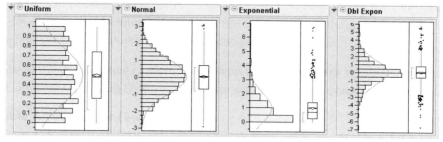

🖑    Get the hand tool from the **Tools** menu or toolbar.

🖑    Click on the histogram and drag to the right, then back to the left.

The histogram bars get narrower and wider (see **Figure 7.7**). Make them wider and then drag up and down to change the position of the bars.

**Figure 7.7**    The Hand Tool Adjusts Histogram Bar Widths

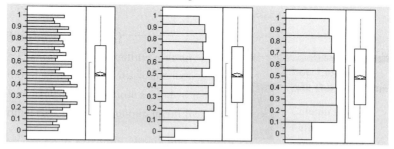

# Outlier and Quantile Box Plots: Optional

*Box plots* are schematics that help show how your data are distributed. JMP's Distribution platform offers two varieties of box plots that you can turn on or off with options accessed by the popup menu icon on the report title bar, as shown here.

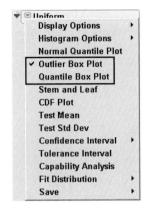

**Figure 7.8** shows these box plots for the simulated distributions. The box part within each plot surrounds the middle half of the data. The lower edge of the rectangle represents the lower quartile, the higher edge represents the upper quartile, and the line in the middle of the rectangle is the median. The distance between the two edges of the rectangle is called the *interquartile range*. The lines extending from the box show the tails of the distribution, points that the data occupy outside the quartiles. These lines are sometimes called *whiskers*.

**Figure 7.8**   Quantile and Outlier Box Plots

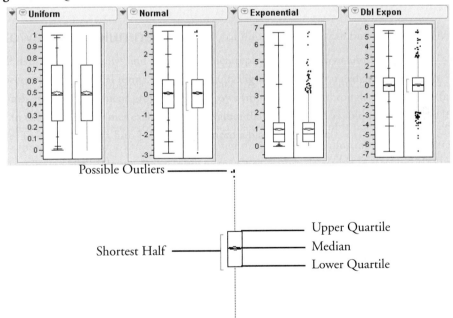

In the outlier box plots, shown on the right of each panel in **Figure 7.8**, the tail extends to the farthest point that is still within 1.5 interquartile ranges from the quartiles. Points farther away are shown individually as outliers.

In the quantile box plots (the left-hand plot in each panel) the tails are marked at certain quantiles. The quantiles are chosen so that if the distribution is Normal, the marks appear approximately equidistant, like the figure on the right. The spacing of the marks in these box plots give you a clue about the Normality of the underlying distribution.

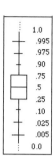

Look again at the boxes in the four distributions in **Figure 7.8**, and examine the middle half of the data in each graph. The middle half of the data are wide in the uniform, thin in the double exponential, and very one-sided in the exponential distribution.

In the outlier box plot, the shortest half (the shortest interval containing half the data) is shown by a red bracket on the side of the box plot. The shortest half is at the center for the symmetric distributions, but off-center for non-symmetric ones. Look at the exponential distribution to see an example of a non-symmetric distribution.

In the quantile box plot, the mean and its 95% confidence interval are shown by a diamond. Since this experiment was created with 1000 observations, the mean is estimated with great precision, giving a very short confidence interval. Confidence intervals are discussed in the following sections.

## Normal Quantile Plots: Optional

*Normal quantile plots* show all the values of the data as points in a plot. If the data are Normal, the points tend to follow a straight line.

🖑    From the popup menu on the title bar, select **Normal Quantile Plot**.

The histograms and Normal quantile plots for the four simulated distributions are shown in **Figure 7.10** to **Figure 7.13**.

The $y$ (vertical) coordinate is the actual value of each data point. The $x$ (horizontal) coordinate is the Normal quantile associated with the rank of the value after sorting the data.

If you are interested in the details, the precise formula used for the Normal quantile values is

$$\Phi^{-1}\left(\frac{r_i}{N+1}\right)$$

where $r_i$ is the rank of the observation being scored, $N$ is the number of observations, and $\Phi^{-1}$ is the function that returns the Normal quantile associated with the probability argument

$$\frac{r_i}{N+1}$$

The Normal quantile is the value on the *x*-axis of the Normal density that has the portion *p* of the area below it, where *p* is the probability argument. For example, the quantile for 0.5 (the probability of being less than the median) is zero, because half (50%) of the density of the standard Normal is below zero. The technical name for the quantiles JMP uses is the *van der Waerden* Normal scores; they are computationally cheap (but good) approximations to the more expensive, exact expected Normal order statistics.

- A red straight line, with confidence limits, shows where the points would tend to lie if the data were Normal. This line is purely a function of the mean and standard deviation of the sample. The line crosses the mean of the data at the Normal quantile of 0. The slope of the line is the standard deviation of the data.

- Dashed lines surrounding the straight line form a confidence interval for the Normal distribution. If the points fall outside these dashed lines, you are seeing a significant departure from Normality. For more details see the section "Testing for Normality" on page 151.

- If the slope of the points is small (relative to the Normal) then you are crossing a lot of (ranked) data with very little variation in the real values, and therefore encounter a dense cluster. If the slope of the points is large, then you are crossing a lot of space that has only a few (ranked) points. Dense clusters make flat sections, and thinly populated regions make steep sections. The overall slope is the standard deviation.

**Figure 7.9**   Normal Quantile Plot Explanation

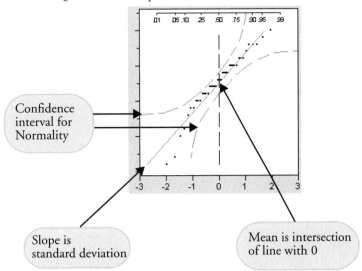

The middle portion of the uniform distribution (**Figure 7.10**) is steeper (less dense) than the Normal. In the tails, the uniform is flatter (more dense) than the Normal. In fact, the tails are truncated at the end of the range, where the Normal tails extend infinitely.

**Figure 7.10**   Uniform Distribution

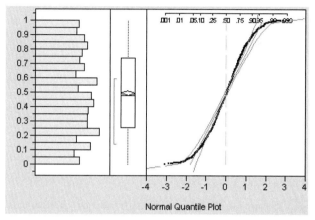

The Normal distribution (**Figure 7.11**) has a Normal quantile plot that follows a straight line. Points at the tails usually have the highest variance and are most likely to fall farther from the line. Because of this, the confidence limits flair near the ends.

**Figure 7.11**   Normal Distribution

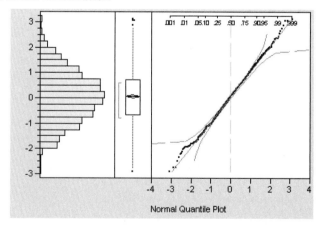

The exponential distribution (**Figure 7.12**) is skewed—that is, one-sided. The top tail runs steeply past the Normal line; it is spread out more than the Normal. The bottom tail is shallow and much denser than the Normal.

**Figure 7.12**  Exponential Distribution

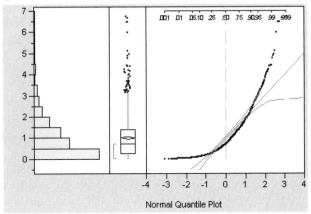

The middle portion of the double exponential (**Figure 7.13**) is denser (more shallow) than the Normal. In the tails, the double exponential spreads out more (is steeper) than the Normal.

**Figure 7.13**  Double Exponential Distribution

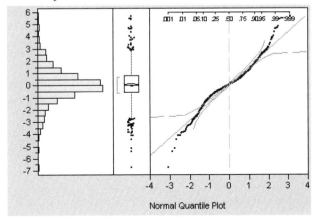

## Stem-and-Leaf Plot

A *stem-and-leaf plot* is a variation on the histogram. It was developed for tallying data in the days when computers were rare and histograms took a lot of time to make. Each line of the plot has a stem value that is the leading digits of a range of column values. The leaf values are made from other digits of the values.

To see two examples, open the Big Class.jmp and the Automess.jmp tables.

🖰 For each table choose **Analyze > Distribution**. The Y variables are weight from the Big Class table and Auto theft from the Automess table.

🖰 When the histograms appear, select **Stem and Leaf** from the options popup menu next to the histogram name.

This option appends stem-and-leaf plots to the end of the text reports.

**Figure 7.14** shows the plot for Weight on the left and the plot for Auto theft on the right. The values in the stem column of the plot are chosen as a function of the range of values to be plotted, with the scale factor at the bottom of the plot.

You can reconstruct the data values by joining the stem and leaf as indicated by the legend on the bottom of the plot. For example, on the bottom line of the Weight plot, you can read the values 64 and 67 (6 from the stem, 4 and 7 from the leaf). At the top, the weight is 172 (17 from the stem, 2 from the leaf).

The leaves respond to mouse clicks.

🖰 Click on the two 5s on the bottom stem of the Auto theft plot.

This highlights the corresponding rows in the data table, which are "California" with the value 154 and the "District of Columbia" with value of 149.

**Figure 7.14**   Examples of Stem-and-Leaf Plots

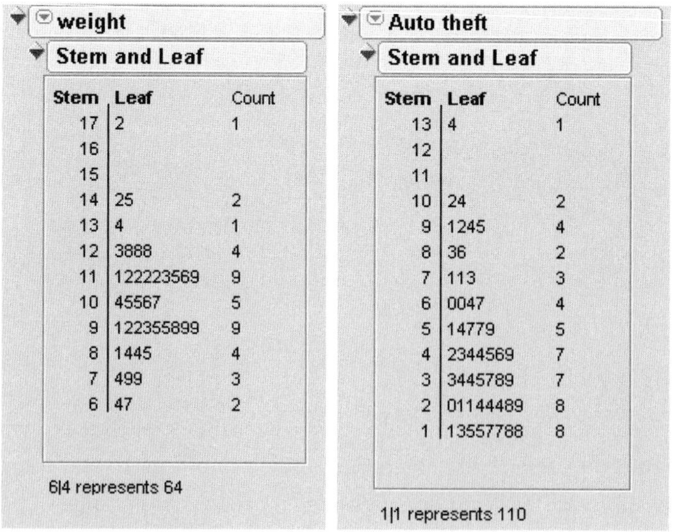

| weight | | |
|---|---|---|
| **Stem and Leaf** | | |
| Stem | Leaf | Count |
| 17 | 2 | 1 |
| 16 | | |
| 15 | | |
| 14 | 25 | 2 |
| 13 | 4 | 1 |
| 12 | 3888 | 4 |
| 11 | 122223569 | 9 |
| 10 | 45567 | 5 |
| 9 | 122355899 | 9 |
| 8 | 1445 | 4 |
| 7 | 499 | 3 |
| 6 | 47 | 2 |

6|4 represents 64

| Auto theft | | |
|---|---|---|
| **Stem and Leaf** | | |
| Stem | Leaf | Count |
| 13 | 4 | 1 |
| 12 | | |
| 11 | | |
| 10 | 24 | 2 |
| 9 | 1245 | 4 |
| 8 | 36 | 2 |
| 7 | 113 | 3 |
| 6 | 0047 | 4 |
| 5 | 14779 | 5 |
| 4 | 2344569 | 7 |
| 3 | 3445789 | 7 |
| 2 | 01144489 | 8 |
| 1 | 13557788 | 8 |

1|1 represents 110

# Mean and Standard Deviation

The *mean* of a collection of values is its average value, computed as the sum of the values divided by the number of values in the sum.

$$\bar{x} = \frac{x_1 + x_1 + \dots + x_n}{n} = \sum \frac{x_i}{n}$$

The sample mean has these properties:

- It is the balance point. The sum of deviations of each sample value from the sample mean is zero.

- It is the least squares estimate. The sum of squared deviations of the values from the mean is minimized. That sum is less than would be computed from any estimate other than the sample mean.

- It is the maximum likelihood estimator of the true mean when the distribution is Normal. It is the estimate that makes the data you collected more likely than any other estimate of the true mean would.

The sample *variance* (denoted $s^2$) is the average squared deviation from the sample mean, which is shown as the expression

$$s^2 = \sum \frac{(x_i - \bar{x})^2}{n-1}$$

The sample *standard deviation* is the square root of the sample variance. The standard deviation is preferred in reports because (among other reasons) it is in the same units as the original data (rather than squares of units).

If you assume a distribution is Normal, you can completely characterize the distribution by its mean and standard deviation.

When you say "mean" and "standard deviation," you are allowed to be ambiguous as to whether you are referring to the true (and usually unknown) parameters of the distribution, or the sample statistics you use to estimate the true values.

# Median and Other Quantiles

Half the data are above and half are below the sample *median*. It estimates the 50th quantile of the distribution. A sample quantile can be defined for any percentage between 0% and 100%;

the 100% quantile is the maximum value, where 100% of the data values are at or below. The 75% quantile is the upper quartile, the value for which 75% of the data values are at or below.

There is an interesting indeterminacy about how to report the median and other quantiles. If you have an even number of observations, there may be several values where half the data are above, half below. There are about a dozen different ways for reporting medians in the statistical literature, many of which are only different if you have tied points on either or both sides of the middle. You can take one side, the other, the midpoint, or a weighted average of the middle values, with a number of weighting options. For example, if the sample values are {1, 2, 3, 4, 4, 5, 5, 5, 7, 8}, the median can be defined anywhere between 4 and 5, including one side or the other, or half way, or two-thirds of the way into the interval. The halfway point is the most common value chosen.

Another interesting property of the median is that it is the least-absolute-values estimator. That is, it is the number that minimizes the sum of the absolute differences between itself and each value in the sample. Least-absolute-values estimators are also called *L1 estimators*, or *MAD estimators* (minimum absolute deviation).

## Mean versus Median

If the distribution is symmetric, the mean and median are estimates of both the expected value of the underlying distribution and its 50% quantile. If the distribution is Normal, the mean is a "better" estimate, in terms of variance, than the median, by a ratio of 2 to 3.1416 (2: $\pi$). In other words, the mean has only 63% of the variance of the median.

If an outlier contaminates the data, the median is not greatly affected, but the mean could be greatly influenced, especially if the outlier is extreme. The median is said to be *outlier-resistant*, or *robust*.

Suppose that you have a skewed distribution, like household income in the United States. This set of data has lots of extreme points on the high end, but is limited to zero on the low end. If you want to know the income of a typical person, then it makes more sense to report the median than the mean. However, if you want to track per-capita income as an aggregating measure, then the mean income might be better to report.

## Higher Moments: Skewness and Kurtosis

*Moment statistics* are those that are formed from sums of powers of the data's values. The first four moments are defined as follows:

- The first moment is the mean, which is calculated from a sum of values to the power 1. The mean measures the center of the distribution.

- The second moment is the variance (and, consequently, the standard deviation), which is calculated from sums of the values to the second power. Variance measure the spread of the distribution.

- The third moment is *skewness*, which is calculated from sums of values to the third power. Skewness measures the asymmetry of the distribution.

- The fourth moment is *kurtosis*, which is calculated from sums of the values to the fourth power. Kurtosis measures the relative shape of the middle and tails of the distribution.

One use of skewness and kurtosis is to help determine if a distribution is Normal and, if not, what the distribution might be. A problem with the higher order moments is that the statistics have higher variance and are more sensitive to outliers.

🖱 To get the skewness and kurtosis, use the red popup menu beside the title of the histogram and select **Display Options > More Moments** from the drop-down list next to the histogram's title.

## Extremes, Tail Detail

The extremes (the minimum and maximum) are the 0% and 100% quantiles.

At first glance, the most interesting aspect of a distribution appears to be where its center lies. However, statisticians often look first at the outlying points—they can carry useful information. That's where the unusual values are, the possible contaminants, the rogues, and the potential discoveries.

In the Normal distribution (with infinite tails), the extremes tend to extend farther as you collect more data. However, this is not necessarily the case with other distributions. For data that is uniformly distributed across an interval, the extremes change less and less as more data is collected. Sometimes this is not helpful, since the extremes are often the most informative statistics on the distribution.

# Statistical Inference on the Mean

The previous sections talked about descriptive graphs and statistics. This section moves on to the real business of statistics: inference. We want to form confidence intervals for the mean and test hypotheses about the mean.

# Standard Error of the Mean

Suppose there exists some true (but unknown) population mean that you estimate with the sample mean. The sample mean comes from a random process, so there is variability associated with it.

The mean is the arithmetic average—the sum of $n$ values divided by $n$. The variance of the mean is $1/n$ of the variance of the original data. Since the standard deviation is the square root of the variance, the standard deviation of the sample mean is $1/\sqrt{n}$ of the standard deviation of the original data.

Substituting in the estimate of the standard deviation of the data, we now define the *standard error of the mean*, which estimates the standard deviation of the sample mean. It is the standard deviation of the data divided by the square root of $n$.

Symbolically, this is written

$$s_{\bar{y}} = \frac{s_y}{\sqrt{n}}$$

where $s_y$ is the sample standard deviation.

The mean and its standard error are the key quantities involved in statistical inference concerning the mean.

# Confidence Intervals for the Mean

The sample mean is sometimes called a *point estimate*, because it's only a single number. The true mean is not this point, but rather this point is an estimate of the true mean.

Instead of this single number, it would be more useful to have an interval that you are pretty sure contains the true mean (say, 95% sure). This interval is called a *95% confidence interval* for the true mean.

To construct a confidence interval, first make some assumptions. Assume

- the data are Normal, and
- the true standard deviation is the sample standard deviation. (This assumption will be revised later.)

Then, the exact distribution of the mean estimate is known, except for its location (because you don't know the true mean).

If you knew the true mean and had to forecast a sample mean, then you could construct an interval around the true mean that would contain the sample mean with probability 0.95. To do this, first obtain the quantiles of the standard Normal distribution that have 5% of the area in its tails. These quantiles are −1.96 and +1.96.

Then, scale this interval by the standard deviation and add in the true mean. Symbolically, compute

$$\mu \pm 1.96 s_{\bar{y}}$$

However, our present example is the reverse of this situation. Instead of a forecast, you already have the sample mean; instead of an interval for the sample mean, you need an interval to capture the true mean. If the sample mean is 95% likely to be within this distance of the true mean, then the true mean is 95% likely to be within this distance of the sample mean. Therefore, the interval is centered at the sample mean. The formula for the approximate 95% confidence interval is

$$95\% \text{ C.I. for the mean} = \bar{x} \pm 1.96 s_{\bar{y}}$$

**Figure 7.15** illustrates the construction of confidence intervals. This is not exactly the confidence interval that JMP calculates. Instead of using the quantile of 1.96 (from the Normal distribution), it uses a quantile from Student's $t$ distribution, discussed later. It is necessary to use this slightly modified version of the Normal distribution because of the extra uncertainty that results from estimating the standard error of the mean. So the formula for the confidence interval is

$$(1-\alpha) \text{ C.I. for the mean} = \bar{x} \pm \left( t_{1-\frac{\alpha}{2}} \cdot s_{\bar{y}} \right)$$

The alpha ($\alpha$) in the formula is the probability that the interval does not capture the true mean. That probability is 0.05 for a 95% interval. The confidence interval is reported on the Distribution platform in the Moments report as the Upper 95% Mean and Lower 95% Mean. It is represented in the quantile box plot by the ends of a diamond (see **Figure 7.16**).

**Figure 7.15**   Illustration of Confidence Interval

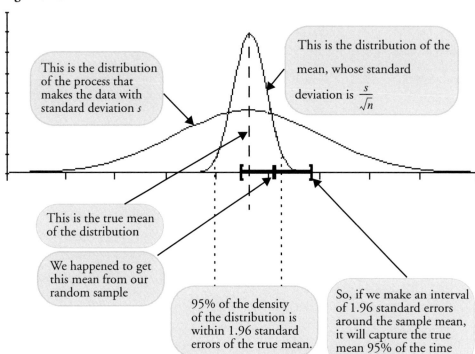

This is the distribution of the process that makes the data with standard deviation $s$

This is the distribution of the mean, whose standard deviation is $\frac{s}{\sqrt{n}}$

This is the true mean of the distribution

We happened to get this mean from our random sample

95% of the density of the distribution is within 1.96 standard errors of the true mean.

So, if we make an interval of 1.96 standard errors around the sample mean, it will capture the true mean 95% of the time

**Figure 7.16**   Moments Report and Quantile Box Plot

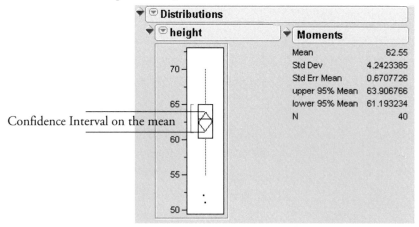

Confidence Interval on the mean

| Moments | |
|---|---|
| Mean | 62.55 |
| Std Dev | 4.2423385 |
| Std Err Mean | 0.6707726 |
| upper 95% Mean | 63.906766 |
| lower 95% Mean | 61.193234 |
| N | 40 |

# Testing Hypotheses: Terminology

Suppose that you want to test whether the mean of a collection of sample values is significantly different from some hypothesized value. The strategy is to calculate a statistic, so that if the true mean were the hypothesized value, getting such a large computed statistic value would be an extremely unlikely event. You would rather believe the hypothesis to be false than to believe this rare coincidence happened. This is a probabilistic version of *proof-by-contradiction*.

The way that you see that event to be rare is to see that its probability is past a point in the tail of the probability distribution of the hypothesis. Often, researchers use 0.05 as a significance indicator, which means that you believe that the mean is different from the hypothesized value if the chance of being wrong is only 5% (one in twenty).

Statisticians have a precise and formal terminology for hypothesis testing:

- The possibility of the true mean being the hypothesized value is called the *null hypothesis*, which you want to reject (refute). Said another way, the null hypothesis is the possibility that the hypothesized value is not different from the true mean. The *alternative hypothesis* is that the mean is different from the hypothesized value, which can be phrased as greater than, less than, or unequal. The latter is called a *two-sided alternative*.

- The situation where you reject the null hypothesis when it happens to be true is called a *Type I error*. This declares that some difference is nonzero when it is really zero. The opposite mistake (not detecting a difference when there is a difference) is called a *Type II error*.

- The probability of getting a Type I error in a test is called the *alpha-level* ($\alpha$-level) of the test. This is the probability that you are wrong if you say that there is a difference. The *beta-level* ($\beta$-level) or *power* of the test is the probability of being right when you say that there is a difference. $1 - \beta$ is the probability of a Type II error.

- Statistics and tests are constructed so that the power is maximized subject to the $\alpha$-level being maintained.

In the past, people obtained critical values for $\alpha$-levels and ended with an accept/reject decision based on whether the statistic was bigger or smaller than the critical value. For example, a researcher would declare that his experiment was significant if his test statistic fell in the region of the distribution corresponding to an $\alpha$-level of 0.05. This $\alpha$-level was specified in advance, before the study was conducted.

Computers have changed this strategy. Now, the $\alpha$-level isn't pre-determined, but rather is produced by the computer after the analysis is complete. In this context, it is called a *p-value* or *significance level*. The definition of a *p*-value can be said in many ways.

- The *p*-value is the α-level at which the statistic would be significant.

- The *p*-value is how unlikely getting so large a statistic would be if the true mean were the hypothesized value.

- The *p*-value is the probability of being wrong if you rejected the null hypothesis. It is the probability of a Type I error.

- The *p*-value is the area in the tail of the distribution of the test statistic under the null hypothesis.

The *p*-value is the number you want to be very small, certainly below 0.05, so that you can say that the mean is significantly different from the hypothesized value. The *p*-values in JMP are labeled according the test statistic's distribution. The label "Prob >|t|" is read as the "probability of getting an even greater absolute *t* statistic, given that the null hypothesis is true."

## The Normal *z*-Test for the Mean

The Central Limit Theorem tells us that if the original response data are Normally distributed, then when many samples are drawn, the means of the samples are Normally distributed. Even if the original response data are not Normally distributed, the sample mean still has an approximate Normal distribution if the sample size is large enough. So the Normal distribution provides a reference to use to compare the sample mean to an hypothesized value.

The standard Normal distribution has a mean of zero and a standard deviation of one. You can center any variable to mean zero by subtracting the mean (at least the hypothesized mean). You can standardize any variable to have standard deviation 1 by dividing by the true standard deviation, assuming for now that you know what it is. If the hypothesis were true, the test statistic you construct should have this standard distribution. Tests using the Normal distribution constructed like this (hypothesized mean and known standard deviation) are called *z-test*s. The formula for a *z*-statistic is

$$z\text{statistic} = \frac{\text{estimate - hypothesized value}}{\text{standard deviation}}$$

You want to find out how unusual your computed *z*-value is from the point of view of believing the hypothesis. If the value is too improbable, then you doubt the null hypothesis.

To get a significance probability, you take the computed *z*-value and find the probability of getting an even greater absolute value. This involves finding the area in the tails of the Normal distribution that are greater than absolute *z* and less than negative absolute *z*. **Figure 7.17** illustrates a two-tailed *z*-test.

**Figure 7.17** Illustration of Two-Tailed $z$-test

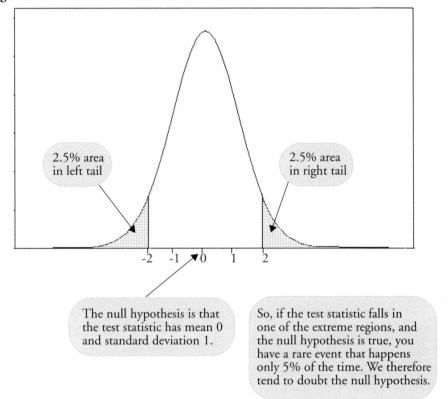

## Case Study: The Earth's Ecliptic

In 1738, the Paris observatory determined with high accuracy that the angle of the earth's spin was 23.472 degrees. But someone suggested that the angle changes over time. Historical accounts were examined and 5 measurements were found dating from 1460 to 1570. These measurements were somewhat different than the Paris measurement, but they were done using much less precise methods. The question is whether the differences in the measurements can be attributed to the errors in measurement of the earlier observations or whether the angle of the earth's rotation actually changed. We need to test the hypothesis that the earth's angle has actually changed.

- Open Cassub.jmp (Stigler, 1986).

- Choose **Analyze > Distribution** and select Obliquity as the Y variable.

- Click **OK**.

The Distribution report shows a histogram of the 5 values.

**Figure 7.18**   Report of observed ecliptic values

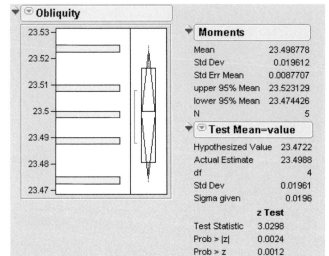

✍ To see the histogram in **Figure 7.18**, get the hand tool from the **Tools** menu (or toolbar) and stretch the bars to the right.

We now want to test that the mean of these values is different than the value from the Paris observatory.

✍ Select **Test Mean** from the popup menu at the top of the histogram.

✍ In the dialog that appears, enter the hypothesized value of 23.47222 (the value measured by the Paris observatory), and enter the standard deviation of 0.0196 found in the Moments table (see **Figure 7.18**).

✍ Click **OK**.

The z-test statistic has the value 3.0298. The area under the Normal curve to the right of this value is reported as Prob > z, which is the probability (*p*-value) of getting an even greater z-value if there was no difference. In this case, the *p*-value is 0.001.

Notice that we are only interested in whether the mean is greater than the hypothesized value. We therefore look at the value of Prob > z, a one-sided test. To test that the mean is different in either direction, the area in both tails is needed. This statistic is two-sided and listed as Prob >|z|, in this case 0.002. The one-sided test Prob < z has a *p*-value of 0.999, indicating that you are not going to prove that the mean is less than the hypothesized value. The two-sided *p*-value is always twice the smaller of the one-sided *p*-values.

## Student's *t*-Test

The *z*-test has a requirement. It requires the value of the standard deviation of the response, and thus the standard deviation of the mean estimate, be known. Usually this true standard deviation value is unknown and you have to estimate the standard deviation.

Using the estimate in the denominator of the statistical test computation requires an adjustment to the distribution that was used for the test. Instead of using a Normal distribution, statisticians use a *Student's t-distribution*. The statistic is called the *Student's t-statistic* and is computed by the formula shown to the right, where $x_0$ is the hypothesized mean and $s$ is the sample standard deviation of the sample data. In words, you can say

$$t = \frac{\bar{x} - x_0}{\frac{s}{\sqrt{n}}}$$

$$t\text{-statistic} = \frac{\text{sample mean} - \text{hypothesized value}}{\text{standard error of the mean}}$$

A large sample estimates the standard deviation very well, and the Student's *t*-distribution is not much different than the Normal distribution, as illustrated in **Figure 7.19**. However, in this example there were only 5 observations.

There is a different *t*-distribution for each number of observations, indexed by a value called *degrees of freedom*, which is the number of observations minus the number of parameters estimated in fitting the model. In this case, five observations minus one parameter (the mean) yields $5 - 1 = 4$ degrees of freedom. As you can see in **Figure 7.19**, the quantiles for the *t*-distribution spread out farther than the Normal when there are few degrees of freedom.

**Figure 7.19**   Comparison of Normal And Student's t Distributions

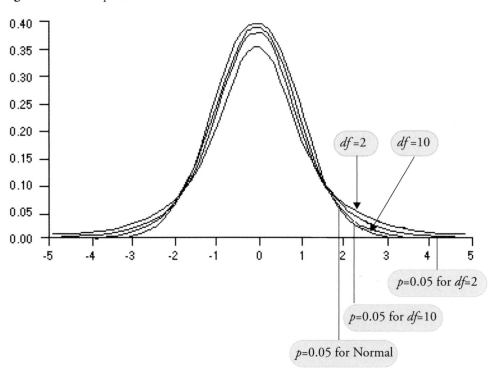

## Comparing the Normal and Student's t Distributions

JMP can produce an animation to show you the relationships in **Figure 7.19**. This demonstration uses the Normal vs. t.JSL script. For more information on opening and running scripts, see "Working With Scripts" on page 53.

🖱   Open the Normal vs t.JSL script.

🖱   Choose **Edit > Run Script**.

You should see the window shown in **Figure 7.19**.

**Figure 7.20**   Normal vs t comparison

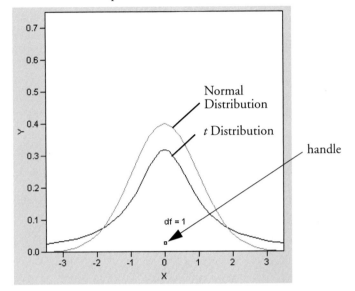

The small square located just above 0 is called a *handle*. It is draggable, and adjusts the degrees of freedom associated with the black *t*-distribution as it moves. The Normal distribution is drawn in red.

    Click and drag the handle up and down to adjust the degrees of freedom of the *t*-distribution.

Notice both the height and the tails of the *t*-distribution. At what number of degrees of freedom do you feel that the two distributions are close to identical?

## Testing the Mean

We now return to the ecliptic case study. It turns out that for a 5% two-tailed test, which uses a *p*-value of a 0.975, the *t*-quantile for 4 degrees of freedom is 2.7764, which is far greater than the corresponding *z*-quantile of 1.96. Let's do the same test again, using this different value.

    Select **Test Mean** and again enter 23.47222 for the hypothesized mean value. This time, do not fill in the standard deviation.

    Click **OK**.

The Test Mean table (shown here) now displays a *t*-test instead of a *z*-test.

When you don't specify a standard deviation, JMP uses the sample estimate of the standard deviation, but this is only an estimate because you didn't enter it yourself. The significance looks questionable, but the *p*-value of 0.039 still looks somewhat convincing, so you can conclude that the angle has changed. When you have a significant result, the idea is that under the null hypothesis, the expected value of the *t*-statistic is zero. It is highly unlikely (probability less than α) for the *t*-statistic to be so far out in the tails. Therefore, you don't put much belief in the null hypothesis.

| Test Mean=value | |
|---|---|
| Hypothesized Value | 23.4722 |
| Actual Estimate | 23.4988 |
| df | 4 |
| Std Dev | 0.01961 |
| **t Test** | |
| Test Statistic | 3.0280 |
| Prob > \|t\| | 0.0389 |
| Prob > t | 0.0194 |
| Prob < t | 0.9806 |

**Note**: You may have noticed that the test dialog offers the options of a Wilcoxon signed-rank nonparametric test. Some statisticians favor nonparametric tests because the results don't depend on the response having a Normal distribution. Nonparametric tests are covered in more detail in the chapter "Comparing Many Means: One-Way Analysis of Variance" on page 199.

## The *p*-Value Animation

Figure 7.17 on page 139 illustrates the relationship between the a two-tailed test and the Normal distribution. Some questions may arise after looking at this picture.

- How would the *p*-value change if the difference between the truth and my observation were different?

- How would the *p*-value change if my test were one-sided instead of two sided?

- How would the *p*-value change if my sample size were different?

To answer these questions, JMP provides an animated demonstration, written in JMP's scripting language. Often, these scripts are stored as separate files (samples are included in the JMP IN Scripts folder). However, some scripts are built into JMP. This *p*-value animation is an example of a built-in script.

☞  Select **PValue Animation** from the drop-down menu on the **Test Mean** outline bar, which produces the window in **Figure 7.21**.

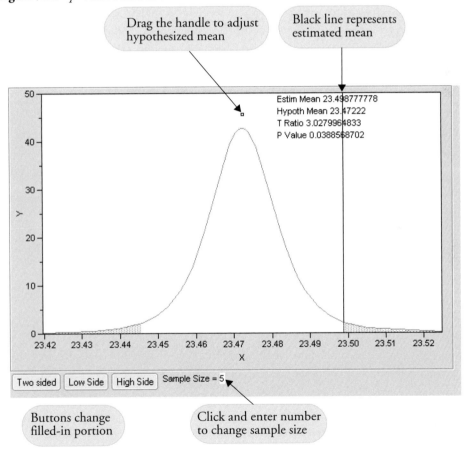

**Figure 7.21**  *p*-Value Animation Window for **Cassub** Data

The black vertical line represents the mean estimated by the **Cassub** data set. The handle can be dragged around the window with the mouse. In this case, the handle represents the true mean under the null hypothesis. To reject this true mean, there must be a significant difference between it and the mean estimated by the data.

The *p*-value calculated by JMP is affected by the difference between this true mean and the estimated mean, and you can see the effect of changing the true mean by dragging the handle.

🖑 Use the mouse to drag the handle left and right. Observe the changes in the *p*-value as the true mean is adjusted.

As expected, the *p*-value decreases as the difference between the true and hypothesized mean increases.

The effect of changing this mean is also illustrated geometrically. As illustrated previously in **Figure 7.17**, the shaded area represents the region where the null hypothesis is rejected. As the area of this region increases, the *p*-value of the test also increases. This demonstrates that the closer your estimated mean is to the true mean under the null hypothesis, the less likely you are to reject the null hypothesis.

This demonstration can also be used to extract other information about the data. For example, you can determine the smallest difference that your data would be able to detect for specific *p*-values. To determine the difference for $p = 0.10$:

🖑 Drag the handle until the *p*-value is as close to 0.10 as possible.

You can then read the estimated mean and hypothesized mean from the text display. The difference between these two numbers is the smallest difference that would be significant at the 0.10 level. Anything smaller would not be significant.

To see the difference between *p*-values for two and one sided tests, use the buttons at the bottom of the window.

🖑 Press the **High Side** button to change the test to a one-sided *t*-test.

The *p*-value decreases because the region where the null hypothesis is rejected has become larger—it is all piled up on one side of the distribution, so smaller differences between the true mean and the estimated mean become significant.

🖑 Repeatedly press the **Two Sided** and **High Side** buttons.

What is the relationship between the *p*-values when the test is one- and two-sided?

🖑 Click on the values for sample size beneath the plot to edit it and see the effect of different sample sizes.

## Power of the *t*-test

As discussed in the section "Testing Hypotheses: Terminology" on page 137, there are two types of error that a statistician is concerned with in conducting a statistical test—Type I and Type II. JMP contains a built-in script to graphically demonstrate the quantities involved in computing the power of a *t*-test.

🖰 In the same popup menu where you found the **Pvalue animation**, select **Power animation** to display the window shown in **Figure 7.22**.

**Figure 7.22**   Power Animation Window

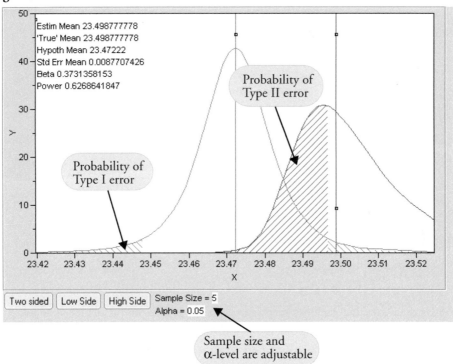

The probability of committing a Type I error (reject the null hypothesis when it is true), often represented by α, is shaded in red. The probability of committing a Type II error (not detecting a difference when there is a difference), often represented as β, is shaded in blue. Power is 1 – β, which is the probability of detecting a difference. The case where the difference is zero is examined below.

There are three handles in this window, one each for the estimated mean (calculated from the data), the true mean (an unknowable quantity that the data estimates), and the hypothesized mean (the mean assumed under the null hypothesis). These handles can be dragged around with your mouse to see how their positions affect power.

**Note:** Click on the values for sample size and alpha beneath the plot to edit them.

🖱 Drag the 'True' mean until it coincides with the hypothesized mean.

This simulates the situation where the true mean is the hypothesized mean in a test where $\alpha=0.05$. What is the power of the test?

🖱 Continue dragging the 'True' mean around the graph.

Can you make the probability of committing a Type II error smaller than the case above, where the two means coincide?

🖱 Drag the 'True' mean so that it is far away from the hypothesized mean.

Notice the shape of the blue distribution (around the 'True' mean) is no longer symmetrical. This is an example of a *non-central t-distribution*.

Finally, as with the *p*-value animation, these same situations can be further explored for one-sided tests using the buttons along the bottom of the window.

🖱 Explore different values for sample size and alpha.

# Practical Significance vs. Statistical Significance

Here are some data that demonstrate that a *statistically* significant difference can be quite different than a *practically* significant difference. Dr. Quick and Dr. Quack are both in the business of selling diets, and they have claims that appear contradictory. Dr. Quack studied 500 dieters and claims,

"A statistical analysis of my dieters shows a statistically significant weight loss for my Quack diet."

The Quick diet, by contrast, shows no significant weight loss by its dieters. Dr. Quick followed the progress of 20 dieters and claims,

"A statistical study shows that on average my dieters lose over three times as much weight on the Quick diet as on the Quack diet."

So which claim is right?

🖰   To compare the Quick and Quack diets, open the Diet.jmp sample data table.

**Figure 7.23** shows a partial listing of the Diet data table.

🖰   Choose **Analyze > Distribution** and assign both variables to Y.

🖰   Click **OK**.

🖰   Select **Test Mean** from the popup menu at the top of each plot to compare the mean weight loss to zero.

You should use the one-sided *t*-test because you are only interested in significant weight loss (not gain)

**Figure 7.23**   Diet Data

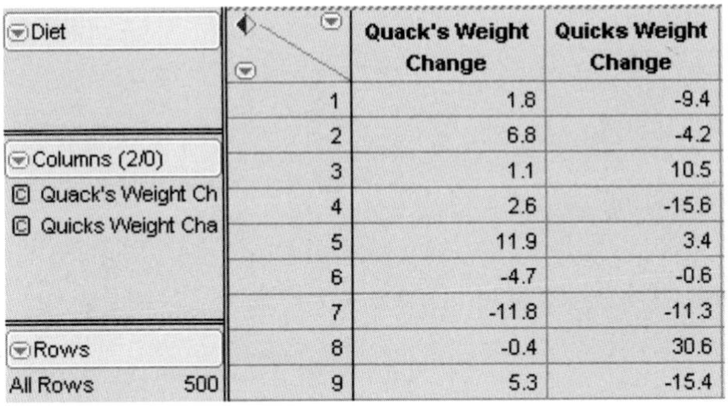

| Diet | | | Quack's Weight Change | Quicks Weight Change |
|---|---|---|---|---|
| | | 1 | 1.8 | -9.4 |
| | | 2 | 6.8 | -4.2 |
| Columns (2/0) | | 3 | 1.1 | 10.5 |
| ◎ Quack's Weight Ch | | 4 | 2.6 | -15.6 |
| ◎ Quicks Weight Cha | | 5 | 11.9 | 3.4 |
| | | 6 | -4.7 | -0.6 |
| | | 7 | -11.8 | -11.3 |
| Rows | | 8 | -0.4 | 30.6 |
| All Rows | 500 | 9 | 5.3 | -15.4 |

If you look closely at the *t*-test results in **Figure 7.24**, you can verify both claims!

**Figure 7.24**  Reports of the Quick and Quack Example

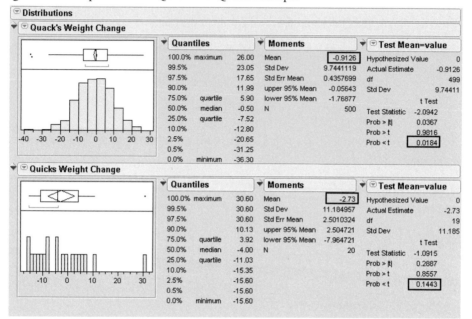

Quick's average weight loss of 2.73 is over three times the 0.91 weight loss reported by Quack. However, Quick's larger mean weight loss was not significantly different from zero, and Quack's small weight loss was significantly different from zero. Quack might not have a better diet, but he has more evidence—500 cases compared with 20 cases. So even though the diet produced a weight loss of less than a pound, it is statistically significant. Significance is about evidence, and having a large sample size can make up for having a small effect.

Dr. Quick needs to collect more cases, and then he can easily dominate the Quack diet (though it seems like even a 2.7-pound loss may not be enough of a practical difference to a customer).

If you have a large enough sample size, even a very small difference can be significant. If your sample size is small, even a large difference may not be significant.

Looking closer at the claims, note that Quick reports on the estimated difference between the two diets, where as Quack reports on the significance probabilities. Both are somewhat empty statements. It is not enough to report an estimate without a measure of variability. It is not enough to report a significance without an estimate of the difference.

The best report is a confidence interval for the estimate, which shows both the statistical and practical significance. The next chapter presents the tools to do a more complete analysis on data like the Quick and Quack diet data.

# Testing for Normality

Sometimes you may want to test whether a set of values has a particular distribution. Perhaps you are verifying assumptions and want to test that the values have a Normal distribution.

A widely used test that the data are from a specific distribution is the *Kolmogorov test* (also called the *Kolmogorov-Smirnov test*). The test statistic is the greatest absolute difference between the hypothesized distribution function and the empirical distribution function of the data. The empirical distribution function goes from 0 to 1 in steps of $1/n$ as it crosses data values. When the Kolmogorov test is applied to the Normal distribution and adapted to use estimates for the mean and standard deviation, it is called the *Lilliefor's test* or the *KSL test*. In JMP, the Lilliefor's quantiles on the cumulative distribution function (cdf) are translated into confidence limits in the Normal quantile plot, so that you can see where the distribution departs from Normality by where it crosses the confidence curves.

Another test of Normality produced by JMP is the *Shapiro-Wilk test* (or the *W-statistic*), which is implemented for samples as large as 2000.

- Look at the Birth Death.jmp data table again, or re-open it if it is closed.

- Choose **Analyze > Distribution** for the variables birth and death. Click **OK**.

- Select **Fit Distribution > Normal** and **Normal Quantile Plot** from the popup menu on the report title bar.

- Select **Goodness of Fit** from the popup menu next to the Fitted Normal report.

Its results are shown in **Figure 7.25**.

**Figure 7.25**  Test Distributions for Normality

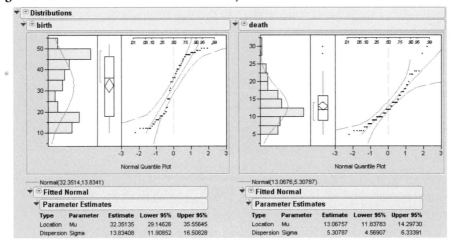

The conclusion is that neither distribution is Normal, though the second is much closer than the first.

This is an example of an unusual situation where you hope the test fails to be significant, since the null hypothesis is that the data are Normal. If you have a large number of observations, you may want to reconsider this tactic because the Normality tests will be sensitive to small departures from Normality, and such small departures would not jeopardize the other analyses you make (because of the central limit theorem), especially because they will also probably be highly significant.

All the distributional tests assume that the data are independent and identically distributed. The most frequent use of Normality tests, however, is for residuals from linear model fits, which have both different variances and are correlated. In reasonably large samples, these two problems are minimal. However, you are bound to learn more from diagnostic plots of the residuals. The distribution of the residuals and their relation to the factors will be more useful information than a significant $p$-value for a Normality test.

So far we have been doing correct statistics, but a few remarks are in order.

1. In most tests, the null hypothesis is something you want to disprove. It is disproven by the contradiction of getting a statistic that would be unlikely if the hypothesis were true. But in Normality tests, you want the null hypothesis to be true. Most testing for Normality is to verify assumptions for other statistical tests.

2. The mechanics for any test where the null hypothesis is desirable are backwards. You can get an undesirable result, but the failure to get it does not prove the opposite—it only says that you have insufficient evidence to prove it isn't true. "Special Topic: Practical Difference" below gives more details on this issue.

3. It is more likely to get a desirable (inconclusive) result if you have very little data. Conversely, if you have thousands of observations, almost any set of data from the real world will be significantly non-Normal.

4. If you have a large sample, the estimate of the mean will be distributed Normally even if the original data is not. This result, from the Central Limit Theorem, is demonstrated in a later section.

5. The test statistic itself doesn't tell you what the nature of the difference from Normality. The Normal quantile plot is better for this. Residuals from regressions can have both these problems.

# Special Topic: Practical Difference

Suppose you really want to show that the mean of some process is some value. Standard statistical tests are of no help, since the failure to show that a mean is *different* from the hypothetical value does not show that it *is* that value. It only says that there is not enough evidence to confirm that it isn't that value.

You can never show that a mean is exactly some hypothesized value, because the mean could be different from that hypothesized value by a very tiny amount. No matter what sample size you have, there is a value that is different from the hypothesized mean by an amount that is so small that it is very unlikely to get a significant difference even if the true difference is zero.

So instead of trying to show that the mean is one hypothesized value, you need to choose an interval around that hypothesized value and try to show that the mean is not outside that interval. This can be done.

There are many situations when you want to control a mean within some specification interval. For example, suppose that you make 20 amp electrical circuit breakers. You need to demonstrate that the mean breaking current for the population of breakers is between 19.9 and 20.1 amps. (Actually, you probably also require that most individual units be in some specification interval, but for now we just focus on the mean.)

The standard way to do this is *TOST method*, the acronym for Two One-Sided Tests [Westlake(1981), Schuirmann(1981), Berger and Hsu (1996)]:

- First you do a one-sided *t*-test that the mean is the low value of the interval, with an upper tail alternative.

- Then you do a one-sided *t*-test that the mean is the high value of the interval, with a lower tail alternative.

- If both tests are significant at some level $\alpha$, then you can conclude that the mean is outside the interval with probability less than or equal to $\alpha$, the significance level. In other words, the mean is not significantly practically different from the hypothesized value, or, in still other words, the mean is practically equivalent to the hypothesized value.

- The test works by a union intersection rule, not described here.

For example,

- Open the Coating.jmp sample data table.

- Select **Analyze > Distribution** and assign weight to the **Y, Columns** role.

- Click **OK**.

When the report appears,

- Select **Test Mean** from the platform drop-down menu and enter 20.2 as the hypothesized value.

- Select **Test Mean** again and enter 20.6 as the hypothesized value.

This tests that the mean weight is between 20.2 and 20.6 with a protection level ($\alpha$) of 0.05.

**Figure 7.26**   Compare Test for Mean at Two Values

| ▼ ⊙ Test Mean=value | | ▼ ⊙ Test Mean=value | |
|---|---|---|---|
| Hypothesized Value | 20.2 | Hypothesized Value | 20.6 |
| Actual Estimate | 20.3969 | Actual Estimate | 20.3969 |
| df | 31 | df | 31 |
| Std Dev | 1.47506 | Std Dev | 1.47506 |
| **t Test** | | **t Test** | |
| Test Statistic | 0.7550 | Test Statistic | -0.7790 |
| Prob > \|t\| | 0.4559 | Prob > \|t\| | 0.4419 |
| Prob > t | 0.2280 | Prob > t | 0.7791 |
| Prob < t | 0.7720 | Prob < t | 0.2209 |

The *p*-value for the hypothesis from below is 0.22, and the *p*-value for the hypothesis from above is also 0.22. Since both of these values are far above the alpha of 0.05 that we were looking for, we declare it not significant. The conclusion is that we have not shown that the mean is practically equivalent to 20.4 ± 0.2 at the 0.05 significance level. We need more data.

# Special Topic: Simulating the Central Limit Theorem

The Central Limit Theorem says that for a very large sample size, the sample mean is very close to Normally distributed, regardless of the shape of the underlying distribution. That is, if you compute means from many samples of a given size, the distribution of those means approaches Normality, even if the underlying population from which the samples were drawn is not.

You can see the Central Limit Theorem in action using the template called Cntrlmt.JMP.

🖱  Open Cntrlmt.JMP.

🖱  Right-click (Control-click on the Macintosh) in the column heading for the N=1 column and select **Formula** from the menu that appears.

🖱  Do this same thing for the rest of the columns, called N=5, N=10, and so on, to look at their formulas.

Looking at the formulas (shown to the right) should give you an idea of what's going on. The expression raising the uniform random number values to the 4th power creates a highly skewed distribution. For each row, the first column, N=1, generates a single uniform random number to the fourth power. For each row in the

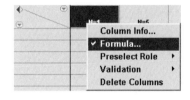

second column, N=5, the formula generates a sample of 5 uniform numbers, takes each to the fourth power, and computes the mean. The next column does the same for a sample size of 10, and the following columns generate means for sample sizes of 50 and 100.

🖱  Add 500 rows to the data table using **Rows > Add Rows**.

You can add as many rows to the table as you want. (But be forewarned that if you have a slow machine, the computations might take some time—each cell in the last column requires 100 random numbers, taken to the fourth power, added together, and divided by 100.) When the computations are complete:

🖱 Choose **Analyze > Distribution**. Select all the variables, assign them as Y variables and click **OK**.

You should see the results in **Figure 7.27**. When the sample size is only 1, the skewed distribution is apparent. As the sample size increases you can clearly see the distributions becoming more and more Normal.

**Figure 7.27**    Example of the Central Limit Theorem in Action

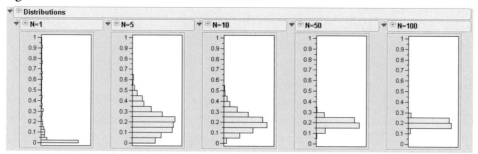

The distributions also become less spread out, since the standard deviation ($\sigma$) of a mean of $n$ items is $\dfrac{\sigma}{\sqrt{n}}$ .

🖱 To see this, select the **Uniform Axes** option in the popup menu beside the uppermost title bar for Distribution.

**Figure 7.28**    Distributions with Uniform Scales

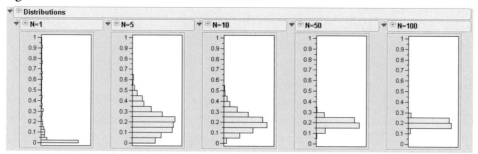

# Special Topic: Seeing Kernel Addition

The idea behind density estimators is not difficult. In essence, a Normal distribution is fit over each data point with a specified standard deviation. Each of these Normal distributions is then summed to produce the overall curve.

To see JMP animate this process for a simple set of data. For details on using scripts, see "Working With Scripts" on page 53.

    ⌐ Open the demoKernel.jsl script, and make sure the check box for **Run this script after opening** is checked.

You should see a window like that in **Figure 7.29**.

**Figure 7.29**   Kernel Addition Demonstration

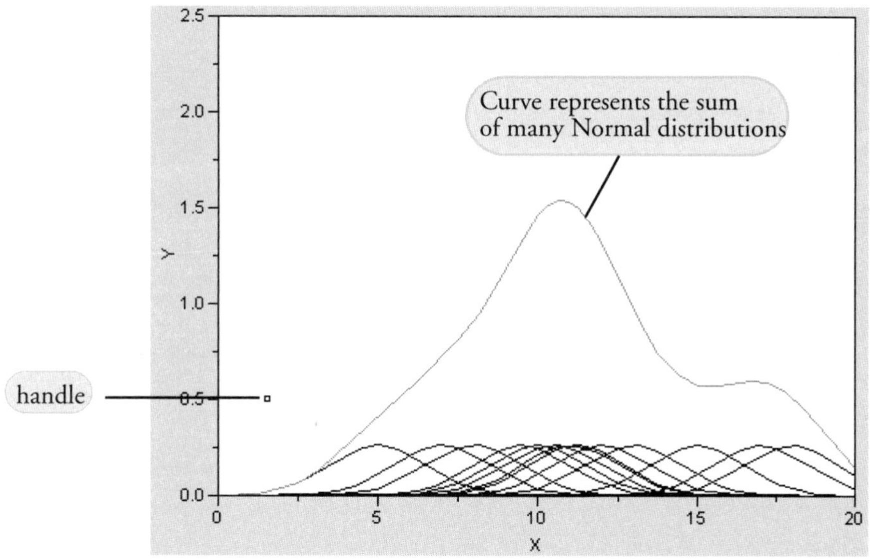

The handle on the left side of the graph can be dragged with the mouse.

    ⌐ Move the handle to adjust the spread of the individual Normal distributions associated with each data point.

The larger red curve is the smoothing spline generated by the sum of the Normal distributions. As you can see, merely adjusting the spread of the small Normal distributions dictates the smoothness of the spline fit.

# Exercises

1.  The file movies.jmp contains a list of the top grossing movies of all time (as of June 2003). It contains data showing the name of a movie, the amount of money it made in the U.S. (Domestic) and in foreign markets (in millions of dollars), its year of release, and the type of movie.

    (a)  Create a histogram of the types of movies in the data. What are the levels of this variable? How many of each level are in the data set?

    (b)  Create a histogram of the domestic gross for each movie. What is the range of values for this variable? What is the average domestic gross of these movies?

    (c)  Consider the histogram you created in part (b) for the domestic gross of these movies. You should notice several outliers in the outlier box plot. Based on your experience, guess what these movies are. Then, move the pointer over each outlier to reveal its name. Were you correct?

    (d)  Create a subset of the data consisting of only drama movies. Create a histogram and find the average domestic and foreign grosses for your subset. Are there outliers in either variable?

2.  The file Analgesics.jmp contains pain ratings from patients after treatments from three different pain relievers. The patients are labeled only by gender in this study. The study was meant to determine if the three pain relievers were different in the amount of pain relief the patients experienced.

    (a)  Create a histogram of the variables gender, drug, and pain. Click on the histogram bars to determine if the distribution of gender is roughly equal among the three analgesics.

    (b)  Create a separate histogram for the variable pain for each of the three different analgesics (Hint: Use the **By** button). Does the mean pain response seem the same for each of the three analgesics?

3.  The file Scores.jmp contains data for the US from the Third International Mathematics and Science Study, conducted in 1995. The variables came from testing over 5000 students for their abilities in Calculus and Physics, and are separated into four regions of the United States. Note that some students took the Calculus test, some took the Physics test, and some took both. Assume that the scores represent a random sample for each of the four regions of the U.S.

    (a)  Produce a histogram and find the mean scores for the U.S. on both tests. By clicking on the bars of the histogram, can you determine that a high calculus score correlates highly with a high Physics score?

(b)   Find the mean scores for the Calculus test for the four regions of the country. Do they appear to be roughly equal?

(c)   Find the mean scores for the Physics tests for the four regions of the country. Do they appear to be roughly equal?

(d)   Suppose that from an equivalent former test, the mean score of U.S. Calculus students was seen to be 450. Does this study show evidence that the score has increased since the last test?

(e)   Construct a 95% confidence interval for the mean calculus score.

(f)   Suppose that Physics teachers say that the overall U.S. score on the Physics test should be higher than 420. Do these data support their claim?

(g)   Construct a 95% confidence interval for the mean Physics score.

4.   The file Cereal.jmp contains nutritional information for 76 kinds of cereal.

(a)   Find the mean number of fat grams for the cereals in this data set. List any unusual observations.

(b)   Use the Distribution platform to find the two kinds of cereal with unusually high fiber content.

(c)   The hot/cold variable is used to specify whether the cereal was meant to be eaten hot or cold. Find the mean amount of sugar contained in the hot cereals and the cold cereals. Construct a 95% confidence interval for each.

5.   Various crime statistics for each of the 50 US states are stored in the file Crime.jmp.

(a)   Examine the distributions of each statistic. Which (if any) do not appear to follow a Normal distribution?

(b)   Which two states are outliers with respect to the robbery variable?

6.   Data for the Brigham Young football team are stored in the Football.jmp data file.

(a)   Find the average height and weight of the players on the team.

(b)   The position variable identifies the primary position of each player. What position has the smallest average weight? What has the highest?

(c)   What position has the largest neck measurements? What position (on average) can bench press the most weight?

7.  The Hot Dogs.jmp data file came from an investigation of the taste and nutritional content of hot dogs.

    (a)  Construct a histogram of the type of hot dogs (beef, meat, and poultry). Are there an equal number of each type considered?

    (b)  The $/oz variable represents the cost per ounce of hot dog. Construct an outlier box plot of this variable and find any outliers.

    (c)  Construct a 95% confidence interval for the caloric content and the protein content of the three types of hot dogs. Which type gives (on average) the highest protein?

    (d)  Test the conjecture tat the mean sodium content of all hot dogs is 410 grams.

8.  Three brands of typewriters were tested for typing speed by having expert typists type identical passages of text. The results are stored in Typing Data.jmp.

    (a)  Are the data for typing speeds Normally distributed?

    (b)  What is the mean typing speed for all typewriters?

    (c)  Find a 95% confidence interval for each of the three typewriter types.

# The Difference Between Two Means

# 8

## Overview

Are the mean responses from two groups different? What evidence would it take to convince you? This question opens the door to many of the issues that pervade statistical inference, and this chapter will explore these issues. Comparing group means also introduces an important statistical distinction regarding how the measurement or sampling process affects the way the resulting data are analyzed. This chapter also talks about validating statistical assumptions.

When two groups are considered, there are two very different situations that lead to two different analyses:

*Independent Groups*—the responses from the two groups are unrelated and statistically independent. For example, the two groups might be two classrooms with two sets of students in them. The responses come from different experimental units or subjects. The responses are uncorrelated and the means from the two groups are uncorrelated.

*Matched Pairs*—the two responses form a pair of measurements coming from the same experimental unit or subject. For example, a matched pair might be a before-and-after blood pressure measurement from the same subject. The responses are correlated, and the statistical method must take that into account.

# Two Independent Groups

For two different groups, the goal might be to estimate the group means and determine if they are significantly different. Along the way, it is certainly advantageous to notice anything else of interest about the data.

## When the Difference Isn't Significant

A study compiled height measurements from 63 children, all age 12. It's safe to say that as they get older, the mean height for males will be greater than for females, but is this the case at age 12? Let's find out:

🖱  Open Htwt12.jmp to see the data shown (partially) below.

There are 63 rows and three columns. This example uses Gender and Height. Gender has the Nominal modeling type, with codes for the two categories, "f" and "m". Gender will be the X variable for the analysis. Height contains the response of interest, and so will be the Y variable.

|  | Gender | Height | Weight |
|---|---|---|---|
| 1 | f | 62.3 | 105 |
| 2 | f | 63.3 | 108 |
| 3 | f | 58.3 | 93 |
| 4 | f | 58.8 | 89 |
| 5 | f | 59.5 | 78.5 |
| 6 | f | 61.3 | 115 |
| 7 | f | 56.3 | 83.5 |
| 8 | f | 64.3 | 110.5 |

Htwt12

Columns (3/0)
- Gender
- Height
- Weight

Rows
All Rows    63

## Check the Data

To check the data, first look at the distributions of both variables graphically with histograms and box plots.

🖱  Choose **Analyze > Distribution** from the menu bar.

🖱  In the launch dialog, select Gender and Height as Y variables.

🖱  Click **OK** to see an analysis window like the one shown in **Figure 8.1**.

Every pilot walks around the plane looking for damage or other problems before starting up. No one would submit an analysis to the FDA without making sure that the data was not confused with data from another study. Do your kids use the same computer that you do? Then check your data. Does your data have so many decimals of precision that it looks like it came from a random number generator? Great detectives let no clue go unnoticed. Great data analysts check their data carefully.

**Figure 8.1**   Histograms and Summary Tables

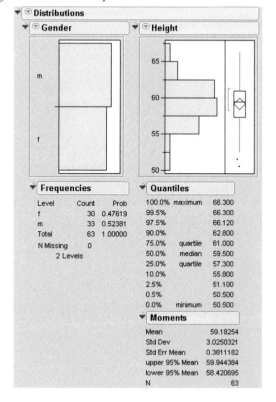

A look at the histograms for **Gender** and **Height** reveals that there are a few more males than females. The overall mean height is about 59, and there are no missing values (N is 63, and there are 63 rows in the table). The box plot indicates that two of the children seem unusually short compared to the rest of the data.

🖰   Move the cursor to the **Gender** histogram, and click on the bar for "f".

Clicking the bar highlights the females in the data table and also highlights the females in the **Height** histogram (See **Figure 8.2**). Now click on the "m" bar, which highlights the males and un-highlights the females.

By alternately clicking on the bars for males and females, you can see the distribution of the subsets highlighted in the **Height** histogram. This gives a preliminary look at the height distribution within each group, and it is these group means we want to compare.

**Figure 8.2**   Interactive Histogram

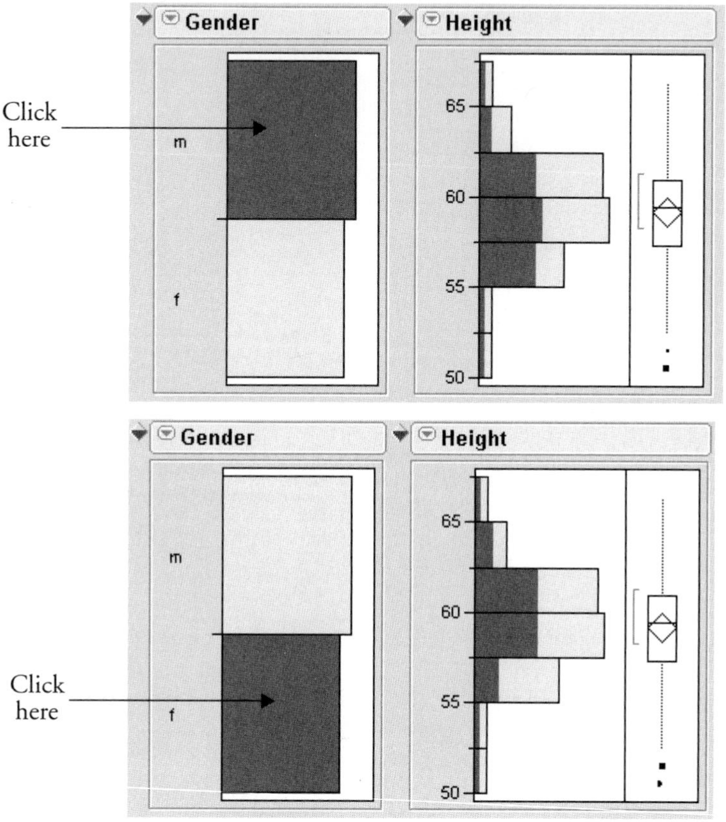

## Launch the Fit Y by X Platform

You can compare group means with the Fit Y by X platform by assigning Height as the continuous Y variable and Gender is the nominal (grouping) X variable. Begin by launching an analysis platform:

🖰   Choose **Analyze > Fit Y by X**.

🖰   In the launch dialog, select Height as Y and Gender as X.

Notice that the role-prompting dialog indicates that you are doing a one-way analysis of variance (ANOVA). Because Height is continuous and Gender is categorical (nominal), the **Fit Y by X** command automatically gives a one-way layout for comparing distributions.

🖰   Click **OK** to see the initial graphs, which are side-by-side vertical dot plots for each group (see the left picture in **Figure 8.3**).

**Figure 8.3**  Plot of the responses, before and after labeling points

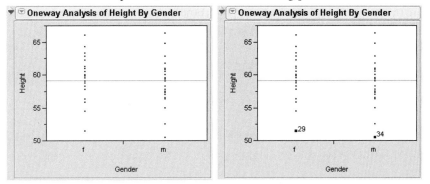

## Examine the Plot

The horizontal line across the middle shows the overall mean of all the observations. To identify possible outliers (students with unusual values):

🖱  Click the lowest point in the "f" vertical scatter, and Shift-click in the lowest point in the "m" sample.

Shift-clicking extends a selection so that the first selection does not un-highlight.

🖱  Choose **Rows > Label/Unlabel** to see the plot on the right in **Figure 8.2**.

Now the points are labeled 29 and 34, the row numbers corresponding to each data point.

## Display and Compare the Means

The next step is to display the group means in the graph, and to obtain an analysis of them.

🖱  Select **t test**.

🖱  Select **Means/Anova/Pooled t** from the red triangle popup menu that shows on the plot's title bar.

This adds analyses that estimates the group means and tests to see if they are different.

**Note:** Normally, you don't select both versions of the *t*-test. We're doing so here for illustration. To determine the correct test for other situations, see "Equal or Unequal Variances?" on page 167.

The **Means/Anova/Pooled t** option automatically displays the *means diamonds* as shown in **Figure 8.4**, with summary tables and statistical test reports.

The center lines of the means diamonds are the group means. The top and bottom of the diamonds form the 95% confidence intervals for the means. You can say the probability is 0.95 that this confidence interval contains the true group mean.

The confidence intervals show whether a mean is significantly different from some hypothesized value, but what can it show regarding whether two means are significantly different? Use the following rule.

> **Interpretation Rule for Means Diamonds:** If the confidence intervals shown by the means diamonds do not overlap, the groups are significantly different (but the reverse is not necessarily true).

It is clear that the means diamonds in this example do overlap, so you need to look closer at the text report beneath the plots to determine if the means are really different. The report, shown in **Figure 8.4**, includes summary statistics, *t*-test reports, an analysis of variance, and means estimates.

Note that the *p*-value of the *t*-test (shown under the label **Prob>|t|** in the **t test** section of the report) table is not significant.

**Figure 8.4**   Diamonds to Compare Group Means.

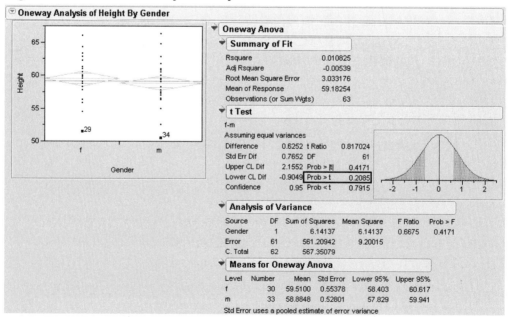

## Inside the Student's *t*-Test

The Student's *t*-test appeared in the last chapter to test whether a mean was significantly different from a hypothesized value. Now the situation is to test whether the difference of two means is significantly different from the hypothesized value of zero. The *t*-ratio is formed by first finding the difference between the estimate and the hypothesized value, and then dividing that quantity by its standard error.

$$t \text{ statistic} = \frac{\text{estimate} - \text{hypothesized value}}{\text{standard error of the estimate}}$$

In the current case, the estimate is the difference in the means for the two groups, and the hypothesized value is zero.

$$t \text{ statistic} = \frac{(\text{mean } 1 - \text{mean } 2) - 0}{\text{standard error of the difference}}$$

For the means of two independent groups, the standard error of the difference is the square root of the sum of squares of the standard errors of the means.

$$\text{standard error of the difference} = \sqrt{s^2_{\text{mean } 1} + s^2_{\text{mean } 2}}$$

JMP does the standard error calculations and forms the tables shown in **Figure 8.4**. Roughly, you look for a *t*-statistic greater than 2 in absolute value to get significance at the 0.05 level. The significance level (the *p*-value) is determined in part by the degrees of freedom (DF) of the *t*-distribution. For this case, DF is the number of observations (63) minus two, because two means are being estimated. With the calculated *t* (0.818) and DF, the *p*-value is determined to be 0.4165. The label Prob>|t| is given to the *p*-value in the test table to indicate that this is the probability of getting an even greater absolute *t* statistic. This is the significance level. Usually a *p*-value less than 0.05 is regarded as significant.

In this example, the *p*-value of 0.4165 isn't small enough to detect a significant difference in the means. Is this to say that the means are the same? Not at all. You just don't have enough evidence to show that they are different. If you collect more data, you might be able to show a significant, albeit small, difference.

## Equal or Unequal Variances?

The report shown in **Figure 8.5** contains two *t*-test reports. The uppermost of the two is labeled **Assuming equal variances**, and is generated with the **Means/Anova/Pooled t** command. The other is labeled **Assuming unequal variances**, and is generated with the **t test** command. Which is the correct report to use?

**Figure 8.5**  *t*-test and ANOVA reports

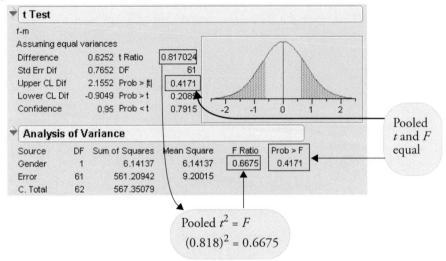

In general, the unequal-variance *t*-test (also known as the *unpooled t*-test) is the preferred test. This is because the unpooled version is quite sensitive to departures from the equal-variance assumption (especially if the number of observations in the two groups is not the same), and often we cannot assume the variances of the two groups are equal. In addition, if the two variances are unequal, the unpooled test maintains the prescribed $\alpha$-level and retains good power. For example, you may think you are conducting a test with $\alpha = 0.05$, but it may in fact be 0.10 or 0.20. What you think is a 95% confidence interval may be, in reality, an 80% confidence interval. (Cryer and Wittmer, 1999) For these reasons, we recommend the unpooled (**t test** command) *t*-test for most situations.

However, the equal-variance variation is included for several reasons.

- For situations with very small sample sizes (for example, having three or fewer observations in each group), the pooled version has slightly more power.

- Pooling the variances is the only option when there are more than two groups, when the *F*-test must be used. Therefore, the pooled *t*-test is a useful analogy for learning the analysis of the more general, multi-group situation. This situation is covered in the next chapter, "Comparing Many Means: One-Way Analysis of Variance" on page 199.

**Rule for *t*-tests:** Unless you have very small sample sizes, or a specific *a priori* reason for assuming the variances are equal, use the *t*-test produced by the **t test** command. When in doubt, use the **t test** command (*i.e.* unpooled) version.

## One-Sided Version of the Test

The Student's *t*-test in the previous example is for a two-sided alternative. In our situation, the difference could go either way, so a two-sided test is appropriate. If you only want to test in one direction, then you can do a little arithmetic on the reported *p*-value, forming one-sided *p* values by using

$$\frac{p}{2} \text{ or } 1 - \frac{p}{2},$$

depending on the direction of the alternative.

**Figure 8.6**  Two- and one-sided $t$-test

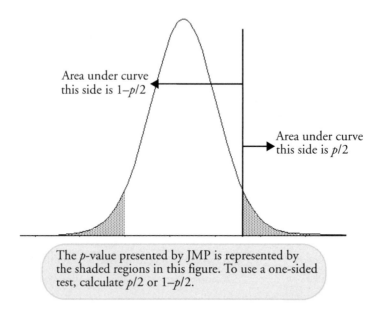

Area under curve
this side is $1-p/2$

Area under curve
this side is $p/2$

> The $p$-value presented by JMP is represented by
> the shaded regions in this figure. To use a one-sided
> test, calculate $p/2$ or $1-p/2$.

In this example, the mean for males was less than the mean for females, so the significance with the alternative to conclude the females higher would be the $p$-value of 0.2083, half the two-tailed $p$-value. Testing the other direction, the $p$-value is 0.7917.

## Analysis of Variance and the All-Purpose $F$-Test

As well as showing the $t$-test, the report in **Figure 8.5** for comparing two groups shows an analysis of variance with its $F$-test. The $F$-test surfaces many times in the next few chapters, so an introduction here is in order. Details will unfold later.

The $F$-test compares variance estimates for two situations, one a special case of the other. Not only is this useful for testing means, but other things, too. Furthermore, when there are only two groups, the $F$-test is equivalent to the pooled (equal variance) $t$-test, and the $F$-ratio is the square of the $t$-ratio. $(0.81)^2 = 0.66$, as you can see in **Figure 8.5**.

To begin, look at the different estimates of variance as reported in the Analysis of Variance table.

First, the analysis of variance procedure pools all responses into one big population and estimates the population mean (the *grand mean*). The variance around that grand mean is estimated by taking the average sum of squared differences of each point from the grand mean.

> The difference between a response value and an estimate such as the mean is called a *residual,* or sometimes the *error.*

In the Analysis of Variance table, this estimate of the variance is in the line labeled C Total (corrected total—corrected for the mean). Both the Sum of Squares and the Mean Square are shown. The Mean Square shown for C Total is the estimate of the variance around the grand mean. Its square root is the estimate of the standard deviation. (Refer back to **Figure 8.4** to see that the standard deviation is 3.025, the square root of the 9.15 reported as the Mean Square for C Total.)

What happens when a separate mean is computed for each group instead of the grand mean for all groups? The variance around these individual means is calculated, and this is shown in the Error line in the Analysis of Variance table. The Mean Square for Error is the estimate of this variance, called *residual variance* (also called $s^2$), and its square root, called the *root mean squared error* (or *s*), is the residual standard deviation estimate.

If the true group means are different, then the separate means give a better fit than the one grand mean. In other words, there will be less variance using the separate means than when using the grand mean. The change in the residual sum of squares from the single-mean model to the separate-means model leads us to the *F*-test shown in the Model line of the Analysis of Variance table. The Mean Square for Model also estimates the residual variance if the hypothesis that the means are the same is true.

The *F*-ratio is the Model Mean Square divided by the Error Mean Square:

$$F\text{-Ratio} = \frac{\text{Mean Square for the Model}}{\text{Mean Square for the Error}} = \frac{6.141}{9.200} = 0.6675$$

The *F*-ratio is a measure of improvement in fit when separate means are considered. If there is no difference between fitting the grand mean and individual means, then both numerator and denominator estimate the same variance (the grand mean residual variance), so the *F*-ratio is around 1. However, if the separate-means model does fit better, the numerator (the model mean square) contains more than just the grand mean residual variance, and the value of the *F*-test increases.

If the two mean squares in the *F*-ratio are statistically independent (and they are in this kind of analysis) then you can use the *F*-distribution associated with the *F*-ratio to get a *p*-value, which tells how likely you are to see the *F*-ratio given by the analysis if there was really no difference in the means.

If the tail probability ($p$-value) associated with the $F$-ratio in the $F$-distribution is smaller than 0.05 (or the $\alpha$-level of your choice), you can conclude that the variance estimates are different, and thus that the means are different.

In this example, the total mean square and the error mean square are not much different; in fact the $F$-ratio is actually less than one, and the $p$-value of 0.4171 (exactly the same as seen for the pooled $t$-test) is far from significant (much greater that 0.05).

The $F$-test can be viewed as whether the variance around the group means (the histogram on the left in **Figure 8.7**) is significantly less than the variance around the grand mean (the histogram on the right). In this case, the variance isn't much different.

In this way, a test of variances is also a test on means. The $F$-test turns up again and again because it is oriented to comparing the variation around two models. Most statistical tests can be constituted this way.

**Figure 8.7** shows histograms that compare residuals from a group means model and a grand mean model. If the effect were significant, the variation showing on the left would have been much less than that on the right.

**Figure 8.7**   Residuals for Group Means Model (left) and Grand Mean Model (right)

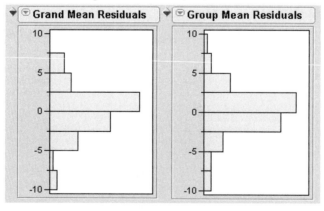

**Terminology for Sums of Squares:** All disciplines that use statistics use analysis of variance in some form. However, you may find different names used for its components. For example, the following are different names for the same kinds of sums of squares (SS):

SS(model) = SS(regression) = SS (between)

SS(error) = SS(residual) = SS(within)

## How Sensitive Is the Test?
## How Many More Observations Are Needed?

So far, in this example, there is no conclusion to report because the analysis failed to show anything. This is an uncomfortable state of affairs. It is tempting to state that we have shown no significant difference, but in statistics this claim is the same as saying inconclusive. Our conclusions can be attributed to not having enough data as easily as to there being a very small true effect.

When a test is not significant, there is additional information you can find out that might be useful:

• How many more observations would make the reported difference become significant? This is called the *Least Significant Number* (LSN).

• How small a difference could the significance test in this example detect? This is called the *Least Significant Value* (LSV). Least significant values are illustrated in conjunction with the *p*-value animation on page 144.

Here is how to address these questions:

🖰 Select **Power** from the popup menu on the Oneway Anova title bar. The analysis first displays the Power Details Dialog shown in **Figure 8.8**.

🖰 In the **Power Details** dialog, check the **Solve for Least Significant Number** and **Solve for Least Significant Value** boxes.

🖰 Click **Done**.

A report titled "Power Details" appears, as shown to the right in **Figure 8.8**. (The concept of Power is discussed in a Chapter 7, "Comparing Many Means: One-Way Analysis of Variance" on page 199.)

**Figure 8.8**  Power Details Dialog and Power Details Report

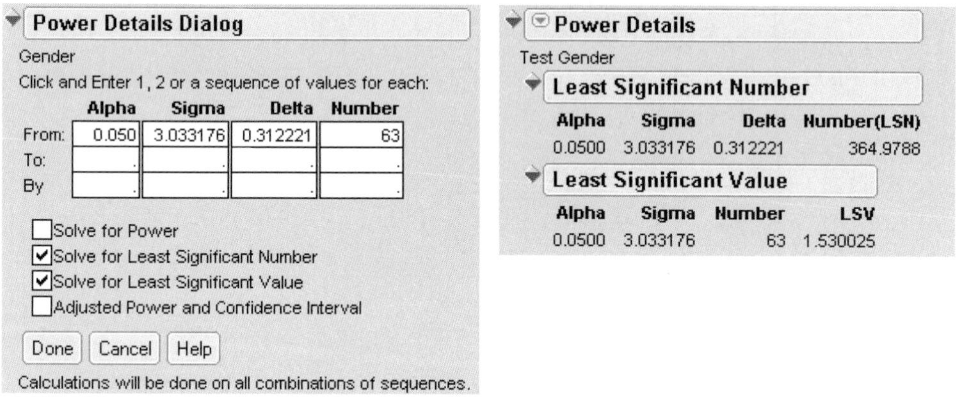

The least significant number (LSN) refers to the sample size. In this case, the actual sample size was 63. But with the variances and differences encountered, the report says you need a sample size of 365 (instead of only 63) to detect a significant difference. If you are planning to collect more data, you should get many more observations than the 365 suggested by the LSN, because the probability of finding a significant difference with LSN observations can be as low as 50%.

The least significant value (LSV) refers to the difference in means. The actual mean height difference in this example is 59.51–58.88 = 0.63 inches (The values 59.51 and 58.88 came from the means table, above the power details report). However, the Power Details report shows that with the existing sample sizes and variances, you need to measure differences as large as 1.53 to declare that the means are significantly different. This is the *sensitivity* of the test. The sensitivity of this test is that you would not have been able to detect differences less than 1.53 inches.

Knowing the sensitivity converts the inconclusive result into one that says that if there is a real difference, it is (95%) likely to be less than 1.53 inches. This converts the inconclusive result into an option to collect more data if the question is still important with this small an effect.

Note, however, that there is no new information in the LSN and LSV. They are merely showing the same results on a different scale.

## When the Difference Is Significant

The 12-year-olds in the previous example don't have significantly different average heights, but let's take a look at the 15-year-olds.

To start, open the sample table called Htwt15.jmp.

Then, proceed as before:

🖱 Choose **Analyze > Fit Y by X**, with Gender as X and Height as Y.

🖱 Click **OK.**

🖱 Select **Means/Anova/t test** from the red triangle popup menu showing beside the report title to add the analysis that estimates group means and tests to see if they are different.

You see the plot and tables shown in **Figure 8.9**.

**Figure 8.9** Analysis for mean heights of 15-year-olds

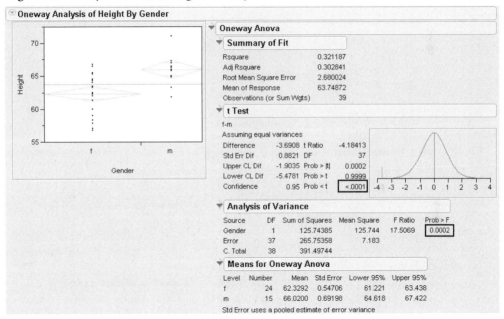

**Note:** We would normally recommend the unpooled (**t test** command) version of the test. We're using the pooled version as a basis for comparison.

The results for the analysis of the 15-year-old heights are completely different than the results for 12-year-olds. Here, the males are significantly taller than the females. You can see this because the confidence intervals shown by the means diamonds do not overlap. You can also see that the $p$-values for the $t$-test ($< 0.001$) and $F$-test (0.0002) are highly significant.

The *F*-test results say that the variance around the group means is significantly less than the variance around the grand mean. These two variances are shown in the histograms of **Figure 8.10**.

**Figure 8.10** Histograms of grand means variance and Group Mean Variance

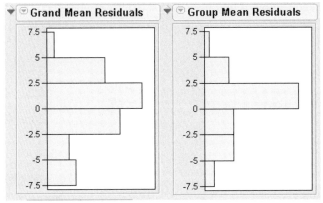

Let's look at the power details for this analysis, as we did in the previous one.

- 🖰 Select **Power** from the red triangle menu next to the Oneway Anova table name. Request the LSV and LSN by clicking the appropriate boxes.

- 🖰 Click **Done**, and the power analysis shown to the right appears.

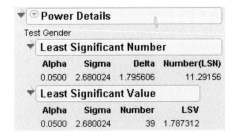

The difference for our data, from the means table, of 66.02–62.33 = 3.69 is much greater than the difference that could have been detected as significant at 0.05, which is the LSV of 1.78. There were 39 observations, but only 12 (the LSN) might have been needed given the differences and variances in this example.

# Normality and Normal Quantile Plots

The *t*-tests (and *F*-tests) used in this chapter assume that the sampling distribution for the group means is the Normal distribution. With sample sizes of at least 30 for each group, Normality is probably a safe assumption. The Central Limit Theorem says that means approach a Normal distribution as the sample size increases even if the original data are not Normal.

If you suspect non-Normality (due to small samples, or outliers, or a non-Normal distribution), consider using nonparametric methods, covered at the end of this chapter.

To assess Normality, use a Normal quantile plot. This is particularly useful when overlaid for several groups, because so many attributes of the distributions are visible in one plot.

🖰  Return to the Fit Y by X platform showing **Height** by **Gender** for the 12-year-olds and select **Normal Quantile Plot > Plot Actual by Quantile in** the popup menu on the report title bar.

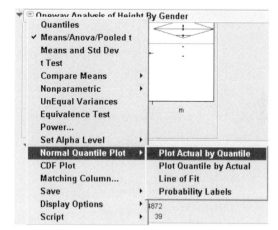

🖰  Do the same for the 15-year-olds.

The resulting plot (**Figure 8.11**) shows the data compared to the Normal distribution. The Normality is judged by how well the points follow a straight line. In addition, the Normal Quantile plot gives other useful information:

- The standard deviations are the slopes of the straight lines. Lines with steep slopes represent the distributions with the greater variances.

- The vertical separation of the lines at the middle shows the difference in the means. The separation of other quantiles shows at other points on the *x*-axis.

The first graph shown in **Figure 8.11** confirms that heights of 12-year-old males and females have nearly the same mean and variance—the slopes (standard deviations) are the same and the positions (means) are only slightly different.

The second graph in **Figure 8.11** shows 15-year-old males and females have different means and different variances—the slope (standard deviation) is higher for the females, but the position (mean) is higher for the males.

The distributions for all groups look reasonably Normal since the points (generally) cluster around their corresponding line.

**Figure 8.11**   Normal Quantile Plots for 12-year-olds and 15-year-olds

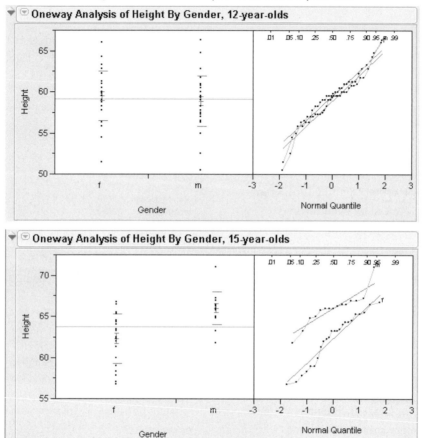

# Testing Means for Matched Pairs

Consider a situation where two responses form a pair of measurements coming from the same experimental unit. A typical situation is a before-and-after measurement on the same subject. The responses are correlated, and if only the group means are compared—ignoring the fact that the groups have a pairing —information is lost. The statistical method called the *paired t-test* allows you to compare the group means, while taking advantage of the information gained from the pairings.

In general, if the responses are positively correlated, the paired *t*-test gives a more significant *p*-value than the *t*-test for independent means (grouped *t*-test) discussed in the previous sections. If responses are negatively correlated, then the paired *t*-test is less significant than the grouped *t*-test. In most cases where the pair of measurements are taken from the same individual at different times, they are positively correlated, but be aware that it is possible for the correlation to be negative.

## Thermometer Tests

A health care center suspected that temperature readings from a new ear drum probe thermometer were consistently higher than readings from the standard oral mercury thermometer. To test this hypothesis, two temperature readings were taken, once with the ear-drum probe, and the other with the oral thermometer. Of course, there was variability among the readings, so they were not expected to be exactly the same. However, the suspicion was that there was a systematic difference—that the ear probe was reading too high.

   For this example, open the Therm.jmp data file.

A partial listing of the data table appears in **Figure 8.12**. The Therm.jmp data table has 20 observations and 4 variables. The two responses are the temperatures taken orally and tympanically (by ear) on the same person on the same visit.

**Figure 8.12**   Comparing Paired Scores

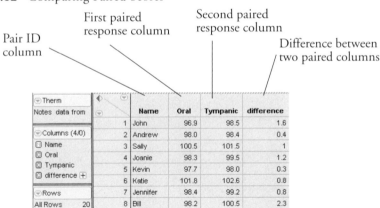

For paired comparisons, the two responses need to be arranged in two columns, each with a continuous modeling type. This is because JMP assumes each row represents a single experimental unit. Since the two measurements are taken from the same person, they belong in the same row. It is also useful to create a new column with a formula to calculate the difference between the two responses. (If your data table is arranged with the two responses in different rows, then use the **Tables > Split** command to rearrange it. For more information, see "Juggling Data Tables" on page 47.)

## Look at the Data

Start by inspecting the distribution of the data. To do this:

- ☞ Choose **Analyze > Distribution** with Oral and Tympanic as Y variables.

- ☞ When the results appear, select **Uniform Scaling** from the popup menu on the Distribution title bar to display the plots on the same scale.

The histograms (in **Figure 8.13**) show the temperatures to have different distributions. The mean looks higher for the Tympanic temperatures. However, as you will see later, this side-by-side picture of each distribution can be very misleading if you try to judge the significance of the difference from this perspective.

What about the outliers at the top end of the Oral temperature distribution? Are they of concern? Can you expect the distribution to be Normal? Not really. *It is not the temperatures that are of interest, but the difference in the temperatures.* So there is no concern about the distribution so far. If the plots showed temperature readings of 110 or 90, there would be concern, because that would be suspicious data for human temperatures.

**Figure 8.13**   Plots and Summary Statistics for Temperature

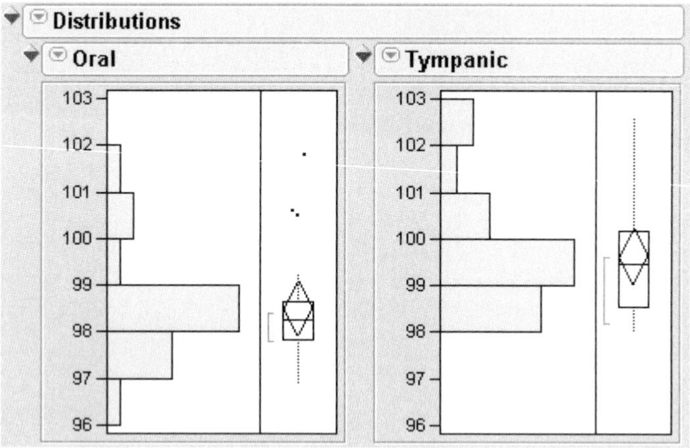

## Look at the Distribution of the Difference

The comparison of the two means is actually a comparison of the difference between them. Inspect the distribution of the differences:

- ☞ Choose **Analyze > Distribution** with difference as the Y variable.

The results (shown in **Figure 8.14**) show a distribution that seems to be above zero. In the Moments table, the lower 95% limit for the mean is 0.828—greater than zero. The Student's *t*-test will show the mean to be significantly above zero.

**Figure 8.14**   Histogram and Moments of the Difference Score

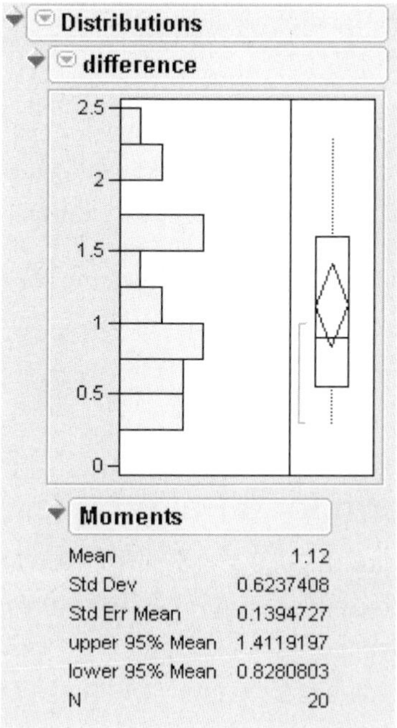

## Student's *t*-test

🖱  Choose **Test Mean** from the popup menu on the report title bar for the scatterplot of the difference variable. When prompted for a hypothesized value, click **OK** to accept the default value of zero.

🖱  If the box for the Wilcoxon's signed-rank test is checked, uncheck it—this is be covered later.

🖱  Click **OK**.

Now you have the *t*-test for testing that the mean over the matched pairs is the same. In this case, the results in the Test Mean table, shown to the right, show a *p*-value of less than 0.0001, which supports our visual guess that there is a significant difference between methods of temperature taking; the tympanic temperatures are significantly higher than the oral temperatures.

| Test Mean=value | |
| --- | --- |
| Hypothesized Value | 0 |
| Actual Estimate | 1.12 |
| df | 19 |
| Std Dev | 0.62374 |
| **t Test** | |
| Test Statistic | 8.0302 |
| Prob > \|t\| | <.0001 |
| Prob > t | <.0001 |
| Prob < t | 1.0000 |

There is also a nonparametric test, the Wilcoxon signed-rank test, described at the end of this chapter, that tests the difference between two means. This test is produced by checking the appropriate box on the test mean dialog.

🖱  Choose **Test Mean** from the popup menu next on the **difference** report title bar and accept zero as the hypothesized value. This time, make sure the box for Wilcoxon's signed-rank test is checked.

🖱  Click **OK**.

## The Matched Pairs Platform for a Paired *t*-Test

JMP offers a special platform for the analysis of paired data. The Matched Pairs platform compares means between two response columns using a paired *t*-test. The primary plot in the platform is a plot of the difference of the two responses on the *y*-axis, and the mean of the two responses on the *x*-axis. This graph is the same as a scatterplot of the two original variables, but turned 45°. A 45° rotation turns the original coordinates into a difference and a sum. By rescaling, this plot can show a difference and a mean, as illustrated in **Figure 8.15**.

**Figure 8.15**   Transforming to Difference by Sum is a Rotation by 45°

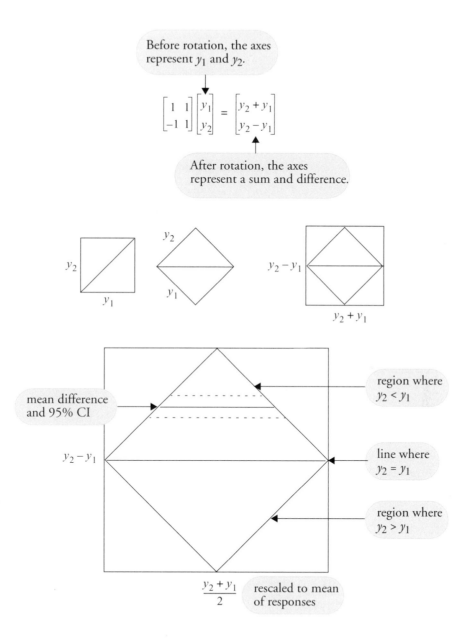

Before rotation, the axes represent $y_1$ and $y_2$.

$$\begin{bmatrix} 1 & 1 \\ -1 & 1 \end{bmatrix} \begin{bmatrix} y_1 \\ y_2 \end{bmatrix} = \begin{bmatrix} y_2 + y_1 \\ y_2 - y_1 \end{bmatrix}$$

After rotation, the axes represent a sum and difference.

region where $y_2 < y_1$

mean difference and 95% CI

line where $y_2 = y_1$

region where $y_2 > y_1$

$\dfrac{y_2 + y_1}{2}$   rescaled to mean of responses

There is a horizontal line at zero, and a confidence interval is plotted using dashed lines. If the confidence interval does not contain the horizontal zero line, the test detects a significant difference.

Seeing the platform in use reveals its usefulness.

☞ Choose **Analyze > Matched Pairs** and use Oral and Tympanic as the paired responses.

☞ Click **OK** to see a scatterplot of Tympanic and Oral as a matched pair.

To see the rotation of the scatterplot in **Figure 8.16** more clearly:

☞ Select the **Reference Frame** option from the popup menu on the Matched Pairs title bar.

☞ Right-click on the vertical axis and choose **Revert Axis** from the menu that appears.

**Figure 8.16**   Scatterplot of Matched Pairs Analysis

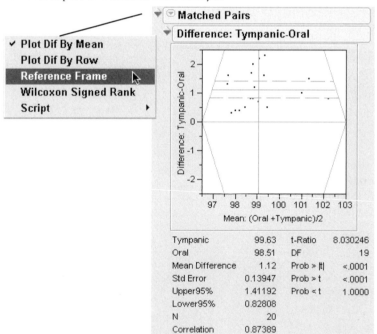

The analysis first draws a reference line where the difference is equal to zero. This is the line where the two columns are equal. If the means are equal, then the points should be evenly distributed along this line; you should see about as many points above this line as below. If a point is above the reference line, it means that the difference is greater than zero. In this example, points above the line show the situation where the Tympanic temperature is greater than the Oral temperature.

Parallel to the reference line at zero is a solid red line that is displaced from zero by an amount equal to the difference in means between the two responses. This red line is the line of fit for the sample. The test of the means is equivalent to asking if the red line through the points is significantly separated from the reference line at zero.

The dashed lines around the red line of fit show the 95% confidence interval for the difference in means.

This scatterplot gives you a good idea of each variable's distribution, as well as the distribution of the difference.

> **Interpretation rule for the Paired *t*-test scatterplot:** If the confidence interval (represented by the dashed lines around the red line) contains the reference line at zero, then the two means are not significantly different.

Another feature of the scatterplot is that you can see the correlation structure. If the two variables are positively correlated, they lie closer to the line of fit, and the variance of the difference is small. If the variables are negatively correlated, then most of the variation is perpendicular to the line of fit, and the variance of the difference is large. It is this variance of the difference that scales the difference in a *t*-test and determines whether the difference is significant.

The paired *t*-test table beneath the scatterplot of **Figure 8.16** gives the statistical details of the test. The results should be identical to those shown earlier in the Distribution platform. The table shows that the observed difference in temperature readings of 1.12 degrees is significantly different from zero.

## Optional Topic:
## An Equivalent Test for Stacked Data

There is a third approach to the paired *t*-test. Sometimes, you receive grouped data with the response values stacked into a single column instead of having a column for each group. Here is how to rearrange the Therm.jmp data table and see what a stacked table looks like:

- Choose **Tables > Stack**.

- When the Stack dialog appears, select Oral and Tympanic as the columns to be stacked. Name the _Stacked_ column Temperature and the _ID_ column Type.

- Click **OK** to see the data shown here.

The response values (temperatures) are in the Temperature column, identified as "Oral" or "Tympanic" by the Type column.

| | Name | difference | Type | Temperature |
|---|---|---|---|---|
| 1 | John | 1.6 | Oral | 96.9 |
| 2 | John | 1.6 | Tympanic | 98.5 |
| 3 | Andrew | 0.4 | Oral | 98.0 |
| 4 | Andrew | 0.4 | Tympanic | 98.4 |
| 5 | Sally | 1 | Oral | 100.5 |
| 6 | Sally | 1 | Tympanic | 101.5 |
| 7 | Joanie | 1.2 | Oral | 98.3 |
| 8 | Joanie | 1.2 | Tympanic | 99.5 |

(Side panel: Untitled, Source, Columns (4/0): Name, difference, Type, Temperature; Rows, All Rows 40)

If you do

**Analyze > Fit Y by X**

with Temperature (the response of both temperatures) as Y and Type (the classification) as X, then select **t test** from the popup menu, you get the *t*-test designed for independent groups. But this is inappropriate for paired data.

However, fitting a model that includes an adjustment for each person fixes the independence problem because the correlation is due to temperature differences from person to person. To do this, you need to use the Fit Model command, covered in detail in Chapter 12, "Fitting Models." The response is modeled as a function of both the category of interest (Type—Oral or Tympanic) and the Name category that identifies the person.

- Choose **Analyze > Fit Model**.

- When the Fit Model dialog appears, add Temperature as Y, and both Type and Name as Model Effects.

- Click **Run Model**.

When you get the results, the *p*-value for the category effect is identical to the *p*-value that you would get from the ordinary paired *t*-test; in fact the *F*-ratio in the effect test is exactly the square of the *t*-test value in the paired *t*-test. In this case the formula is

(Paired *t*-test statistic)$^2$ = $8.032^2$ = 64.4848 = (stacked *F*-test statistic)

The Fit Model platform gives you a plethora of information, but for this example you need only open the Effect Test table (**Figure 8.17**). It shows an *F*-ratio of 64.48, which is exactly the square of the *t*-ratio of 8.03 found with the previous approach. It's just another way of doing the same test.

**Figure 8.17**   Equivalent *F*-test on Stacked Data

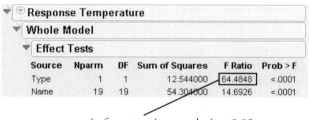

t-ratio from previous analysis = 8.03
*F*-ratio = square of *t*-ratio = 64.48

The alternative formulation for the paired means covered in the this section is important for cases in which there are more than two related responses. Having many related responses is a *repeated-measures* or *longitudinal* situation. The generalization of the paired *t*-test is called the *multivariate* or $T^2$ approach, whereas the generalization of the stacked formulation is called the *mixed-model* or *split-plot* approach.

# Special Topic: The Normality Assumption

The paired *t*-test assumes the differences are Normally distributed. With 30 pairs or more, this is probably a safe assumption—the results are reliable even if the distribution is not very Normal. The temperatures example only has 20 observations, so some people may like to check the Normality. To do this:

⟡   Use the Therm.jmp data table and do **Analyze > Distribution** on the variable difference.

⟡   Select **Normal Quantile Plot** and **Quantile Box Plot** from the popup menu on the Distribution platform title bar.

🖰 Select **Fit Distribution > Normal**
from the same menu as shown here.

🖰 Scroll down to the Fitted Normal
report, and select **Goodness of Fit**
from the red triangle popup menu
found on its title bar (See **Figure
8.18**).

The quantile plot, outlier box plot, and the
S-shaped Normal quantile plot are all
indicative of a slightly skewed distribution.
The Goodness of Fit table in **Figure 8.18**,
with a *p*-value of 0.1326, also indicates that
the distribution might not be Normal.

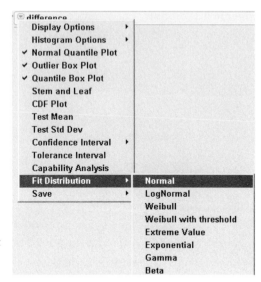

If you are concerned with non-Normality, nonparametric methods or a data transformation
should be considered.

**Figure 8.18**    Looking at the Normality of the Difference Score

# Two Extremes of Neglecting the Pairing Situation: A Dramatization

What happens if you do the wrong test? What happens if you do a *t*-test for independent groups on highly correlated paired data?

Consider the following two data tables:

🖱 Open the sample data table called Bptime.jmp to see the left-hand table in **Figure 8.25**.

This table represents blood pressure measured for ten people in the morning and again in the afternoon. The hypothesis is that, on average, the blood pressure in the morning is the same as it is in the afternoon.

🖱 Open the sample data table called BabySleep.jmp to see the right-hand table in **Figure 8.19**.

In this table, a researcher examined ten two-month-old infants at 10 minute intervals over a day to count the intervals in which a baby was asleep or awake. The hypothesis is that at two months old, the asleep time is equal to the awake time.

**Figure 8.19**   The Bptime and Babysleep data tables

| bpTime | | | BP AM | BP PM | Dif | | babySleep | | | Awake | Asleep | Dif |
|---|---|---|---|---|---|---|---|---|---|---|---|---|
| side by side | | | | | | | side by side | | | | | |
| univ. test Dif | × | 1 | 70 | 94 | 24 | | univ. test Dif | × | 1 | 110 | 131 | 21 |
| paired t-test | × | 2 | 85 | 100 | 15 | | paired t-test | × | 2 | 126 | 113 | -13 |
| bivariate | × | 3 | 92 | 106 | 14 | | bivariate | × | 3 | 85 | 156 | 71 |
| | × | 4 | 97 | 113 | 16 | | | × | 4 | 140 | 100 | -40 |
| Columns (3/0) | × | 5 | 110 | 130 | 20 | | Columns (3/0) | × | 5 | 92 | 149 | 57 |
| ☐ BP AM | × | 6 | 110 | 131 | 21 | | ☐ Awake | × | 6 | 70 | 170 | 100 |
| ☐ BP PM | × | 7 | 126 | 142 | 16 | | ☐ Asleep | × | 7 | 148 | 94 | -54 |
| ☐ Dif ⊞ | × | 8 | 137 | 149 | 12 | | ☐ Dif ⊞ | × | 8 | 97 | 142 | 45 |
| Rows | × | 9 | 140 | 156 | 16 | | Rows | × | 9 | 137 | 106 | -31 |
| All Rows        10 | × | 10 | 148 | 170 | 22 | | All Rows        10 | × | 10 | 110 | 130 | 20 |

Let's do the incorrect *t*-test—the *t*-test for independent groups. Actually doing this involves reorganizing the data using the **Stack** Columns.

🖱 In two separate tables, stack Awake and Asleep to form a single column, and BP AM and BP PM to form a single column.

🖱 Select **Analyze > Fit Y by X** on both new tables, using the Stack column as Y and the ID column as X.

🖑    Choose **t test** from the popup menu on the title bar of each plot.

The results for the two analyses are shown in **Figure 8.20**. The conclusions are that there is no significant difference between Awake and Asleep time, nor is there a difference between time of blood pressure measurement. The summary statistics are the same in both analysis and the probability is the same, showing no significance ($p = 0.1426$).

**Figure 8.20**    Results of $t$-test for Independent Means

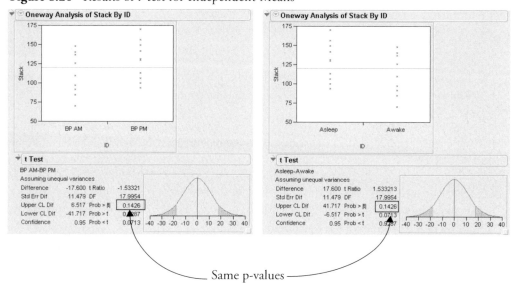

Now do the proper test, the paired $t$-test.

🖑    Using the **Distribution** command, examine a distribution of the Dif variable in each table.

🖑    Double click on the axis of the blood pressure histogram and make its scale match the scale of the baby sleep axis.

🖑    Then, test that each mean is zero (see **Figure 8.21**).

In this case the analysis of the differences leads to very different conclusions.

•    The mean difference between time of blood pressure measurement is highly significant because the variance is very small (Std Dev=3.89).

•    The mean difference between awake and asleep time is not significant because the variance of that difference is large (Std Dev=51.32).

So don't judge the mean of the difference by the difference in the means without noting that the variance of the difference is the measuring stick, and that the measuring stick depends on the correlation between the two responses.

**Figure 8.21**   Histograms and Summary Statistics Show the Problem

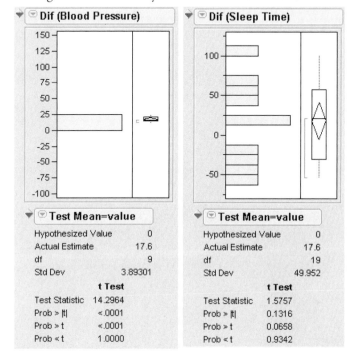

The scatterplots produced by the Bivariate platform (**Figure 8.22**) and the Matched Pairs platform (**Figure 8.23**) shows what is happening. The first pair is highly positively correlated, leading to a smaller variance for the difference. The second pair is highly negatively correlated, leading to a very large variance for the difference.

**Figure 8.22**   Bivariate scatterplots of blood pressure data and baby sleep data

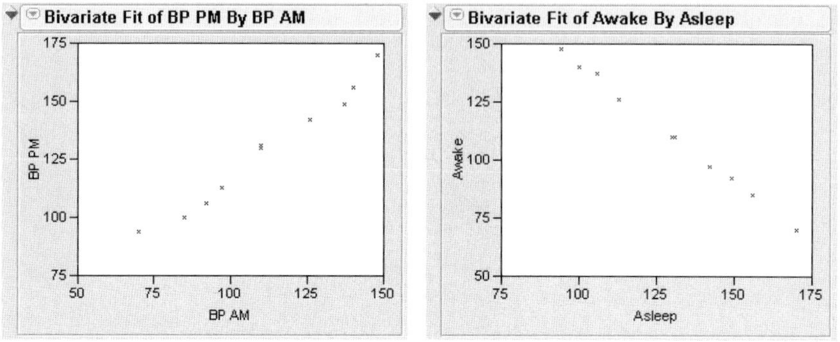

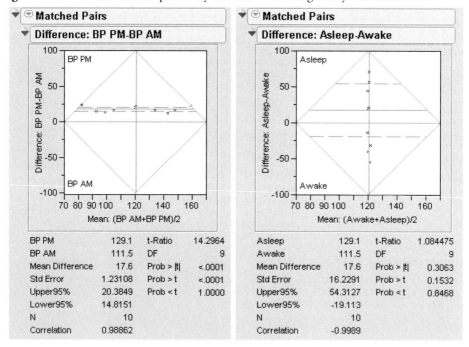

**Figure 8.23**   Paired *t*-test for positively correlated and negatively correlated data

To review, make sure you can answer the following question:

What is the reason that you use a different *t*-test for matched pairs?

a. Because the statistical assumptions for the *t*-test for groups aren't satisfied with correlated data.

b. Because you can detect the difference much better with a paired *t*-test—the paired *t*-test is much more sensitive to a given difference.

c. Because you might be overstating the significance if you used a group *t*-test rather than a paired *t*-test.

d. Because you are testing a different thing.

Answer: All of the above.

a. The grouped *t*-test assumes that the data are uncorrelated and paired data are correlated. So you would violate assumptions using the grouped *t*-test.

b. Most of the time the data are positively correlated, so the difference has a smaller variance than you would attribute if they were independent. So the paired *t*-test is more powerful—that is, more sensitive.

c. There may be a situation in which the pairs are negatively correlated, and if so, the variance of the difference would be greater than you expect from independent responses, and the difference would have greater variance. The grouped *t*-test would overstate the significance.

d. You are testing the same thing in that the mean of the difference is the same as the difference in the means. But you are testing a different thing in that the variance of the mean difference is different than the variance of the differences in the means (ignoring correlation), and the significance for means is measured with respect to the variance.

**Mouse Mystery**

Comparing two means is not always straightforward. Consider this story.

A food additive showed promise as a dieting drug. An experiment was run on mice to see if it helped control their weight gain. If it proved effective, then it could be sold to millions of people trying to control their weight.

After the experiment was over, the average weight gain for the treatment group was significantly less than for the control group, as hoped for. Then someone noticed that the treatment group had fewer observations than the control group. It seems that the food additive caused the obese mice in that group to tend to die young, so the thinner mice had a better survival rate for the final weighing.

These tables are set up such that the values are identical for the two responses, as a marginal distribution, but the values are paired differently so that the bpTime difference is highly significant and the babySleep difference is non-significant. This illustrates that it is the distribution of the difference that is important, not the distribution of the original values. If you don't look at the data correctly, the data can appear the same even when they are dramatically different.

# A Nonparametric Approach

## Introduction to Nonparametric Methods

Nonparametric methods provide ways to analyze and test data that do not depend on assumptions about the distribution of the data. In order to ignore Normality assumptions, nonparametric methods disregard some of the information in your data. Typically, instead of using actual response values, you use the *rank ordering* of the response.

Most of the time you don't really throw away much relevant information, but you avoid information that might be misleading. A nonparametric approach creates a statistical test that it ignores all the spacing information between response values. This protects the test against distributions that have very non-Normal shapes, and can also provide insulation from data contaminated by rogue values.

In many cases, the nonparametric test has almost as much power as the corresponding parametric test and in some cases has more power. For example, if a batch of values is Normally distributed, the rank-scored test for the mean has 95% efficiency relative to the most powerful Normal-theory test.

The most popular nonparametric techniques are based on functions (scores) of the ranks:

- The rank itself, called a *Wilcoxon score*.
- Whether the value is greater than the median; whether the rank is more than $\frac{n+1}{2}$, called the *Median test*.
- A Normal quantile, computed as in Normal quantile plots, called the *van der Waerden* score.

Nonparametric methods are not contained in a single platform in JMP, but are available through many platforms according to the context where that test naturally occurs.

## Paired Means: The Wilcoxon Signed-Rank Test

The Wilcoxon signed-rank test is the nonparametric analog to the paired *t*-test. You do a signed-rank test by testing the distribution of the difference of matched pairs, as discussed previously. The following example shows the advantage of using the signed-rank test when data are non-Normal.

🖰 Open the Chamber.jmp table.

The data represent electrical measurements on 24 wiring boards. Each board is measured first when soldering is complete, and again after three weeks in a chamber with a controlled environment of high temperature and humidity (Iman 1995).

- 🖱 Examine the diff variable (difference between the outside and inside chamber measurements) with **Analyze > Distribution**.

- 🖱 Select the **Fit Distribution > Normal** from the popup menu on the title bar of the diff histogram.

- 🖱 Select **Goodness of Fit** from the popup menu on the Fitted Normal Report.

The Shapiro-Wilk $W$-test that results tests the assumption that the data are Normal. The probability of 0.0076 given by the Normality test indicates that the data are significantly non-Normal. In this situation, it might be better to use signed ranks for comparing the mean of diff to zero.

**Figure 8.24**   The Chamber data and test for Normality

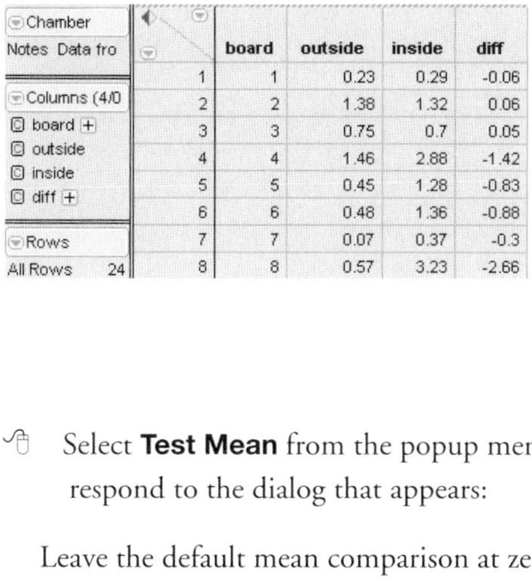

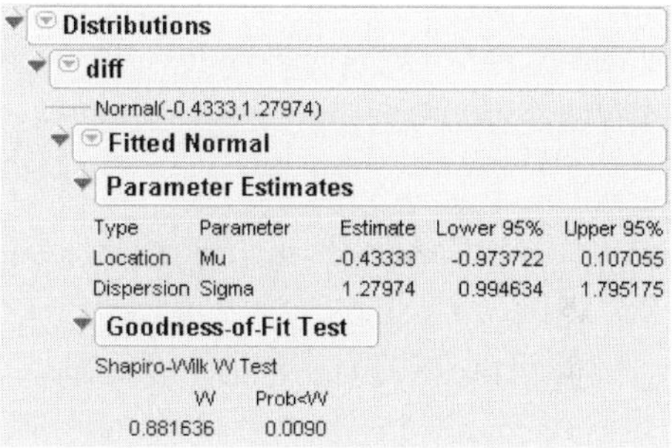

🖱 Select **Test Mean** from the popup menu on the diff histogram title bar. When you respond to the dialog that appears:

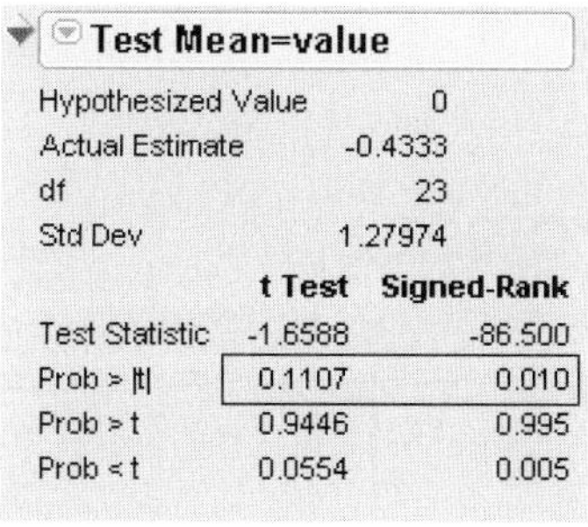

- Leave the default mean comparison at zero.

- Leave the standard deviation blank, to be computed from the sample.

- Make sure the **Wilcoxon Signed-Rank** check box is checked.

- Click **OK**.

| Test Mean=value | | |
|---|---|---|
| Hypothesized Value | 0 | |
| Actual Estimate | -0.4333 | |
| df | 23 | |
| Std Dev | 1.27974 | |
| | t Test | Signed-Rank |
| Test Statistic | -1.6588 | -86.500 |
| Prob > \|t\| | 0.1107 | 0.010 |
| Prob > t | 0.9446 | 0.995 |
| Prob < t | 0.0554 | 0.005 |

Note that the standard $t$-test probability is insignificant ($p = 0.1107$). However, in this example, the signed-rank test detects a difference between the groups with a $p$-value of 0.01.

## Independent Means: The Wilcoxon Rank Sum Test

If you want to test the means of two independent groups, as in the *t*-test, but nonparametrically, then you can rank the responses and analyze the ranks instead of the original data. This is the *Wilcoxon rank sum test*. It is also known as the *Mann-Whitney U test* because there is a different formulation of it that was not discovered to be equivalent to the Wilcoxon rank sum test until after it had become widely used.

> 🖱 Open Htwt15 again, and choose **Analyze > Fit Y by X** with Height as Y and Gender as X.

This is the same platform that gave the *t*-test.

> 🖱 Select the **Nonparametric-Wilcoxon** command from the popup menu on the title bar at the top of the report.

The result is the report in **Figure 8.25**. This table shows the sum and mean ranks for each group, then the Wilcoxon statistic along with an approximate *p*-value based on the large-sample distribution of the statistic. In this case, the difference in the mean heights is declared significant, with a *p*-value of 0.0002. If you have small samples, you should consider also checking the tables of the Wilcoxon to obtain a more exact test, because the Normal approximation is not very precise in small samples.

**Figure 8.25**   Wilcoxon rank sum test for independent groups

### Wilcoxon / Kruskal-Wallis Tests (Rank Sums)

| Level | Count | Score Sum | Score Mean | (Mean-Mean0)/Std0 |
|-------|-------|-----------|------------|-------------------|
| f | 24 | 350.5 | 14.6042 | -3.728 |
| m | 15 | 429.5 | 28.6333 | 3.728 |

### 2-Sample Test, Normal Approximation

| S | Z | Prob>\|Z\| |
|---|---|-----------|
| 429.5 | 3.72806 | 0.0002 |

### 1-way Test, ChiSquare Approximation

| ChiSquare | DF | Prob>ChiSq |
|-----------|----|-----------| 
| 14.0064 | 1 | 0.0002 |

# Exercises

1. The file On-Time Arrivals.jmp (*Aviation Consumer Home Page,* 1999) contains the percentage of airlines' planes that arrived on time in 29 airports (those that the Department of Transportation designates "reportable"). You are interested in seeing if there are differences between certain months.

(a) Suppose you want to examine the differences between March and June. Is this a situation where a grouped test of two means is appropriate, or would a matched pairs test be a better choice?

(b) Based on your answer in (a), determine if there is a difference in on-time arrivals between the two months.

(c) Similarly, determine if there is a significant difference between the months June and August, and also between March and August.

2. William Gossett was a pioneer in statistics. In one famous experiment, he wanted to investigate the yield from corn planted from two different types of seeds. One type of seed was dried in the normal way, while the other was kiln-dried. Gossett planted one of each seed in eleven different plots and measured the yield for each one. The drying methods are represented by the columns Regular or Kiln in the data file Gosset's Corn (Gosset 1908).

(a) This is a matched pairs experiment. Explain why it is inappropriate to use the grouped-means method of determining the difference between the two plots.

(b) Using the matched pairs platform, determine if there is a difference in yield between kiln-dried corn and regular-dried corn.

3. The data file Companies.jmp (*Fortune* Magazine, 1990) contains data on sales, profits, and employees for two different industries (Computers and Pharmaceutical). This exercise is interested in detecting differences between the two types of companies.

(a) Suppose you wanted to test for differences in sales amounts for the two business types. First, examine histograms of the variables Type and Sales, and comment on the output.

(b) In comparing sales for the two types of companies, should you use grouped means or matched pairs for the test?

(c) Using your answer in part (b), determine if there is a difference between the sales amount of the two types of companies.

(d) Should you throw out any outliers in your analysis of part (c)? Comment on why this would or would not be appropriate in this situation.

4. Automobile tire manufacturing companies are obviously interested in the quality of their tires. One of their measures of quality is tire treadwear. In fact, all major manufacturers regularly sample their production and conduct tire treadwear tests. There are two accepted methods of measuring tread-wear—one based on weight loss during use, the other based on groove wear. A scientist at one of

these manufacturers decided to see if the two methods gave different results for the wear on tires. He set up an experiment that measured 16 tires, each by the two methods. His data is assembled in the file Tire Tread Measurement.jmp (Stichler, Richey, and Mandel, 1953).

    (a)   Determine if there is a difference in the two methods by using the matched pairs platform.

    (b)   Now, determine if there is a difference in the two methods by using group means. To do this, you will need to "stack" the data. Select **Tables**, then **Stack**, and select both columns to be stacked. After pressing **OK**, you will have a data table with two columns, one with the Weight/Groove identifier, the other with the measurement. At this point you are ready to carry out your analysis with the Fit Y By X platform.

    (c)   Which of the two methods (matched pairs or grouped means) is correct?

    (d)   Would the scientist have had different results by using the wrong method?

5.   The data table Cars.jmp (Henderson and Velleman, 1981) contains information on several different brands of cars, including number of doors and impact compression for various parts of the body during crash tests.

    (a)   Is there a difference between two- and four-door cars when it comes to impact compression on left legs?

    (b)   Is there a difference between two- and four-door cars when it comes to compression on right legs?

    (c)   Is there a difference between two- and four-door cars when it comes to head impact compression?

6.   The data in Chamber.jmp represent electrical measurements on 24 electrical boards (this is the same data used in "Paired Means: The Wilcoxon Signed-Rank Test" on page 194). Each measurement was taken when soldering was complete, then again three weeks later after sitting in a temperature- and humidity-controlled chamber. The investigator wants to know if there is a difference between the measurements.

    (a)   Why is this a situation that calls for a matched pairs analysis?

    (b)   Determine if there is a significant difference between the means before and after the time in the chamber.

# Comparing Many Means: One-Way Analysis of Variance

## Overview

In Chapter 6, "The Difference Between Two Means," the $t$-test was the tool needed to compare the means of two groups. However, if you need to test the means of more than two groups, the $t$-test can't handle the job. This chapter shows how to compare more than two means using the one-way *analysis of variance*, or ANOVA for short. The $F$-test, which has made brief appearances in previous chapters, is the key element in an ANOVA. It is the statistical tool necessary to compare many groups, just as the $t$-test compares two groups. This chapter also introduces multiple comparisons and power calculations, reviews the topic of unequal variances, and extends nonparametric methods to the one-way layout.

# What Is a One-Way Layout?

A one-way layout is the organization of data when a response is measured across a number of groups, and the distribution of the response may be different across the groups. The groups are labeled by a classification variable, which is a column in the JMP data table with the nominal or ordinal modeling type.

Usually, one-way layouts are used to compare group means. **Figure 9.1** shows a schematic that compares two models. The model on the left fits a different mean for each group, and model on the right indicates a single grand mean (a single-mean model).

**Figure 9.1**    Different Mean for Each Group Versus a Single Overall Mean

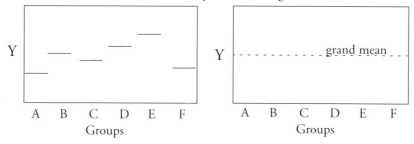

The previous chapter showed how to use the $t$-test and the $F$-test to compare two means. When there are more than two means, the $t$-test is no longer applicable; the $F$-test must be used.

An $F$-test has the following features:

- An $F$-test compares two models, one constrained and the other unconstrained. The constrained model fits one grand mean. The unconstrained model for the one-way layout fits a mean for each group.

- The measurement of fit is done by accumulating the sum of the squares of *residuals*, where the residual is the difference between the actual response and the fitted response.

- A Mean Square is calculated by dividing a sum of squares by its degrees of freedom. Mean Squares are estimates of variance, sometimes under the assumption that certain hypotheses are true.

- Degrees of Freedom (DF) are numbers, based on the number of parameters and number of data points in the model, that you divide by to get an unbiased estimate of the variance (see the chapter "What Are Statistics?" on page 91 for a definition of bias).

- An *F*-statistic is a ratio of Mean Squares (MS) that are independent and have the same expected value. In our discussion, this ratio is

$$\frac{\text{Model MS}}{\text{Total MS}}$$

- If the null hypothesis that there is no difference between the means is true, this *F*-statistic has an *F distribution*.

- If the hypothesis is not true (if there is a difference between the means), the mean square for model in the numerator of the *F*-ratio has some effect in it besides the error variance. This numerator produces a large (and significant) *F* if there is enough data.

- When there is only one comparison, the *F*-test is equivalent to the pooled (equal-variance) *t*-test. In fact, when there is only one comparison, the *F*-statistic is the square of the pooled *t*-statistic. This is true despite the fact that the *t*-statistic is derived from the distribution of the estimates, whereas the *F*-test is thought of in terms of the comparison of variances of residuals from two different models.

# Comparing and Testing Means

The file DrugLBI.jmp contains the results of a study that measured the response of 30 subjects to treatment by one of three drugs (Snedecor and Cochran, 1967). To begin,

 Open DrugLBI.jmp.

The three drug types are called "a", "d", and "placebo." The LBS column is the response measurement. (The LBI column is used in a more complex model, covered in Chapter 12, "Fitting Models.")

| Drug | LBI | LBS |
|------|-----|-----|
| a | 6 | 4 |
| a | 10 | 13 |
| a | 6 | 1 |
| a | 11 | 8 |
| a | 3 | 0 |
| d | 6 | 0 |
| d | 6 | 2 |
| d | 7 | 3 |

 For a quick look at the data,
choose **Analyze > Distribution** and select Drug and LBS as Y variables.

Note in the histogram on the left in **Figure 9.2** that the number of observations is the same in each of the three drug groups; that is what is meant by a *balanced design*.

**Figure 9.2**    Distributions of Model Variables

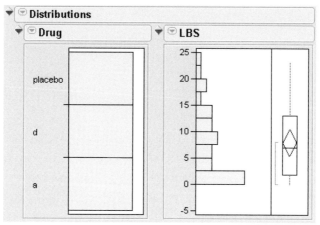

🖰    Next, choose **Analyze > Fit Y by X** with Drug as X and LBS as Y.

Notice that the Launch dialog displays the message that you are requesting a one-way analysis.

The launch dialog shows the appropriate analysis for the selected variables.

🖰    Click **OK.**

The results window in **Figure 9.3** appears. The initial plot on the left shows the distribution of the response in each drug group. The line across the middle is at the grand mean.

**Figure 9.3**   Distributions of drug groups

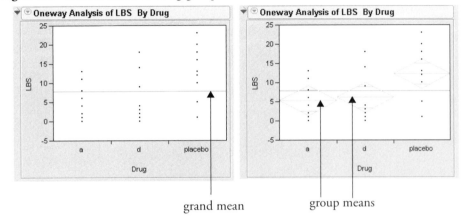

grand mean                        group means

# Means Diamonds: A Graphical Description of Group Means

🖑   Select **Means/Anova** from the popup menu showing on the title bar of the plot.

This adds means diamonds to the plot and also adds a set of text reports. The plot on the right in **Figure 9.3** shows means diamonds:

- The middle line in the diamond is the response group mean for the group.
- The vertical endpoints form the 95% confidence interval for the mean.
- The x-axis is divided proportionally by group sample size.

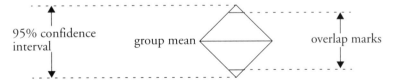

If the means are not much different, they will be close to the grand mean. If the confidence intervals (the points of the diamonds) don't overlap, the means are significantly different.

See the section "Display and Compare the Means" on page 165 for details on means diamonds.

# Statistical Tests to Compare Means

The **Means/Anova** command produces a report composed of the three tables shown in **Figure 9.4**:

- The **Summary of Fit** table gives an overall summary of how well the model fits.

- The **Analysis of Variance** table gives sums of squares and an *F*-test on the means.

- The **Means for Oneway Anova** table shows the group means, standard error, and upper and lower 95% confidence limits.

**Figure 9.4**  One-way ANOVA Report

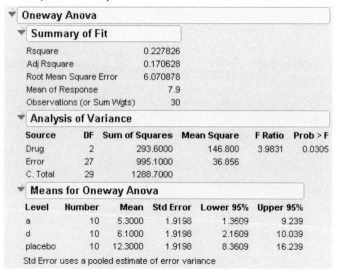

The Summary of Fit and the Analysis of Variance tables may look like a hodgepodge of numbers, but they are all derived by a few simple rules. **Figure 9.5** illustrates how the statistics relate.

**Figure 9.5**  Summary of Fit and ANOVA tables

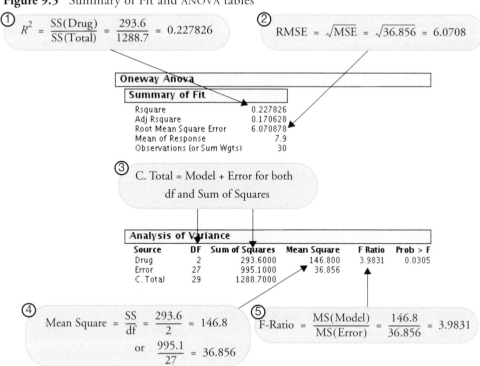

The Analysis of Variance table (**Figure 9.4** and **Figure 9.5**) describes three source components:

### C. Total

The C. Total Sum of Squares (SS) is the sum of the squares of residuals around the grand mean. "C. Total" stands for *corrected total* because it is corrected for the mean. The C Total degrees of freedom is the total number of observations in the sample minus 1.

### Error

After you fit the group means, the remaining variation is described in the Error line. The Sum of Squares is the sum of squared residuals from the individual means. The remaining unexplained variation is C Total minus Model and is called Error for both the sum of squares and the degrees of freedom. The Error Mean Square estimates the variance.

### Model

The Sum of Squares for the Model line is the difference of C Total and Error. It is a measure of how much the residuals' sum of squares is accounted for by fitting the model rather than fitting only the grand mean. The degrees of freedom in the drug example is the number of parameters in the model (the number of groups, 3) minus 1.

Everything else in the Analysis of Variance table and the Summary of Fit table is derived from these quantities.

## Mean Square

*Mean Squares* are the sums of squares divided by their respective degrees of freedom.

## F-ratio

The *F-ratio* is the model mean square divided by the error mean square. The *p*-value for this *F*-ratio comes from the *F*-distribution.

## RSquare

The *RSquare* ($R^2$) is the proportion of variation explained by the model. In other words, it is the model sum of squares divided by the total sum of squares.

## Adjusted RSquare

The *Adjusted RSquare* is more comparable over models with different numbers of parameters (degrees of freedom). It is the error mean square divided by the total mean square, subtracted from 1:

$$1 - \frac{\text{Error MS}}{\text{Total MS}}.$$

## Root Mean Square Error

The Root Mean Square Error is the square root of the Mean Square for Error in the Analysis of Variance table. It estimates the standard deviation of the error.

So what's the verdict for the drugs? The *F*-value of 3.98 is significant with a *p*-value of 0.03, which confirms that there is a significant difference in the means. The *F*-test does not give any specifics about which means are different, only that there is at least one pair of means that is statistically different.

The *F*-test shows whether the variance of residuals from the model is better than the variances of the residuals from only fitting a grand mean. In this case, the answer is yes, but just barely. The histograms shown to the right compare the residuals from the grand means (left) with the group mean residuals (right).

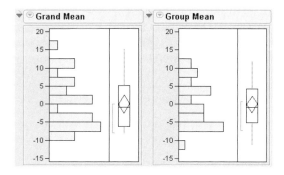

# Means Comparisons for Balanced Data

Which means are significantly different from which other means? It looks like the mean for the drug "placebo" separates from the other two. However, since all the confidence intervals for the means intersect, it takes further digging to see significance.

If two means were composed from the same number of observations, then you can use the overlap marks to get a more precise graphical measure of which means are significantly different. Two means are significantly different when their overlap marks don't overlap. The overlap marks are placed into the confidence interval at a distance of $1/(\sqrt{2})$, a distance given by the Student's *t*-test of separation.

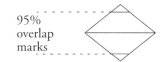

For balanced data, to be significantly different, two means must not overlap their overlap marks.

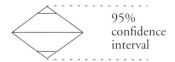

When two means do not have the same number of observations, then the design is unbalanced and the overlap marks no longer apply. Another technique using *comparison circles* can be used instead. The next section describes comparison circles and shows you how to interpret them.

# Means Comparisons for Unbalanced Data

Suppose, for the sake of this example, the drug data are unbalanced. That is, there are not the same number of observations in each group. The following steps will unbalance the

DrugLBl.jmp data in an extreme way to illustrate an apparent paradox, as well as introduce a new graphical technique.

🖰  Change **Drug** in rows 1, 4–7, and 9 from "a" to "placebo". Change **Drug** in rows 2 and 3 to "d". Change **LBS** in row 10 to "4." (Be careful not to save this modified table over the original copy in your sample data.)

Now drug "a" has only 2 observations, whereas "d" has 12 and "placebo" has 16. The mean for "a" will have a very high standard error because it is supported by so few observations compared with the other two levels.

Again, use the **Fit Y by X** command to look at the data:

🖰  Choose **Analyze > Fit Y by X** for the modified data, and as before, select the **Means/ Anova** option from the Oneway menu.

🖰  Select **Compare Means > Each Pair, Student's t** from the platform menu on the scatterplot title bar.

The modified data should give results like those illustrated in **Figure 9.6**. The *x*-axis divisions are proportional to the group sample size, which causes drug "a" to be very thin, because it has fewer observations. The confidence interval on its mean is large compared with the others. Comparison circles for Student's *t*-tests appear to the right of the means diamonds.

**Figure 9.6**   Comparison Circles to Compare Group Means

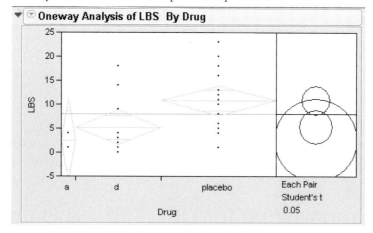

Comparison circles are a graphical technique that let you see significant separation among means in terms of how circles intersect. This is the only graphical technique that works in general with both equal and unequal sample sizes. The plot displays a circle for each group, with the centers lined up vertically. The center of each circle is aligned with its corresponding group mean. The radius of a circle is the 95% confidence interval for its group mean, as you can see by comparing a circle with its corresponding means diamond. The non-overlapping confidence intervals shown by the diamonds for groups that are significantly different correspond directly to the case of non-intersecting comparison circles.

When the circles intersect, the angle of intersection is the key to seeing if the means are significantly different. If the angle of intersection is exactly a right angle, 90°, then the means are on the borderline of being significantly different. To see why, imagine the radii of the circles as legs of a right triangle, where the hypotenuse's length is the confidence limit of the difference of the two means. Then, the actual difference in means is the same as the confidence interval length for the difference, called the *Least Significant Difference* (the *LSD*).

If the circles are farther apart than the right angle case, then the outside angle is more acute and the means are significantly different. If the circles are closer together, the angle is larger than a right angle, and the means are not significantly different. **Figure 9.7** illustrates these angles of intersection.

**Figure 9.7**   Diagram of How to Interpret Comparison Circles

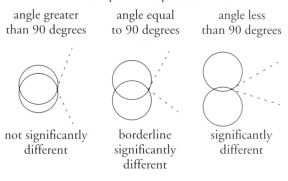

So what are the conclusions for the drug example shown in **Figure 9.6**?

First, be assured that you don't need to hunt down a protractor to figure out the size of the angles of intersection. Click on a circle and see what happens—the circle highlights and becomes red. Groups that are not different from it also show in red. All groups that are significantly different remain black.

🖑   Click on the "placebo" circle and use the circles to compare group means.

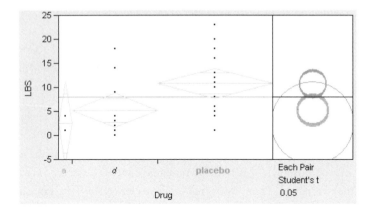

- The "placebo" and "d" means are represented by the smaller circles. The circles are farther separated than would occur with a right angle. The angle is acute, so these two means are significantly different.

- The circle for the "d" mean is completely nested in the circle for "a", so they are not significantly different.

- The "a" mean is well below the "d" mean, which is significantly below "placebo." By transitivity, one might expect "a" to be significantly different than "placebo." The problem with this logic is that the standard error around the "a" mean is so large that it is not significantly different from "placebo", even though it is farther away than "d." The angle of intersection is greater than a right angle, so they are not significantly different.

This complexity in relationships when the data are unbalanced is the reason a more complex graphic is needed to show relationships. The Fit Y by X platform lets you see the difference with the comparison circles and verify group differences statistically with the Means Comparisons tables as shown in **Figure 9.8**.

The Means Comparisons table uses the concept of Least Significant Difference (LSD). In the balanced case, this is the separation that any two means must have from each other to be significantly different. In the unbalanced case, there is a different LSD for each pair of means.

The Means Comparison report shows all the comparisons of means ordered from high to low. The elements of the table show the absolute value of the difference in two means minus the LSD. If the means are farther apart than the LSD, then they are significantly different and the element is positive. For example the element that compares "placebo" and "d" is +0.88, which says the means are 0.88 more separate than needed to be significantly different. If the means are not significantly different, then the LSD is greater than the difference, so the element in the table is negative. The elements for the other two comparisons are negative, showing no significant difference.

**Figure 9.8**   Statistical Text Reports to Compare Groups

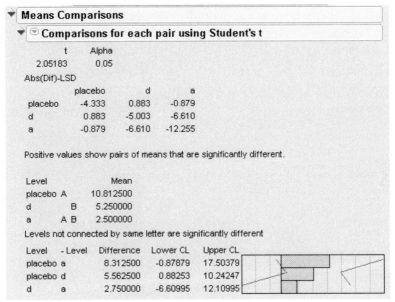

In addition, a table shows the classic SAS-style means comparison with letters. Levels that share a letter are not significantly different from each other. For example, both levels d and a share the letter B, so d and a are not significantly different from each other.

An Ordered Differences report, closed by default, appears below the text reports. It lists the differences between groups in decreasing order, with confidence limits of the difference. A bar chart displays the differences with blue lines representing the confidence limits.

The last thing to do in this example is to restore your copy of the DrugLBI.jmp table to its original state so it can be used in other examples. To do this,

🖑   Choose **File > Revert,** or reopen the data table.

# Special Topic: Adjusting for Multiple Comparisons

Making multiple comparisons, such as comparing many pairs of means, increases the possibility of committing a Type I error. Remember, a Type I error is the error of declaring a difference significant (based on statistical test results) that is actually not significant. The more tests you do, the more likely you are to happen upon a significant difference occurring by chance alone. If you are comparing all possible pairs of means in a large one-way layout, there are many possible tests, and a Type I error becomes very likely.

There are many methods that modify tests to control for an overall error rate. This section covers one of the most basic, the *Tukey-Kramer Honestly Significant Difference* (HSD). The Tukey-Kramer HSD uses the distribution of the maximum range among a set of random variables.

🖰   After reverting to the original copy of DrugLBI.jmp, again choose
   **Analyze > Fit Y by X** for the variables LBS as Y Drug as X. To expedite this, you may hit the **Recall** button in the Fit Y by X dialog.

🖰   Next, select the following three commands from the popup menu on the title bar:
   **Means/Anova, Compare Means > Each Pair, Student's t**, and
   **Compare Means > All Pairs, Tukey HSD**.

These commands should give you the results shown in **Figure 9.9**.

**Figure 9.9**   *t*-tests and Tukey-Kramer Adjusted *t*-tests for One-Way ANOVA

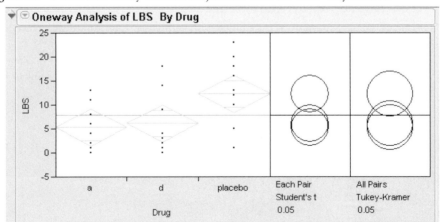

The comparison circles work as before, but have different kinds of error rates.

The Tukey-Kramer comparison circles are larger than the Student's $t$ circles. This protects more tests from falsely declaring significance, but this protection makes it harder to declare two means significantly different.

If you click on the top circle, you see that the conclusion is different between the Student's $t$ and Tukey-Kramer's HSD for the comparison of "placebo" and "d." This comparison is significant for Student's $t$-test but not for Tukey's test.

The difference in significance occurs because the quantile that is multiplied into the standard errors to create a Least Significant Difference has grown from 2.05 to 2.48 between Student's $t$-test and the Tukey-Kramer test.

The only positive element in the Tukey table is the one for the "a" versus "placebo" comparison (**Figure 9.10**).

**Figure 9.10**  Means Comparisons Table for One-Way ANOVA

**Means Comparisons**

**Comparisons for each pair using Student's t**

|   | t | Alpha |
|---|---|---|
|   | 2.05183 | 0.05 |

Abs(Dif)-LSD

|   | placebo | d | a |
|---|---|---|---|
| placebo | -5.5707 | 0.6293 | 1.4293 |
| d | 0.6293 | -5.5707 | -4.7707 |
| a | 1.4293 | -4.7707 | -5.5707 |

Positive values show pairs of means that are significantly different.

| Level |   | Mean |
|---|---|---|
| placebo | A | 12.300000 |
| d | B | 6.100000 |
| a | B | 5.300000 |

Levels not connected by same letter are significantly different

| Level | - Level | Difference | Lower CL | Upper CL |   |
|---|---|---|---|---|---|
| placebo | a | 7.000000 | 1.42932 | 12.57068 |   |
| placebo | d | 6.200000 | 0.62932 | 11.77068 |   |
| d | a | 0.800000 | -4.77068 | 6.37068 |   |

**Comparisons for all pairs using Tukey-Kramer HSD**

|   | q* | Alpha |
|---|---|---|
|   | 2.47942 | 0.05 |

Abs(Dif)-LSD

|   | placebo | d | a |
|---|---|---|---|
| placebo | -6.7316 | -0.5316 | 0.2684 |
| d | -0.5316 | -6.7316 | -5.9316 |
| a | 0.2684 | -5.9316 | -6.7316 |

Positive values show pairs of means that are significantly different.

| Level |   | Mean |
|---|---|---|
| placebo | A | 12.300000 |
| d | A B | 6.100000 |
| a | B | 5.300000 |

Levels not connected by same letter are significantly different

| Level | - Level | Difference | Lower CL | Upper CL |   |
|---|---|---|---|---|---|
| placebo | a | 7.000000 | 0.26843 | 13.73157 |   |
| placebo | d | 6.200000 | -0.53157 | 12.93157 |   |
| d | a | 0.800000 | -5.93157 | 7.53157 |   |

# Are the Variances Equal Across the Groups?

The one-way ANOVA assumes that the variance is the same within each group. The Analysis of Variance table shows the note "Std Error uses a pooled estimate of error variance." When testing the difference between two means, as in this chapter, JMP provides separate reports for both equal and unequal variance assumptions. This is why, when there are only two groups, the command is **Means/Anova/Pooled t**, because ANOVA pools the variances like the pooled *t*-test does.

Before you get too concerned about the equal-variance issue, be aware that there is always a list of issues to worry about—it is not usually useful to be overly concerned about this one.

🖱 Select **Quantiles** from the red triangle popup menu showing beside the report title.

This command displays quantile box plots for each group as shown in **Figure 9.11**. Note that the interquartile range (the height of the boxes) is not much different for drugs a and b, but is somewhat different for placebos. The placebo group seems to have a slightly larger interquartile range.

**Figure 9.11** Quantile Box Plots

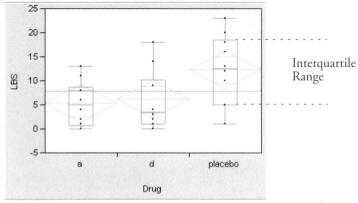

🖱 Select **Quantiles** again to turn the box plots off.

A more effective graphical tool to check the variance assumption is the Normal Quantile plot.

🖱 Select **Normal Quantile Plot > Plot Actual By Quantile** from the popup menu on the title bar.

This option displays a plot next to the Means Diamonds as shown in **Figure 9.12**. The Normal Quantile plot compares mean, variance, and shape of the group distributions.

There is a line on the Normal Quantile plot for each group. The height of the line shows the location of the group. The slope of the line shows the group's standard deviation. So, lines that appear to be parallel have similar standard deviations. The straightness of the line segments connecting the points shows how close the shape of the distribution is to the Normal distribution. Note that the "placebo" group is both higher and has a greater slope, which indicates a higher mean and a higher variance, respectively.

**Figure 9.12**   Normal Quantile Plot

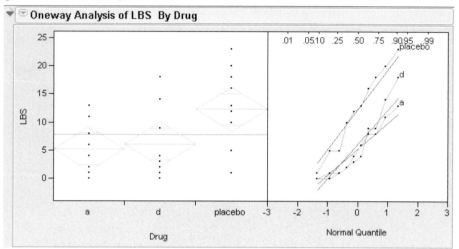

It's easy to get estimates of the standard deviation within each group:

🖱 Select **Means and Std Dev** from the red triangle popup menu showing on the report title to see the reports in **Figure 9.13**.

**Figure 9.13**   Mean and Standard Deviation report

▼ **Means and Std Deviations**

| Level | Number | Mean | Std Dev | Std Err Mean | Lower 95% | Upper 95% |
|---|---|---|---|---|---|---|
| a | 10 | 5.3000 | 4.64399 | 1.4686 | 1.9779 | 8.622 |
| d | 10 | 6.1000 | 6.15449 | 1.9462 | 1.6973 | 10.503 |
| placebo | 10 | 12.3000 | 7.14998 | 2.2610 | 7.1852 | 17.415 |

You can conduct a statistical test the equality of the variances as follows:

🖱 Select **UnEqual Variance** from the popup menu on the title bar to see the Tests the Variances are Equal tables in **Figure 9.14**.

**Figure 9.14**   Tests the Variances are Equal Report

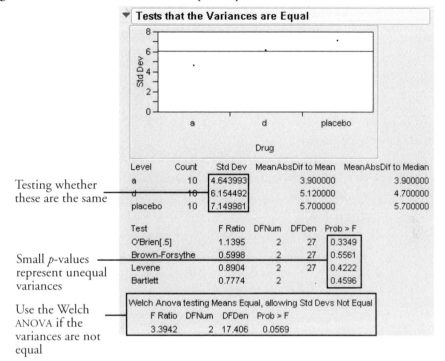

Testing whether these are the same

Small *p*-values represent unequal variances

Use the Welch ANOVA if the variances are not equal

Tests that the Variances are Equal

| Level | Count | Std Dev | MeanAbsDif to Mean | MeanAbsDif to Median |
|---|---|---|---|---|
| a | 10 | 4.643993 | 3.900000 | 3.900000 |
| d | 10 | 6.154492 | 5.120000 | 4.700000 |
| placebo | 10 | 7.149981 | 5.700000 | 5.700000 |

| Test | F Ratio | DFNum | DFDen | Prob > F |
|---|---|---|---|---|
| O'Brien[.5] | 1.1395 | 2 | 27 | 0.3349 |
| Brown-Forsythe | 0.5998 | 2 | 27 | 0.5561 |
| Levene | 0.8904 | 2 | 27 | 0.4222 |
| Bartlett | 0.7774 | 2 | . | 0.4596 |

Welch Anova testing Means Equal, allowing Std Devs Not Equal

| F Ratio | DFNum | DFDen | Prob > F |
|---|---|---|---|
| 3.3942 | 2 | 17.406 | 0.0569 |

To interpret these reports, note that the **Std Dev** column lists the estimates you are testing to be the same. Then note the results listed under **Prob>F**. These numbers are *p*-values from testing the assumption that the variances are equal. Therefore, small *p*-values suggest that variances are not equal.

As expected, there is no evidence here that the variances are unequal. None of the *p*-values are small.

Each of the four tests in **Figure 9.14** (O'Brien, Brown-Forsythe, Levene, and Bartlett) tests whether the variances are equal, but each uses a different method for measuring variability.

One way to evaluate dispersion is to take the absolute value of the difference of each response from its group mean. Mathematically, this means to look at $|x_i - \bar{x}|$ for each response.

- *Levene's Test* estimates the mean of these absolute differences for each group (shown in the table as **MeanAbsDif to Mean**), and then does a *t*-test (or equivalently, an *F*-test) on these estimates.

- The *Brown-Forsythe Test* measures the differences from the median instead of the mean and then tests these differences.

- *O'Brien's Test* tricks the *t*-test by telling it that the means were really variances.

- *Bartlett's Test* derives the test mathematically, using an assumption that the data are Normal. Though powerful, Bartlett's test is sensitive to departures from the Normal distribution.

Statisticians have no apologies for offering different tests with different results. Each test has its advantages and disadvantages.

## Testing Means with Unequal Variances

If you think the variances are different, then you should consider avoiding the standard *t*-and *F*-tests that assume the variances are equal. Instead, use the Welch ANOVA *F*-test that appears with the unequal variance tests (**Figure 9.14**). The test can be interpreted as an *F*-test in which the observations are weighted by an amount inversely proportional to the variance estimates. This has the effect of making the variances comparable.

The *p*-values may disagree slightly with those obtained from other software providing unequal-variance tests. These differences arise because some methods round or truncate the denominator degrees of freedom for computational convenience. JMP uses the more accurate fractional degrees of freedom.

In practice, the hope is that there are not conflicting results from different tests of the same hypothesis. However, conflicting results do occasionally occur, and there is an obligation to report the results from all reasonable perspectives.

# Special Topic: Power

In the drug example, the *F*-test was significant with a *p*-value of 0.03. What if you wanted to be very cautious and achieve a *p*-value of less than 0.01, a 1% chance of falsely declaring significance? To do this, more data is needed. One solution is to use the *Least Significant Number* (LSN) as a guide to choosing the sample size. Even with more data, however, it is still up to chance whether you achieve the *p*-value you want. *Power* is the probability of achieving a certain significance when the true means and variances are specified. You can use the power concept to help choose a sample size that is likely to give significance for certain effect sizes and variances.

Power has the following ingredients:

- the effect size—that is, the separation of the means

- the standard deviation of the error, often called *sigma*

- Alpha, the significance level

- the number of observations, the sample size.

Similarly, there are four ways to increase power.

## Increase the Effect Size

Larger differences are easier to detect. For example, when designing an experiment to test a drug, administer as large a difference in doses as possible. Also, use balanced designs.

How do you increase the effect size? You could look for ways to determine when big differences occur. The problem with this is these bigger differences might not reflect what is going on in the process of interest.

Another way to increase the observed differences is to be bold in your selection of widely separated groups. If you are studying children whose ages range from 10 to 11, the chance of finding an age-related difference is probably small. Comparing 10-year-olds to 15-year-olds is more likely to produce a difference. *Be bold*.

## Decrease Residual Variance

If you have less noise it is easier to find differences. Sometimes blocking, or testing within subjects, or the selection of a more homogeneous sample can reduce noise.

The quest of total quality management is to get rid of variability. Of course, this is fundamentally impossible—noise happens. One way to reduce noise is to control extraneous factors. Another way is to estimate the noise effect by including variables in the model to control for sources of nuisance variability. Either way reduces error variance. The down side of this is that it makes the study more difficult to perform and more complicated to analyze. *Life is not easy*.

## Increase the Sample Size

With larger samples, the standard error of the estimate of effect size is smaller. The effect is estimated with more precision as the sample size increases. Roughly, the precision increases in proportion to the square root of the sample size.

Experimental units can be expensive, but more data are usually desirable. Another approach to increase power is to see whether it is advantageous to allocate the experimental units to a better design. Experimental design and sample size determination are the main tools to increase power. The down side is that it takes effort and expertise on your part to design good studies. *Life is not simple*.

### Accept Less Protection

Increase $\alpha$. There is nothing magic about $\alpha = 0.05$. A larger $\alpha$ lowers the cut-off value. A statistical test with $\alpha = 0.20$ declares significant differences more often (and also leads to false conclusions more often).

What if you set $\alpha$ to 0.5? This means that when there is no difference, you still declare a significant difference half the time. This will reduce your experimental costs dramatically and increase your conclusiveness. Good luck, though, in selling this to your boss or to a journal editor. You'll also need to hire a good lawyer to handle the damage claims and a psychiatrist to help you sleep at night. *Protect yourself.*

## Power in JMP

Many JMP analyses give you the option to conduct a power analysis. Wherever the **Power** option appears in a popup menu, you can request a power analysis of the statistical test you are performing. When you select **Power**, a dialog similar to the one shown on the right appears.

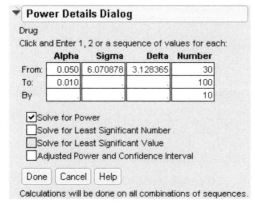

In the Power Details dialog, you set up the hypothetical scenario, specifying what you think might be true values for the effect size and the residual variance. Then, the analysis tells you how likely you are to detect the effect at different sample sizes and alpha levels.

You can try out this powerful tool on the analysis of variance for the Drug data covered in the previous sections.

☞    Select **Power** from the menu next to the Oneway Anova table title to see the Power Details dialog at the bottom of the text report.

Complete the editable power fields as shown in the Power Dialog above.

### Alpha

defaults to 0.05. Click an empty alpha field to type a second value of 0.01.

### Sigma

defaults to the Root Mean Square Error from the Summary table in the Analysis of Variance report (see **Figure 9.4**), since Root Mean Square Error estimates the standard deviation of the error. You can edit Sigma if you want to find the power for other error variances.

## Delta

is a general estimate of the effect size. It is the sum of squares for the hypothesis being tested divided by the total sample size. Note that some books use a standardized effect size estimate rather than a raw effect size estimate. JMP keeps the effect size estimate apart from the error estimate so that they can be looked at separately.

## Number

is the total number of observations in the table.

- Type a range of sample sizes from 30 to 100 in increments of 10.

- Check the **Solve for Power** box in the dialog and click **Done**.

By typing in a range of numbers, power calculations are completed for each combination of values you entered in the dialog. The table of power calculations shown in **Figure 9.15** is appended to the analysis tables.

- Embellish the analysis with a plot using the command accessed by the triangular popup command beneath the power table, which gives the plot of power by sample size shown to the right in **Figure 9.15**.

To read the plot, pick values you would like to have and note the corresponding information. For example, if you want 90% probability (power) of achieving a significance of 0.01, then the sample size needs to be slightly above 70. For the same power at 0.05 level significance, the sample size only needs to be 50.

**Figure 9.15**   Power analysis results

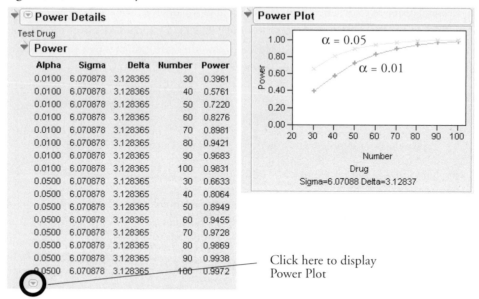

Click here to display Power Plot

# Nonparametric Methods

JMP also offers nonparametric methods in the Fit Y by X platform. Nonparametric methods, introduced in the previous chapter, use only the rank order of the data and ignore the spacing information between data points. The nonparametric tests do not assume the data have a Normal distribution. This section first reviews the rank-based methods, then generalizes the Wilcoxon rank-sum method to the $k$ groups of the one-way layout.

## Review of Rank-Based Nonparametric Methods

The nonparametric tests are useful to test whether means or medians are the same across groups. However, the usual assumption of Normality is not made. Nonparametric tests use functions of the response ranks, called rank scores (Hajek 1969).

JMP offers the following nonparametric tests for testing whether distributions across factor levels are centered at the same location. Each is the most powerful rank test for a certain distribution, as indicated in Table 9.1.

- Wilcoxon rank scores are the ranks of the data.

- Median rank scores are either 1 or 0 depending on whether a rank is above or below the median rank.

- Van der Waerden rank scores are the quantiles of the standard Normal distribution for the probability argument formed by the rank divided by $n-1$. This is the same score that is used in the Normal quantile plots.

**Table 9.1.** Guide for Using Nonparametric Tests

| Fit Y By X Analysis Option | Two Levels | Two or more levels | Most powerful for errors distributed as |
|---|---|---|---|
| **Nonpar-Wilcoxon** | Wilcoxon rank-sum (Mann-Whitney $U$) | Kruskal-Wallis | Logistic |
| **Nonpar-Median** | Two-Sample Median | $k$-Sample Median (Brown-Mood) | Double Exponential |
| **Nonpar-VW** | Van der Waerden | $k$-sample Van der Waerden | Normal |

## The Three Rank Tests in JMP

As an example, use the DrugLBI.jmp example and request nonparametric tests to compare the Drug group means of LBS:

⌐ Choose **Analyze > Fit Y by X** with LBS as Y and Drug as X.

The one-way analysis of variance platform appears showing the distributions of the three groups, as seen previously in **Figure 9.3**.

⌐ In the popup menu on the title bar of the scatterplot, select the four tests that compare groups:

- **Means/Anova**, producing the $F$-test from the standard parametric approach

- **Nonparametric > Wilcoxon Test**, also known as the Kruskal-Wallis test when there are more than two groups

- **Nonparametric > Median Test** for the median test

- **Nonparametric > Van der Waerden Test** for the Van der Waerden test.

**Figure 9.16** shows the results of the four tests that compare groups. In this example, the Wilcoxon and the Van der Waerden agree with the parametric $F$-test in the ANOVA and show borderline significance for a 0.05 $\alpha$-level, despite a fairly small sample and the possibility that the data are not Normal.

The median test is much less powerful than the others and doesn't detect a difference in this example.

**Figure 9.16** Parametric and Non-Parametric tests for drug example

**Oneway Analysis of LBS By Drug**

**Oneway Anova**

**Analysis of Variance**

| Source | DF | Sum of Squares | Mean Square | F Ratio | Prob > F |
|---|---|---|---|---|---|
| Drug | 2 | 293.6000 | 146.800 | 3.9831 | 0.0305 — Significant |
| Error | 27 | 995.1000 | 36.856 | | |
| C. Total | 29 | 1288.7000 | | | |

**Wilcoxon / Kruskal-Wallis Tests (Rank Sums)**

| Level | Count | Score Sum | Score Mean | (Mean-Mean0)/Std0 |
|---|---|---|---|---|
| a | 10 | 122 | 12.2000 | -1.433 |
| d | 10 | 132.5 | 13.2500 | -0.970 |
| placebo | 10 | 210.5 | 21.0500 | 2.425 |

**1-way Test, ChiSquare Approximation**

| ChiSquare | DF | Prob>ChiSq |
|---|---|---|
| 6.0612 | 2 | 0.0483 — Significant |

**Median Test (Number of Points Above Median)**

| Level | Count | Score Sum | Score Mean | (Mean-Mean0)/Std0 |
|---|---|---|---|---|
| a | 10 | 4 | 0.400000 | -0.762 |
| d | 10 | 4 | 0.400000 | -0.762 |
| placebo | 10 | 7 | 0.700000 | 1.523 |

**1-way Test, ChiSquare Approximation**

| ChiSquare | DF | Prob>ChiSq |
|---|---|---|
| 2.3200 | 2 | 0.3135 — Not Significant |

**Van der Waerden Test (Normal Quantiles)**

| Level | Count | Score Sum | Score Mean | (Mean-Mean0)/Std0 |
|---|---|---|---|---|
| a | 10 | -3.69293 | -0.36929 | -1.568 |
| d | 10 | -2.24452 | -0.22445 | -0.953 |
| placebo | 10 | 5.93744 | 0.59374 | 2.521 |

**1-way Test, ChiSquare Approximation**

| ChiSquare | DF | Prob>ChiSq |
|---|---|---|
| 6.4804 | 2 | 0.0392 — Significant |

# Exercises

1. This exercise uses the Movies.jmp data set. You are interested in discovering if there is a difference in earnings between the different classifications of movies.

   (a) Use the Distribution platform to examine the variable Type. How many levels are there? Are there roughly equal numbers of movies of each type?

   (b) Use the Fit Y by X platform to perform an ANOVA, with Type as X and Worldwide $ as Y. Does the test show differences among the different types?

(c) Use Comparison Circles to explore the differences. Does it appear that Action and Drama are significantly different than all other types?

(d) Examine a Normal Quantile Plot of this data and comment on the equality of the variances of the groups. Are they different enough to require a Welch ANOVA? If so, conduct one and comment on its results.

2. The National Institute of Standards and Technology (NIST) references research involving the doping of silicon wafers with phosphorus. The wafers were doped with phosphorous by neutron transmutation doping in order to have nominal resistivities of 200 ohm/cm. Each data point is the average of 6 measurements at the center of each wafer. Measurements of bulk resistivity of silicon wafers were made at NIST with 5 probing instruments on each of 5 days, with the data stored in the table Doped Wafers.jmp (see Ehrstein and Croarkin). The experimenters are interested in testing differences among the instruments.

(a) Examine a histogram of the resistances reported for the instruments. Do the data appear to be Normal? Conduct a statistical test to test for Normality.

(b) Conduct an ANOVA to determine if there is a difference between the probes in measuring resistance.

(c) Comment on the sample sizes involved in this investigation. Do you feel confident with your results?

3. The data table Michelson.jmp contains data (as reported by Stigler, 1977) collected to determine the speed of light in air. Five separate collections of data were made in 1879 by Michelson, and the speed of light was recorded in km/sec. The values for velocity in this table have had 299,000 subtracted from them.

(a) The true value (accepted today) for the speed of light is 299,792.5 km/sec. What is the mean of Michelson's responses?

(b) Is there a significant statistical difference between the trials? Use an ANOVA or a Welch ANOVA—whichever is appropriate—to justify your answer.

(c) Using Student's $t$ Comparison Circles, find the group of observations that is statistically different from all the other groups.

(d) Does excluding the result in part (c) improve Michelson's prediction?

4. *Run-Up* is a term used in textile manufacturing to denote waste. Manufacturers often use computers to lay out designs on cloth in order to minimize waste, and the percentage difference between human layouts and computer-modeled layouts is the run-up. There are some cases where humans

get better results than the computers, so don't be surprised if there are a few negative values for run-up in the table Levi Strauss Run-Up.jmp (Koopmans, 1987). The data was gathered from five different supplier plants in order to determine if there were differences among the plants.

(a)  Produce histograms for the values of Run Up for each of the five plants.

(b)  Test for differences between supplier plants by using the three non-parametric tests provided by JMP. Do they give similar results?

(c)  Compare these results to results given by an ANOVA and comment on the differences. Would you trust the parametric or non-parametric tests more?

5.  The data table Scores.jmp contains a subset of results from the *Third International Mathematics and Science Study*, conducted in 1995. The data contain information on scores for Calculus and Physics, divided into four regions of the country.

(a)  Is there a difference among regions in Calculus scores?

(b)  Is there a difference among regions for Physics Scores?

(c)  Do the data fall into easily definable groups? Use comparison circles to explore their groupings.

6.  Three brands of typewriters were tested for typing speed by having expert typists type identical passages of text. The results are stored in Typing Data.jmp.

(a)  Complete an analysis of variance comparing the typing speed of the experts on each brand of typewriter.

(b)  Does one brand of typewriter yield significantly higher speeds than the other two? Defend your answer.

7.  A manufacturer of widgets determined the quality of its product by measuring abrasion on samples of finished products. The manufacturer was concerned that there was a difference in the abrasion measurement for the two shifts of workers that were employed at the factory. Use the data stored in Abrasion.jmp to compute a *t*-test of abrasion comparing the two shifts. Is there statistical evidence for a difference?

8.  The manufacturers of a medication were concerned about adverse reactions in patients that were treated with their drug. Data on adverse reactions is stored in AdverseR.jmp. The duration of the adverse reaction is stored in the ADR DURATION variable.

(a)  Patients given a placebo are noted with PBO listed in the treatment group variable, while those that received the standard drug regimen are designated with ST_DRUG.

Test whether there is a significant difference in adverse reaction times between the two groups.

(b)   Test whether there is a difference in adverse reaction times based on the sex (gender) of the patient.

(c)   Test whether there is a difference in adverse reaction time based on the race of the patient.

(d)   Redo the analyses in parts (a)-(c) using a nonparametric test. Do the results differ?

(e)   A critic of the study claims that the weights of the patients in the placebo group is not the same as that of the treatment group. Do the data support the claim of the critic?

9.   To judge the efficacy of a three pain relievers, a consumer group conducted a study of the amount of relief each patient received. The amount of relief is measured by each participant rating the amount of relief on a scale of 0 (no relief) to 20 (complete relief). Results of the study are stored in the file Analgesics.jmp.

(a)   Is there a difference in relief between the males and females in the study?

(b)   Conduct an analysis of variance comparing the three types of drug. Is there a significantly significant difference among the three types of drug?

(c)   Find the mean amount of pain relief for each of the three types of pain reliever.

(d)   Does the amount of relief differ for males and females? To investigate this question, conduct an analysis of variance on pain relief for each of the two genders (a separate analysis for males and females. HINT: assign gender as a By-variable). Is there a significant difference in relief for the female subset? For the male subset?

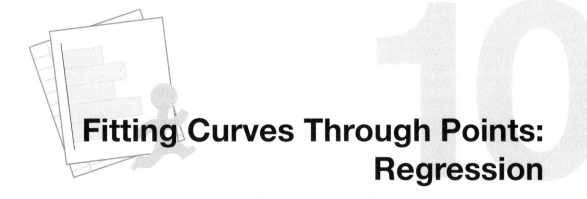

# Fitting Curves Through Points: Regression

## Overview

Regression is a method of fitting curves through data points. It is a straightforward and useful technique, but people new to statistics often ask about its rather strange name—regression.

Sir Francis Galton, in his 1885 Presidential address before the anthropology section of the British Association for the Advancement of Science (Stigler, 1986), described his study of how tall children are compared with the heights of their parents. In this study, Galton defined ranges of parents' heights, and then calculated the mean child height for each range. He drew a straight line that went through the means (as best as he could), and thought he had made a discovery when he found that the child heights tended to be more moderate than the parent heights. For example, if a parent was very tall, the children tended to be tall, but not as tall as the parents. If a parent was very short, the child tended to be short, but not as short as the parent. This discovery he called a regression to the mean, with the regression meaning "to come back to".

Somehow, the term regression became associated with the technique of fitting the line, rather than the process describing inheritance. The term has stuck for 110 years, and is probably going to stay. At least the consequences of Galton giving a misleading term to the statistics profession were milder than the consequences from the idea he gave to anthropology (eugenics). Galton's data are covered later in this chapter.

This chapter covers the case where there is only one factor—the kind of regression situation that you can see on a scatterplot graph.

# Regression

Fitting one mean is easy. Fitting several means is not much harder. How do you fit a mean when it changes as a function of some other variable? In essence, how do you fit a line or a curve through data?

## Least Squares

In regression, you pick an equation type (linear, polynomial, and so forth) and allow the fitting mechanism to determine some of its parameters (coefficients). These parameters are determined by the method of least squares, which finds the parameter values that minimizes the sum of squared distances from each point to the line of fit. **Figure 10.1** illustrates a least-squares regression line.

**Figure 10.1**   Straight-Line Least-Squares Regression

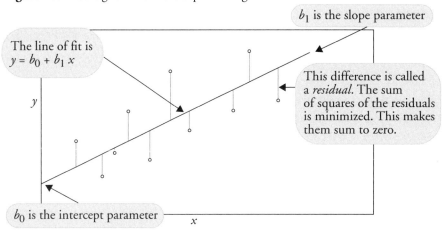

For any regression model, the term *residual* denotes the difference between the actual response value and the value predicted by the line of fit. When talking about the true (unknown) model rather than the estimated one, these differences are called the *errors* or *disturbances*.

Least squares regression is the method of fitting of a model to minimize the sum of squared residuals.

The regression line has interesting balancing properties with regard to the residuals. The sum of the residuals is always zero, which was also true for the simple mean fit. You can think of the fitted line as balancing data in the up-and-down direction. If you add the product of the residuals times the *x*(regressor) values, this sum is also zero. This can be interpreted as the line balancing the data in a rotational sense. Chapter 21, "Machines of Fit," shows how these least squares properties can be visualized in terms of the forces of data acting like springs on the line of fit.

An important special case is when the line of fit is constrained to be horizontal (flat). The equation for this fit is a constant; if you constrain the slope of the line to be zero, the coefficient of the $x$ term (regressor) is zero, and the X term drops out of the model. In this situation, the estimate for the constant is the sample mean. This special case is important because it leads to the statistical test of whether the regressor really affects the response.

## Seeing Least Squares

The principal of least squares can be seen with one of the sample scripts included in the Sample Scripts folder.

    &#9758;   Open and run the demoLeastSquares.jsl script.

Opening and running scripts is covered in "Working With Scripts" on page 53.

**Figure 10.2**   demoLeastSquares Display

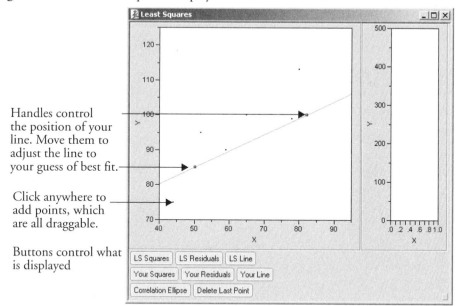

There is a line in the scatterplot with two small squares on it. These two rectangular handles are draggable, and are used in this case to move the line to a position that you think best summarizes the data.

🖰  Press the **Your Residuals** button. Use the handles to move the line around until you think the residuals (in blue) are as small as they can be.

🖰  Press the **Your Squares** button and again try to minimize the total area covered by the blue squares.

To assist you in minimizing the area of the squares, a second graph is displayed to the right of the scatterplot. Think of it as a "thermometer" that measures the sum of the area of the squares in the scatterplot. The least squares criterion selects the line that minimizes this area. To see the least-squares line as calculated by JMP,

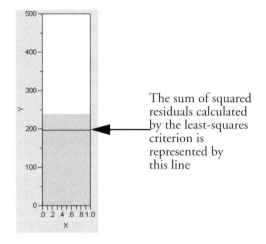

The sum of squared residuals calculated by the least-squares criterion is represented by this line

🖰  Press the button labeled **LS Line** to display the least-squares line.

🖰  Press the buttons **LS Residuals** and **LS Squares** to display the residuals and squares for the least squares line.

You should notice in the graph that displays the sum of the squares, a horizontal line has been added. This represents the sum of the squared residuals from the line calculated by the least squares criterion.

Finally, to illustrate that the least squares criterion performs as it claims to:

🖰  Using the handles, drag your line so that it coincides with the least squares line.

Observe that the sum of squares is now the same as the sum calculated by the least squares criterion.

🖰  Using one of the handles, move your line off of the least squares line, first in one direction, then in the other.

Notice that as your line moves off the line of least squares in any way, the sum of the squares increases. Therefore, the least squares line is truly the line that minimizes the sum of the squared residuals.

## Fitting a Line and Testing the Slope

Eppright *et al.* (1972) as reported in Eubank (1988) measured 72 children from birth to 70 months. You can use regression techniques to examine how the weight to height ratio changes as kids grow up,

🖱  Open the Growth.jmp sample table to see the Eppright data.

🖱  Choose **Analyze > Fit Y by X** and select ratio as Y and age as X. When you click **OK** the result is a scatterplot of ratio by age.

Look for the triangular popup menu icon on the title bar of the plot. Click this icon to see fitting commands for adding regression fits to the scatterplot.

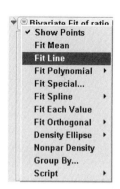

🖱  Select **Fit Mean** and then **Fit Line** from the fitting popup menu.

• **Fit Mean** draws the horizontal line at the mean of ratio.

• **Fit Line** draws the regression line through the data.

These commands also add statistical tables to the regression report. You should see a scatterplot much like the one shown in **Figure 10.3**. The statistical tables are actually displayed beneath the scatterplot in the report window, but have been rearranged here to save space.

Each kind of fit you select has its own popup menu icon that lets you request fitting details.

🖱  Click the popup menu for **Linear Fit**, and select **Confid Curves: Fit**.

This command adds dashed lines around the regression line. The parameter estimates report shows the estimated coefficients of the regression line. The fitted regression is the equation

$$ratio = 0.6656 + 0.005276 \text{ age} + residual$$

**Figure 10.3**   Straight-line least-squares regression

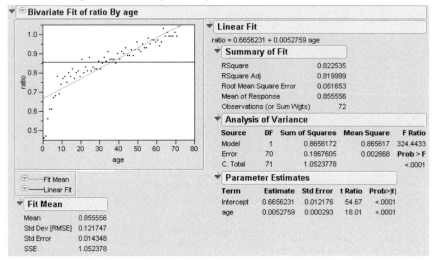

## Testing the Slope by Comparing Models

If we assume that the linear equation is adequate to describe the relationship of the weight to height ratio with growth (which turns out to be incorrect), we have some questions to answer:

- Does the regressor really affect the response?

- Does the ratio of weight to height change as a function of age? Is the true slope of the regression line zero?

- Is the true value for the coefficient of age zero?

- Is the sloped regression line significantly different than the horizontal line at the mean?

Actually, these are all the same question.

"The Difference Between Two Means" on page 161 presented two analysis approaches that turned out to be equivalent. One approach used the distribution of the estimates, which resulted in the *t*-test. The other approach compared the sum of squared residuals from two models where one model was a special case of the other. This model comparison approach resulted in an *F*-test. In regression, there are the same two equivalent approaches: (1) distribution of estimates, and (2) model comparison.

The model comparison is between the regression line and what the line would be if the slope were constrained to be zero; that is, you compare the fitted regression line with the horizontal line at the mean. If the regression line is a better fit than the line at the mean, then the slope of the regression line shows as significantly different from zero. This is often stated negatively: "If the regression line doesn't fit better than the horizontal fit, then the slope of the regression line will not test as significantly different from zero."

The *F*-test in the Analysis of Variance table is the comparison that tests the slope of the fitted line. It compares the sum of squared residuals from the regression fit to the sum of squared residuals from the sample mean. **Figure 10.4** diagrams the relationship between the quantities in the statistical reports and corresponding plot.

Here are descriptions of the quantities in the statistical tables:

## C Total

corresponds to the sum of squares error if you fit only the mean. You can verify this by looking at the Fit Mean table from in the previous example. The C Total sum of squares (SS) is 1.0524 for both the mean fit and the line fit.

## Error

is the sum of squared residuals after fitting the line, 0.1868. This is sometimes casually referred to as the residual, or residual error. You can think of Error as leftover variation— variation that didn't get explained by fitting a model.

## Model

is the difference between the error sum of squares in the two models (the horizontal mean and the sloped regression line). It is the sum of squares resulting from the regression, 0.8656. You can think of Model as a measure of the variation in the data that was explained by fitting a regression line.

## Mean Square

is a sum of squares divided by its respective degrees of freedom. The Mean Square for Error is the estimate of the error variance (0.002668 in this example).

## Root Mean Square Error

is found in the Summary of Fit report. It estimates the error standard deviation of the error and is calculated as the square root of the Mean Square for Error.

If the true regression line has slope zero, then the model isn't explaining any of the variation. The model mean square and the error mean square both estimate the residual error variance, and therefore have the same expected value.

The *F*-statistic is calculated as the model mean square divided by the error mean square. If the model and error both have the same expected value, the *F*-statistic is 1. However, if the model mean square is larger than the mean square for error, you suspect that the slope is not zero and the model is explaining some variation. The *F*-ratio has an *F*-distribution under the hypothesis that (in this example) age has no effect on ratio.

**Figure 10.4**  Diagram to Compare Models

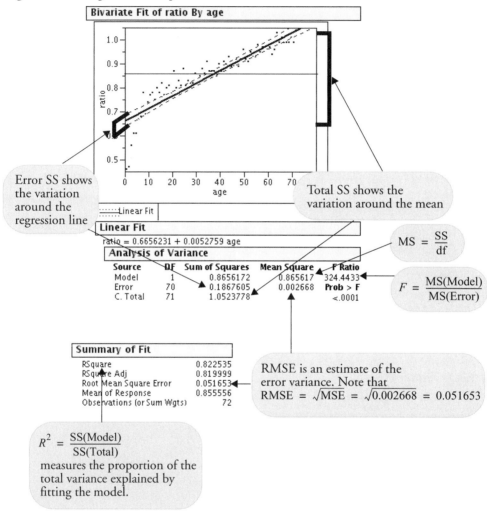

**The Distribution of the Parameter Estimates**

The formula for a simple straight line only has two parameters (the intercept and the slope). For this example, the model can be written as follows:

$$\text{ratio} = b_0 + b_1 \text{ age} + \text{residual}$$

where $b_0$ is the intercept and $b_1$ is the slope.

The Parameter Estimates Table has these quantities:

**Std Error**

is the estimate of the standard deviation attributed to the parameter estimates.

**t-Ratio**

is a test that the true parameter is zero. The *t*-ratio is the ratio of the estimate to its standard error. Generally, you are looking for *t*-ratios that are greater than 2 in absolute value, which usually correspond to significance probabilities of less than 0.05.

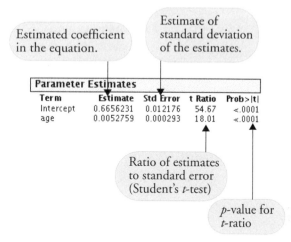

Estimated coefficient in the equation.

Estimate of standard deviation of the estimates.

| Parameter Estimates | | | | |
|---|---|---|---|---|
| Term | Estimate | Std Error | t Ratio | Prob>|t| |
| Intercept | 0.6656231 | 0.012176 | 54.67 | <.0001 |
| age | 0.0052759 | 0.000293 | 18.01 | <.0001 |

Ratio of estimates to standard error (Student's *t*-test)

*p*-value for *t*-ratio

**Prob>|t|**

is the significance probability (*p*-value). You can translate this as "the probability of getting an even greater absolute *t* value by chance alone if the true value of the slope is zero."

Note that the *t*-ratio for the age parameter, 18.01, is the square root of the *F*-Ratio in the Analysis of Variance table, 324.44. You can double-click on the *p*-values in the tables to show more decimal places, and see that the *p*-values are exactly the same. This is not a surprise—the *t*-test for simple regression is testing the same hypothesis as the *F*-test.

## Confidence Intervals on the Estimates

There are several ways to look at the significance of the estimates. The *t*-tests for the parameter estimates, discussed previously, test that the parameters are significantly different from zero. A more revealing way to look at the estimates is to obtain confidence limits that show the range of likely values for the true parameter values.

☞   Beside **Linear Fit**, select the **Confid Curves: Fit** command from the popup menu.

This command adds the confidence curves to the graph, as illustrated previously in **Figure 10.4**.

The 95% confidence interval forms the smallest interval whose range includes the true parameter values with 95% confidence. The upper and lower confidence limits are calculated by adding and subtracting respectively the standard error of the parameter times a quantile value corresponding to a (0.05)/2 Student's *t*-test.

Another way to find the 95% confidence interval is to examine the Parameter Estimates tables. Although the 95% confidence interval values are initially hidden in the report, they can be made visible.

🖱 Right-click on the report and select **Columns > Lower 95%** and **Columns > Upper 95%**, as shown in **Figure 10.5**.

**Figure 10.5**   Add Confidence Intervals to Table

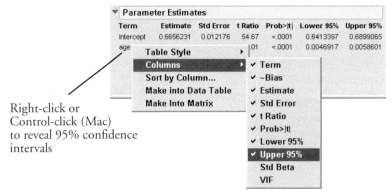

An interesting way to see this concept is to look from the point of view of the sum of squared errors. Imagine the sum of squared errors (SSE) as a function of the parameter values, so that as you vary the parameter values you calculate the SSE at each set of parameter values. The least squares estimates are where this surface is at a minimum.

The left plot in **Figure 10.6** shows a three-dimensional view of this interpretation for the growth data regression problem. The 3-D view shows the curvature of the SSE surface as a function of the parameters, with a minimum in the center. The *x*- and *y*-axes are a grid of parameter values and the *z*-axis is the computed SSE for those values.

The contour plot on the right in **Figure 10.6** shows the elliptical contours corresponding to given SSE values as a function of the parameters. The least squares values for the parameters are the center of the ellipses.

**Figure 10.6**   Representation of Confidence Limit Regions

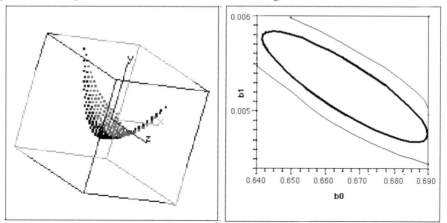

One way to form a 95% confidence interval is to *turn the F-test upside down*. You take an F value that would be the criterion for a 0.05 test (3.97), multiply it by the MSE, and add that to the SSE. This gives a higher SSE of 0.19737 and forms a confidence region for the parameters. Anything that produces a smaller SSE is believable because it corresponds to an F-test with a *p*-value greater than 0.05.

The 95% confidence region is the inside elliptical shape in the plot on the right in **Figure 10.6**. The flatness of the ellipse corresponds to the amount of correlation of the estimates. You can look at the plot to see what parameter values correspond to the extremes of the ellipse in each direction.

- The horizontal scale corresponds to the intercept parameter. The confidence limits are the positions of the vertical tangents to the inner contour line, indicating a low point of 0.6413 and high point of 0.6899.

- The vertical scale corresponds to the slope parameter for **age**. The confidence limits are the positions of the vertical tangents to the inner contour line, indicating a low of 0.00469 and high point of 0.00586. These are the lower and upper 95% confidence limits for the parameters.

## Examine Residuals

It is always a good idea to take a close look at the residuals from a regression (the difference between the actual values and the predicted values):

🖰   Select **Plot Residuals** from the **Linear Fit** popup menu beneath the scatterplot (**Figure 10.7**).

This command appends the residual plot shown in **Figure 10.7** to the bottom of the regression report.

**Figure 10.7**   Scatterplot to Look at Residuals

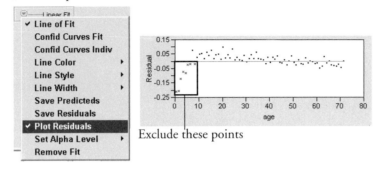

Exclude these points

The picture you usually hope to see is the residuals scattered randomly about a mean of zero. So, in residual plots like the one shown in **Figure 10.7**, you are looking for patterns and for points that violate this random scatter. This plot is suspicious because the left side has a pattern of residuals below the line. These points influence the slope of the regression line (**Figure 10.3**), pulling it down on the left. You can see what the regression would look without these points by excluding them from the analysis.

## Exclusion of Rows

To exclude points (rows) from an analysis, you highlight the rows and assign them the **Exclude** row state characteristic as follows.

- Get the Brush tool from the **Tools** menu or toolbar.

- Shift-drag the brush (which shows on the plot as a stretch rectangle) to highlight the points at the lower left of the plot (indicated by an x marker in **Figure 10.7**).

- Choose **Rows > Exclude / Include**.

You then see a *do not use* sign on those row number area of the data grid.

- Click the original scatterplot window and select **Remove Fit** from the **Fit Mean** popup menu to remove the horizontal line and clean up the plot.

- Again select **Fit Line** from the Bivariate menu to overlay a regression line with the low-age points excluded from the analysis.

The plot in **Figure 10.8** shows the two regression lines. Note that the new line of fit appears to go through the bulk of the points better, ignoring the points at the left that have been excluded.

  ✒ To see the residuals plot for the new regression line, select **Plot Residuals** from the second Linear Fit popup (the options popup for this new fit).

Note in **Figure 10.8** that the residuals no longer have a pattern to them.

**Figure 10.8**  Regression with Extreme Points Included

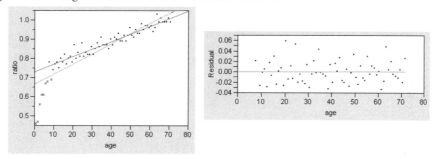

## Time to Clean Up

This scatterplot is needed for the next example, so let's clean it up:

  ✒ Choose **Row > Clear Row States**.

This removes the Excluded row state status from the points so that you can use them in the next steps.

  ✒ To finish the current example, select **Remove Fit** from the second **Linear Fit** popup menu, to remove the example regression that excluded outlying points.

# Polynomial Models

Rather than excluding some points, let's try fitting a curved line through all the points. The simplest curved line is the quadratic curve (a parabola). Its equation contains a term for the squared value of the regressor, age:

$$\text{ratio} = b_0 + b_1 \, \text{age} + b_2 \, \text{age}^2 + \text{residual}$$

To fit a curved line to ratio by age:

⚙ Select the **Fit Polynomial** command on the platform Bivariate menu, with **2, quadratic** from its hierarchical submenu.

The left plot in **Figure 10.9** shows the best fitting polynomial of degree 2 and line fit overlaid on the scatterplot. You can visually compare the straight line and curved line and compare them statistically with the Analysis of Variance reports for both fits that show beneath the plot.

**Figure 10.9**  Comparison of Linear and Second-order Polynomial Fits

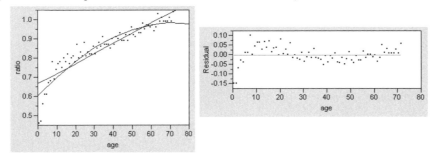

## Look at the Residuals

To examine the residuals,

⚙ Select **Plot Residuals** from the **Polynomial Fit degree=2** popup menu.

You should see a plot similar to the plot on the right in **Figure 10.9**. There still appears to be a pattern in the residuals, so you might want to continue and fit a model with higher order terms.

## Higher Order Polynomials

If you want to give more flexibility to the curve, you can specify higher-order polynomials, adding a term to the third power, to the fourth power, and so forth:

⚙ With the scatterplot active, request a polynomial of degree 4 from the popup menu on the scatterplot title bar.

⚙ Then, select **Plot Residuals** from the **Polynomial Fit degree=4** options menu, which gives the plot on the right in **Figure 10.10**.

Note that the residuals no longer appear to have a pattern to them.

**Figure 10.10**   Comparison of Linear and Fourth-order Polynomial fits

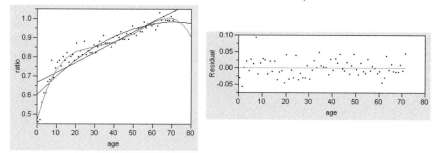

## Distribution of Residuals

It is also informative to look at the shape of the distribution of the residuals. If the distribution departs dramatically from the Normal, then you may be able to find further phenomena in the data.

**Figure 10.11** shows histograms of residuals from the line fit, the polynomial degree-2 fit, and the polynomial degree-4 fit. You can see the distributions evolve toward Normality for the better fitting models with more parameters.

To generate these histograms,

☞   Select the **Save Residuals** command found in the popup menu for each regression fit.

This forms three new columns in the data table.

☞   Choose **Analyze > Distribution** for the three new columns of residual values.

**Figure 10.11**   Histograms for Distribution of Residuals

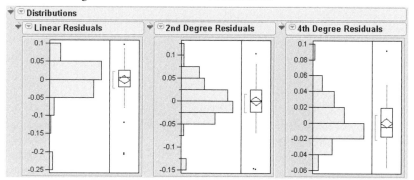

# Transformed Fits

Sometimes, you can fit a curve better if you transform either the Y or X variable (or sometimes both). You can use the Fit Y by X platform to do this:

🖰   Again, choose **Analyze > Fit Y by X** for ratio by age.

🖰   Then, select **Fit Special** from the popup menu on the scatterplot title bar.

The **Fit Special** command displays a dialog that lists natural log, square, square root, exponential, and other transformations, as selections for both the X and Y variables. Try fitting ratio to the log of age:

🖰   Click the **Natural Logarithm: (log x)** radio button for X. When you click **OK**, you should see the left plot in **Figure 10.12**.

**Figure 10.12**   Comparison of Fits

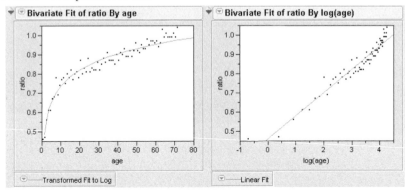

Alternatively, you could create a new column in the data table and compute the log of age, then use **Analyze > Fit Y by X** to do a straight-line regression of ratio on this transformed age. The results are identical except you see the line of fit as a straight line in the log scale, as shown in the plot on the right in **Figure 10.12**.

If you transform the Y variable, then you can't compare the $R^2$ and error sums of squares of the transformed variable fit with the untransformed variable fit—you are fitting a different variable.

## Spline Fit

It would be nice if you could fit a flexible leaf spring through the points. The leaf spring would resist bending somewhat, but would take gentle bends when it needed to fit the data better. A smoothing spline is exactly that kind of fit. With smoothing splines, you can specify how stiff to make the curve. If it is too rigid it looks like a straight line, but if you make the spline too flexible it curves to try to fit each point. Use these commands to see the plots in **Figure 10.13**:

🖰   Choose **Analyze > Fit Y by X** for ratio and age and then select **Fit Spline** from the popup menu with **lambda=10**.

🖰   Select **Fit Spline** again from the popup menu again, but with **lambda=1000**.

**Figure 10.13**   Comparison of Less Flexible and More Flexible Spline Fits

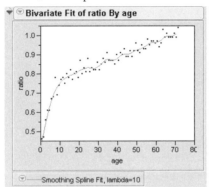 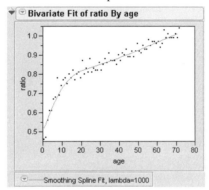

# Special Topic: Why Graphics Are Important

Some statistical packages don't show graphs of the regression, while others require you to make an extra effort to see the graph. The following data table can help you understand the kind of phenomena that you miss if you don't look at the graph.

🖰   Open Anscombe.jmp (Anscombe, 1973).

🖰   Choose **Analyze > Fit Y by X** four times, to fit Y1 by X1, Y2 by X2, Y3 by X3, and Y4 by X4.

🖰   For each pair, select the **Fit Line** command from the Fitting popup menu on the title bar above each scatterplot.

First, look at the text reports for each bivariate analysis, shown in **Figure 10.14**, and compare them. Notice that the reports are nearly identical. The $R^2$ values, the $F$-tests, the parameter estimates and standard errors—they are all the same. Does this mean the situations are the same?

**Figure 10.14** Statistical Reports for Four Analyses

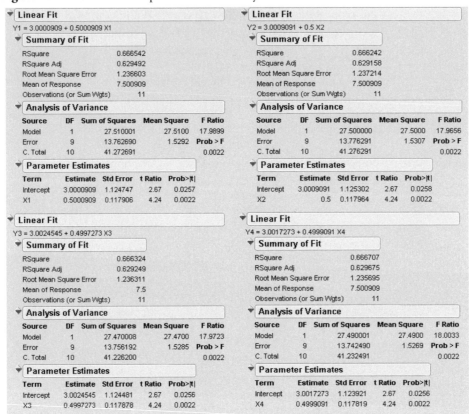

Now look at the graphs of the four relations shown in **Figure 10.15**, and note the following characteristics:

- Y1 by X1 shows a Normal regression situation.

- The points in Y2 by X2 follow a parabola, so a quadratic model is appropriate, with the square of X2 as an additional term in the model.

- There is an extreme outlier in Y3 by X3, which increases the slope of the line that is a perfect fit otherwise. As an exercise, exclude the outlying point and fit another line.

- In Y4 by X4 all the x-values are the same except for one point, which completely determines the slope of the line. This situation is called *leverage*. It is not necessarily bad, but you ought to know about it.

**Figure 10.15**  Regression Lines for Four Analyses

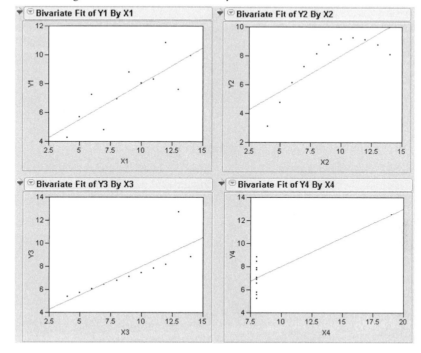

# Special Topic: Why It's Called Regression
# What Happens When *x* and *y* Are Switched?

Remember the story about the study done by Sir Francis Galton mentioned at the beginning of this chapter? He examined the heights of parents and their grown children, perhaps to gain some insight into what degree height is an inherited characteristic. He concluded that the children's heights tended to be more moderate than the parent's heights, and used the term "regression to the mean" to name this phenomenon. For example, if a parent was very tall, the children would be tall, but less so than the parents. If a parent was very short, the child would tend to be short, but less so than the parent.

The case of Galton is interesting not only because it was the first use of regression, but also because Galton failed to notice some properties of regression that would have changed his mind about using regression to draw the conclusions that he drew. To investigate Galton's data:

🖱 Open Galton.jmp and choose **Analyze > Matched Pairs** with child ht and parent ht as the Y paired responses.

The data in the Galton table comes from Galton's published table, but the values are *jittered* by a random amount up to 0.5 in either direction to each point. The jittering was done so that in plots all the points show instead of overlapping. Also, Galton multiplied the women's heights by 1.08 to make them comparable to men's. The parent's height is defined as the average of the two parents.

The scatterplot produced by the Matched Pairs platform is the same as that given by the Fit Y by X platform, but it is rotated by 45°. If the difference between the two variables is zero (the parent and child's heights are, in essence, the same), the points cluster around a horizontal reference line at zero. The mean difference is shown as a horizontal line, with the 95% confidence interval above and below. If the confidence region does not include the horizontal reference line at zero, then the means are not significantly different at the 0.05 level. This represents the *t*-test that Galton could have hypothesized to see if the mean height of the child is the same as the parent.

**Figure 10.16**   Matched Pairs Output of Galton's Data

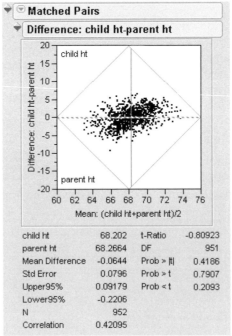

| child ht | 68.202 | t-Ratio | -0.80923 |
|---|---|---|---|
| parent ht | 68.2664 | DF | 951 |
| Mean Difference | -0.0644 | Prob > \|t\| | 0.4186 |
| Std Error | 0.0796 | Prob > t | 0.7907 |
| Upper95% | 0.09179 | Prob < t | 0.2093 |
| Lower95% | -0.2206 | | |
| N | 952 | | |
| Correlation | 0.42095 | | |

However, this is not the test that Galton considered. He invented regression to fit an arbitrary line and then tested to see if the slope of the line was 1 (*i.e.* tested if the line was diagonal). If the line has a slope of one, then the predicted height of the child is the same as that of the parent, except for a generational constant. A slope of less than one indicates regression in the sense that the children tended to have more moderate heights (closer to the mean) than the parents. To look at this regression,

　🖰　Select **Analyze > Fit Y by X**, with parent's height as X and child's height as Y.

　🖰　Select **Fit Line** from the platform popup menu.

When you examine the Parameter Estimates table and the regression line in the left plot of **Figure 10.17**, you see that the least squares regression slope is 0.61—far below 1. This confirms the regression toward the mean.

Is Galton's technique fair to the hypothesis? If the children's heights were more moderate than the parents, shouldn't the parent's heights be more extreme than the children's? To find out, you can reverse the model and try to predict the parent's heights from the children's heights. The analysis on the right in **Figure 10.17** shows the results when the parent's height is Y and children's height is X. Because the previous slope was less than one, symmetry would imply this analysis would give a slope greater than 1. Instead it is 0.28, even less than the first slope.

**Figure 10.17** Child's Height and Parent's Height

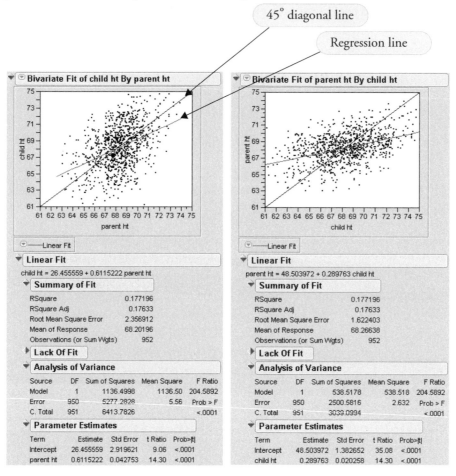

Instead of phrasing the conclusion that children tended to regress to the mean, Galton could have worded his conclusion to say that there is a somewhat weak relationship. With regression, there is no symmetry between the Y and X variables. The slope of Y on X is not the reciprocal of the slope of X on Y; you cannot solve the X by Y fit by taking the Y by X fit and solving for the other variable.

Regression is not symmetric because the error that is minimized is in one direction only—that of the Y variable. So if the roles are switched, a completely different problem is solved.

It always happens that the slope will be smaller than the reciprocal of the inverted variables. However, there is a way to fit the slope symmetrically, so that the role of both variables is the same. This is what is you do when you calculate a *correlation*.

The correlation characterizes the bivariate Normal continuous density. The contours of the Normal density form ellipses as illustrated in **Figure 10.18**. If there is a strong relationship, the ellipses become elongated along a diagonal axis. The line along this axis even has a name—it's called the *first principal component*.

It turns out that least squares does not fit a line to the principal component. Instead, it bisects the contour ellipses at the points of their vertical tangents (see **Figure 10.18**).

If you reverse the direction of finding midpoints or tangents, you describe what the regression line would be if you reversed the role of the Y and X variables. If you draw the X by Y line fit in the Y by X diagram as shown in **Figure 10.18**, it intersects the ellipses at their horizontal tangents.

So Galton's phenomenon of regression to the mean was more an artifact of the method, rather than something to learn about the data.

**Figure 10.18**  Diagram Comparing Regression and Correlation

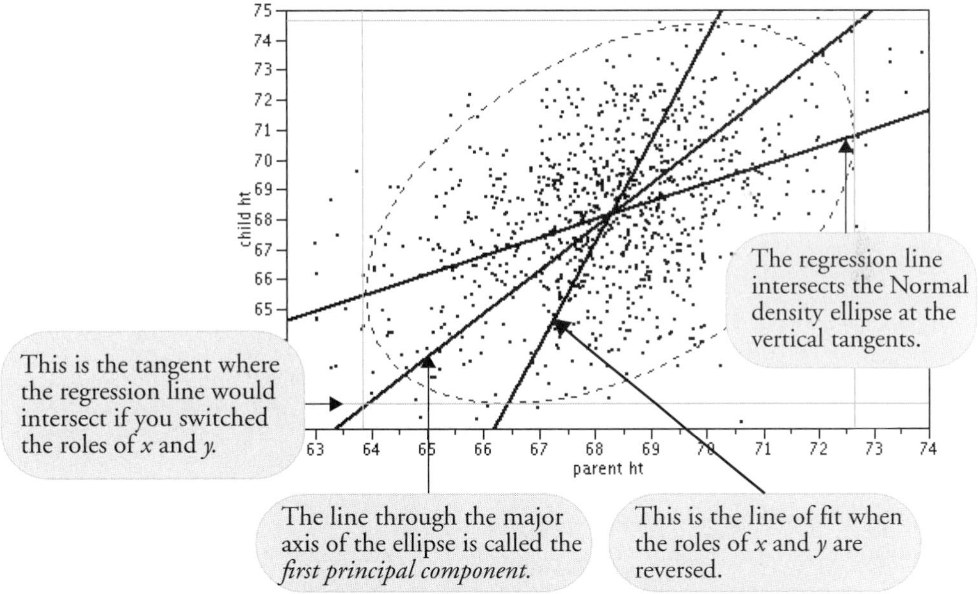

This is the tangent where the regression line would intersect if you switched the roles of *x* and *y*.

The regression line intersects the Normal density ellipse at the vertical tangents.

The line through the major axis of the ellipse is called the *first principal component*.

This is the line of fit when the roles of *x* and *y* are reversed.

**Correlation**

| Variable | Mean | Std Dev | Correlation | Signif. Prob | Number |
|---|---|---|---|---|---|
| parent ht | 68.26638 | 1.787649 | 0.420947 | 0.0000 | 952 |
| child ht | 68.20196 | 2.59697 | | | |

# Curiosities

## Sometimes It's the Picture That Fools You

An experiment by a molecular biologist generated some graphs similar to the scatterplots in **Figure 10.19**. Looking quickly at the plot on the left, where would you guess the least squares regression line lies? Now look at the graph on the right to see where the least-squares fit really appeared.

**Figure 10.19**    Beware of hidden dense clusters

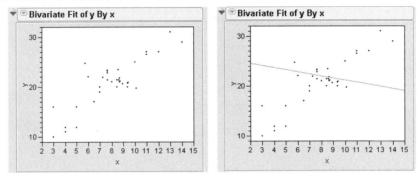

The biologist was perplexed. How did this unlikely looking regression line happen?

It turns out that there is a very dense cluster you can't see. This dense cluster of thousands of points dominated the slope estimate even though the few points farther out had more individual leverage. There was nothing wrong with the computer. It's just that the human eye is sometimes fooled, especially when many points occupy same position. (This example uses the Slope.jmp sample data table)

## High Order Polynomial Pitfall

Suppose you want to develop a prediction equation for predicting ozone based on the population of a city. The lower-order polynomials fit fine, but why not take the "more is better" approach and try a higher order one, say, sixth degree.

  To see the next example, open the Polycity.jmp sample data table. Do **Fit Y by X** with OZONE as Y and POP as X.

  Choose **Fit Polynomial > 6** from the Fitting popup menu on the scatterplot title bar.

As you can see in the bivariate fit shown to the right, the curve fits very well—too well, in fact. How trustworthy is the ozone prediction for a city with a population of 7500?

This overfitting phenomenon, shown to the right, occurs in higher order polynomials when the data are unequally spaced.

More is not always better.

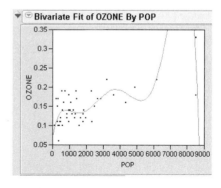

## The Pappus Mystery on the Obliquity of the Ecliptic

Ancient measurements of the angle of the earth's rotation disagree with modern measurements. Is this because modern ones are based on different (and better) technology, or did the angle of rotation actually change?

"Case Study: The Earth's Ecliptic" on page 139 introduced the angle-of-the-ecliptic data. The data that goes back to ancient times is in Cassini.jmp table (Stigler 1986). **Figure 10.20** shows the regression of the obliquity (angle) by time. The regression suggests that the angle has changed over time. The mystery is that the measurement by Pappus is not consistent with the rest of the line. Was Pappus's measurement flawed or did something else happen at that time? We probably will never know.

These kinds of mysteries sometimes lead to detective work that results in great discoveries. Marie Curie discovered radium because of a small discrepancy in measurements made with pitchblende experiments. If she hadn't noticed this discrepancy, the progress of physics might have been delayed.

Outliers are not to be dismissed casually. Moore and McCabe (1989) point out a situation in the 1980s where a satellite was measuring ozone levels over the poles. It automatically rejected a number of measurements because they were very low. Because of this, the ozone holes were not discovered until years later by experiments run from the ground that confirmed the satellite measurements.

**Figure 10.20**   Measurements of the earth's angular rotation

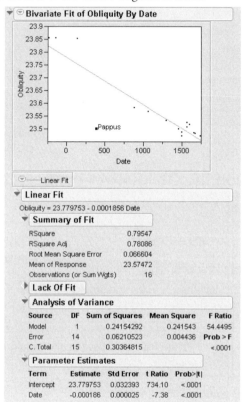

# Exercises

1.  This exercise deals with the data on the top grossing box-office movies (as of June 2003) found in the data table **movies.jmp**. Executives are interested in predicting the amount movies will gross overseas based on the domestic gross.

    (a)  Examine a scatterplot of **Domestic $** vs. **Foreign $**. Do you notice any outliers? Identify them.

    (b)  Fit a line through the mean of the data.

    (c)  Perform a linear regression on **Foreign $** vs. **Domestic $**. Does this linear model describe the data better than the constrained model with only the mean? Justify your answer.

(d)   Exclude any outliers that you found in part (a) and re-run the regression. Describe the differences between this model and the model that included all the points. Which would you trust more?

(e)   Construct a subset of this data consisting of only movies labeled "Drama" or "Comedy". Then, fit a line to this subset. Is it different than the model in part (c)?

(f)   On the subsetted data, fit a line for the "Drama" movies and a separate one for "Comedy" movies. Comment on what you see, and whether you think a single prediction equation for all types of movies will be useful to the executives.

2.   How accurate are past elections in predicting future ones? To answer this question, open the file Presidential Elections.jmp (see Ladd and Carle). This file contains the percent of votes cast for the democratic nominee in three recent elections.

(a)   Produce histograms for the percent of votes in each election, with the three axes having uniform scaling (the **Uniform Scaling** option is in the drop-down menu at the very top of the Distribution platform.) What do you notice about the three means? If you find a difference, explain why it might be there.

(b)   Find the $r^2$ for 1996 vs. 1980 and 1984 vs. 1980. Comment on the associations you see.

(c)   Would the lines generated in these analyses be useful in predicting the percent of voters in the next presidential election? Justify your answer.

3.   Open the file Birth Death.jmp to see data on the birth and death rates of several countries around the world.

(a)   Identify any univariate outliers in the variables birth and death.

(b)   Fit a mean and a line to the data with birth as X and death as Y. Is the linear fit significantly better than the constrained fit using just the mean?

(c)   Produce a residual plot for the linear fit in part (b).

The linear fit produces the following ANOVA table and parameter estimates:

**Analysis of Variance**

| Source | DF | Sum of Squares | Mean Square | F Ratio |
|--------|----|----|----|----|
| Model | 1 | 963.6731 | 963.673 | 63.4814 |
| Error | 72 | 1092.9891 | 15.180 | **Prob > F** |
| C. Total | 73 | 2056.6622 | | <.0001 |

**Parameter Estimates**

| Term | Estimate | Std Error | t Ratio | Prob>|t| |
|------|----|----|----|----|
| Intercept | 4.5709585 | 1.158603 | 3.95 | 0.0002 |
| birth | 0.2626354 | 0.032963 | 7.97 | <.0001 |

The slope of the regression line seems to be highly significant. Why, then, is this model inappropriate for this situation?

(d) Use the techniques of this chapter to fit several transformed, polynomial, or spline models to the data and comment on the best one.

4. The table Solubility.jmp (Koehler and Dunn, 1988) contains data from a chemical experiment that tested the solubility characteristics of seven chemicals.

   (a) Produce scatterplots of all the solutions vs. ethanol. Based on the plots, which chemical has the highest correlation with methane?

   (b) Carbon tetrachloride has solubility characteristics that are highly correlated with hexane. Find a 95% confidence interval for the slope of the regression line of Carbon tetrachloride vs. hexane.

   (c) Suppose the roles of the variables in part (b) were reversed. What can you say about the slope of the new regression line?

# Categorical Distributions

## Overview

When a response is categorical, a different set of tools is needed to analyze the data. This chapter focuses on simple categorical responses and introduces these topics:

- There are two ways to approach categorical data. This chapter refers to them as *choosing* and *counting*. They use different tools and conventions for analysis.

- The concept of variability in categorical responses is more difficult than in continuous responses. Monte Carlo simulation helps demonstrate how categorical variability works.

- The *chi-square test* is the fundamental statistical tool for categorical models. There are two kinds of chi-square tests. They test the same thing in a different way and get similar results.

Fitting models to categorical response data is covered in Chapter 10, "Categorical Models."

# Categorical Situations

A *categorical response* is one in which the response is from a limited number of choices (called *response categories*). There is a probability associated with each of these choices, and these probabilities sum to 1.

Categorical responses are common:

- Consumer preferences are usually categorical: Which do you like the best— tea, coffee, juice, or soft drinks?

- Medical outcomes are often categorical: Did the patient live or die?

- Biological responses are often categorical: Did the seed germinate?

- Mechanical responses can be categorical: Did the fuse blow at 20 amps?

- Any continuous response can be converted to a categorical response: Did the temperature reach 100 degrees?

# Categorical Responses and Count Data: Two Outlooks

It is important to understand that there are two approaches to handling categorical responses. The two approaches generally give the same results, but they use different tools and terms.

First, imagine that each observation represents the response of a chooser. Based on conditions of the observation, the chooser is going to respond with one of the response categories. For example, the chooser might be selecting a dessert from the choices pie, ice cream, or cake. Each response category has some probability of being chosen, and that probability varies depending on other characteristics of the observational unit.

Now reverse the situation and think of yourself as the observation collector for one of the categories. For example, suppose that you sell the pies. The category "Pies" now is a sample category for the vendor, and the response of interest is how many pies can be sold in a day. Given total sales for the day of all desserts, the interest is in the market share of the pies.

**Figure 11.1** diagrams these two ways of looking at categorical distributions.

**Figure 11.1**   Customer or Supplier?

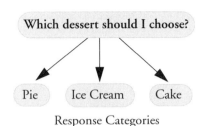

The customer/chooser thinks in terms of *logistic regression*, where the Y variable is which dessert you choose and the X variables affect the probabilities associated with each dessert category. The supplier/counter thinks about *log-linear models*, where the Y is the count, and the X effect is the dessert category. There can be other X's interacting with that X.

The modeling traditions are also different. Customer/chooser-oriented analyses, such as a live/die medical analysis, use continuous X's (like dose, or how many years you smoked). Supplier/counter-oriented analysts, typified by social scientists, are most comfortable with categorical X's (like age, race, gender, and personality type) because that keeps the data count-oriented.

The probability distributions for the two approaches are also different. This book won't go into the details of the distributions, but you can be aware of distribution names. Customer/chooser oriented analysts refer to the *Bernoulli distribution* of the choice. Supplier/counter oriented analysts refer to the *Poisson distribution* of counts in each category. However, both approaches refer to the *multinomial distribution.*

- To the customer/chooser analysts, the multinomial counts are aggregation statistics.

- To the supplier /counter analysts, the multinomial counts are the count distribution within a fixed total count.

The customer/chooser analyst thinks the basic analysis is fitting response category probabilities. The supplier/counter analyst thinks that basic analysis is a one-way analysis of variance on the counts and uses weights because the distribution is Poisson instead of Normal.

Both orientations are right—they just have different outlooks on the same statistical phenomenon.

In this book, the emphasis is on the customer/chooser point of view, also known as the *logistic regression* approach. With logistic regression, it is important to distinguish the responses (Y's), which have the random element, from the factors (X's), which are fixed from the point of view of the model. The X's and Y's must be distinguished before the analysis is started.

Let's be clear on what the X's and Y's are for the chooser point of view:

- Responses (Y's) identify a choice or an outcome. They have a random element because the choice is not determined completely by the X factors. Examples of responses are patient outcome (lived or died), or desert preference (Gobi or Sahara).

- Factors (X's) identify a sample population, an experimentally controlled condition, or an adjustment factor. They are not regarded as random even if you randomly assigned them. Examples of factors are gender, age, treatment or block.

**Figure 11.2** illustrates the X and Y variables for both outlooks on categorical models.

**Figure 11.2**   Categories or Counts?

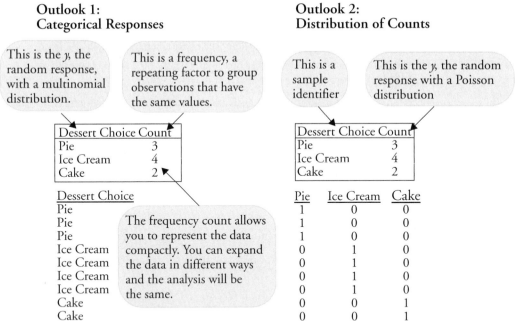

The other point of view is the *log-linear model* approach. The log-linear approach regards the count as the Y variable and all the other variables as X's. After fitting the whole model, the effects that are of interest are identified. Any effect that has no response category variable is discarded, since it is just an artifact of the sampling design. Log-linear modeling uses a technique called *iterative proportional fitting* to obtain test statistics. This process is also called *raking*.

# A Simulated Categorical Response

A good way to learn statistical techniques is to simulate data with known properties, and then analyze the simulation to see if you find the structure that you put into the simulation.

These steps describe the simulation process:

1.  Simulate one batch of data, then analyze.

2.  Simulate more batches, analyze them, and notice how much they vary.

3.  Simulate a larger batch to notice that the estimates have less variance.

4.  Do a batch of batches—simulations that for each run obtain sample statistics over a new batch of data.

5.  Use this last batch of batches to look at the distribution of the test statistics.

## Simulating Some Categorical Response Data

Let's make a world where there are three soft drinks. The most popular ("Sparkle Cola") has a 50% market share and the other two ("Kool Cola" and "Lemonitz") are tied at 25% each. To simulate a sample from this population, create a data table that has one variable (call it Drink Choice), which is drawn as a random categorical variable using the following formula:

$$p = \text{Random Uniform}();$$
$$\text{If} \begin{cases} p < 0.25 \Rightarrow \text{"Kool Cola"} \\ p < 0.5 \;\; \Rightarrow \text{"Lemonitz"} \\ \text{else} \;\;\;\;\; \Rightarrow \text{"Sparkle Cola"} \end{cases};$$

This formula first draws a uniform random number between 0 and 1 using the **Random Uniform** function, and assigns the result to a temporary variable p. Then it compares that random number to the If conditions of the three responses and picks the first response where p is true. Each case returns the character name of the soft drink as the response value.

Note that there are two statements in this formula, which are delimited with semicolons. The Formula Editor operations needed to construct this formula include the following operations: **Variables: New** creates a temporary variable named p, The **If** statement from the **Conditions** functions assigns soft drink names to probability conditions given by the **Random Uniform** function.

This table has already been created.

🖰    Open the data table Cola.jmp, which contains the formula shown previously.

The table is stored with no rows. A data table stored with no rows and columns with formulas is called a *table template*.

🖰    Choose **Rows > Add Rows** to add 50 rows to the table.

🖰    Choose **Analyze > Distribution** with the Drink Choice variable as Y, which gives an analysis like that in **Figure 11.3**.

Don't expect to get the exact same numbers that we show here, because the formula generates random data. Each time the computations are performed, a different set of data is produced.

Note that even though the data are based on the true probabilities of 0.25, 0.25, and 0.50, the estimates came out differently (0.24, 0.28, and 0.48). Your data have random values with somewhat different probabilities.

**Figure 11.3**   Histogram and frequencies of simulated data

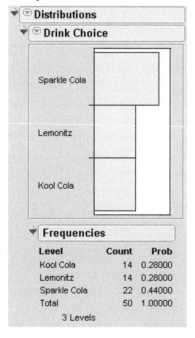

## Variability in the Estimates

The following sections distinguish between $\rho$ (Greek rho), the true value of a probability, and $p$, its estimate. The true value $\rho$ is a fixed number, but its estimate $p$ is an outcome of a random process, so it has variability associated with it.

You cannot compute a standard deviation of the original response—it has character values. However, the variability in the probability estimates is well-defined and computable.

The variability of an estimate is expressed by its variance or its standard deviation. The variance of $p$ is given by the formula

$$\frac{\rho(1-\rho)}{n}$$

For Sparkle Cola, having a $\rho$ of 0.50, the variance of the probability estimate is (0.5•0.5)/50, 0.005. The standard deviation of the estimate is the square root of this variance, 0.07071. **Table 9.1** compares the difference between the true $\rho$ and its estimate $p$. Then, it compares the true standard deviation of the statistic $p$, and the standard error of $p$, which estimates the standard deviation of $p$.

Remember, the term *standard error* is used to label an estimate of the standard deviation of another estimate. Only because this is a simulation with known true values (parameters) can you see both the standard errors and the true standard deviations.

**Table 11.1.** Simulated probabilities and estimates

| Level | $\rho$, the true probability | $p$, the estimate of $\rho$ | True standard deviation of the estimate | Standard Error of the Estimate |
|---|---|---|---|---|
| Kool Cola | 0.25 | 0.18 | 0.06124 | 0.05433 |
| Lemonitz | 0.25 | 0.26 | 0.06124 | 0.06203 |
| Sparkle Cola | 0.50 | 0.56 | 0.07071 | 0.07020 |

This simulation shows a lot of variability. As with Normally distributed data, you can expect to go 2 standard deviations from the true probability about 5% of the time.

Now let's see how the estimates vary with a new set of random responses.

🖰 Right-click on the Drink Choice column heading and select **Formula**.

🖰 Click **Apply** on the calculator to re-evaluate the random formula.

🖰 Again perform **Analyze > Distribution** on Drink Choice.

🖰    Repeat this evaluate/analyze cycle four times.

Each repetition results in a new set of random responses and a new set of estimates of the probabilities. **Table 9.2** gives the estimates from the four Monte Carlo runs.

**Table 11.2.** Estimates from Monte Carlo Runs

| Level | probability | probability | probability | probability |
|---|---|---|---|---|
| Kool Cola | 0.28000 | 0.18000 | 0.26000 | 0.40000 |
| Lemonitz | 0.26000 | 0.32000 | 0.24000 | 0.18000 |
| Sparkle Cola | 0.46000 | 0.50000 | 0.50000 | 0.42000 |

With only 50 observations, there is a lot of variability in the estimates. The "Kool Cola" probability estimate varies between 0.18 and 0.40, the "Lemonitz" estimate varies between 0.18 and 0.32, and the "Sparkle Cola" estimate varies between 0.42 and 0.50.

## Larger Sample Sizes

What happens if the sample size increases from 50 to 500? Remember that the variance of $p$ is

$$\frac{\rho(1 - \rho)}{n}$$

With more data, the probability estimates have a much smaller variance. To see what happens when we add observations,

🖰    Choose **Rows > Add Rows** and enter 450, to get a total of 500 rows.

🖰    Perform **Analyze > Distribution** for the response variable Drink Choice.

Five hundred rows give a smaller variance, 0.0005, and a standard deviation at about $\sqrt{0.005} = 0.02$. The figure to the right shows the first simulation for 500 rows.

**Frequencies**

| Level | Count | Prob | StdErr Prob |
|---|---|---|---|
| Kool Cola | 115 | 0.23000 | 0.01882 |
| Lemonitz | 135 | 0.27000 | 0.01985 |
| Sparkle Cola | 250 | 0.50000 | 0.02236 |
| Total | 500 | 1.00000 | 0.00000 |

3 Levels

🖰    Repeat the evaluate/analyze cycle four times.

**Table 9.3** shows the results of the next 4 simulations.

**Table 11.3.** Estimates from four Monte Carlo Runs

| Level | Probability | Probability | Probability | Probability |
|---|---|---|---|---|
| Kool Cola | 0.28000 | 0.25000 | 0.25600 | 0.23400 |
| Lemonitz | 0.24200 | 0.28200 | 0.23400 | 0.26200 |
| Sparkle Cola | 0.47800 | 0.46800 | 0.51000 | 0.50400 |

Note that the probability estimates are closer to the true values and that the standard errors are smaller.

## Monte Carlo Simulations for the Estimators

What do the distributions of these counts look like? Variances can be easily calculated, but what is the distribution of the estimate? Statisticians often use Monte Carlo simulations to investigate the distribution of the statistics.

To simulate estimating a probability (which has a true value of 0.25 in this case) over a sample size (50 in this case), construct the formula shown here.

The **Random Uniform** function generates a random value distributed uniformly between 0 and 1. The random value is checked to see if it is less than 0.25. The term in the numerator will be 1 or 0 depending on this comparison. It will be 1 about 25% of the time, and 0 about 75% of the time. This random number is generated 50 times, and the sum of them is divided by 50.

$$\frac{\sum_{j=1}^{50} \text{Random Uniform()} < 0.25}{50}$$

This formula is a simulation of a Bernoulli event with 50 samplings. The result estimates the probability of getting a 1. In this case, you happen to know the true value of this probability (0.25) because you generated the data.

Now, it is important to see how well these estimates behave. Theoretically, the mean (expected value) of the estimate, $p$, is 0.25 (the true value), and the standard deviation that is the square root of

$$\frac{p \cdot (1 - p)}{n}$$

which is 0.061237.

## Distribution of the Estimates

The sample data has a table template called Simprob.jmp that is a Monte Carlo simulation for the probability estimates of 0.25 and 0.5, based on 50 and 500 trials. You can add 1000 rows to the data to draw 1000 Monte Carlo trials. (**Note:** If you have a slow computer, be aware that the calculations can be time intensive. Adding 1000 rows requires processing 1,100,000 random numbers.)

To see how the estimates are distributed:

- Open Simprob.jmp.

- Choose **Rows > Add Rows** and enter 1000.

- Next choose **Analyze > Distribution** and use all the columns as Y variables.

- When the histograms appear, select the **Uniform Scaling** option from the check-mark popup menu beside the word "Distributions" on the title bar of the report.

- Get the grabber tool from the **Tools** menu (or palette) and drag the histograms to adjust the bar widths and positions.

**Figure 11.4** and Table 11.4 show the properties as expected:

- The variance narrows as the sample size goes up.

- The distribution of the estimates is approximately Normally distributed, especially as the sample size gets large.

**Figure 11.4**  Histograms for simulations with various $n$ and $p$ values

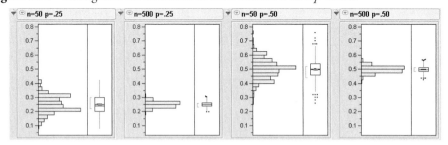

The estimates of the probability $p$ of getting response indicator values of 0 or 1 are a kind of mean. So, as the sample gets larger, the value of $p$ gets closer and closer to 0.50, the mean of 0 and 1. Like the mean for continuous data, the standard error of the estimate relates to the sample size by the factor $1/(\sqrt{n})$.

The Central Limit Theorem applies here. It says that the estimates approach a Normal distribution when there are a large number of observations.

**Table 11.4.** Summary of simulation results

| True Value of $p$ | N used to estimate $p$ | Mean of the trials of the estimates of $p$ | Standard deviation of trials of estimates of $p$ | True mean of estimates | True standard deviation of the estimates |
|---|---|---|---|---|---|
| 0.25 | 50 | 0.24774 | 0.06118 | 0.25 | 0.061237 |
| 0.25 | 500 | 0.24971 | 0.01974 | 0.25 | 0.019365 |
| 0.50 | 50 | 0.49990 | 0.06846 | 0.50 | 0.070711 |
| 0.50 | 500 | 0.50058 | 0.02272 | 0.50 | 0.022361 |

# The $X^2$ Pearson Chi-Square Test Statistic

Because of the Normality of the estimates, it is reasonable to use Normal-theory statistics on categorical response estimates. Remember that the Central Limit Theorem says that the sum of a large number of independent and identically-distributed random values have close to a Normal distribution.

However, there is a big difference between having categorical and continuous responses. With categorical responses, the variances of the differences are known. They are a function of $n$ and the probabilities. The hypothesis specifies the probabilities, so calculations can be made under the null hypothesis. Rather than using an $F$-statistic, this situation calls for the $\chi^2$ *(chi-square) statistic*.

The standard chi-square for this model is the following scaled sum of squares:

$$\chi^2 = \frac{\sum_{j=1}^{n} (\text{Observed}_j - \text{Expected}_j)^2}{\text{Expected}_j}$$

where Observed and Expected refer to cell counts rather than probabilities.

# The $G^2$ Likelihood-Ratio Chi-Square Test Statistic

Whereas the Pearson chi-square assumes Normality of the estimates, another kind of chi-square test is calculated with direct reference to the probability distribution of the response and so does not require Normality.

Define the *maximum likelihood estimator* to be the one that finds the values of the unknown parameters that maximize the probability of the data. In statistical language, it finds parameters that make the data that actually occurred less improbable than they would be with any other parameter values. The term *likelihood* means the probability has been evaluated as a function of the parameters with the data fixed.

It would seem that this requires a lot of guesswork in finding the parameters that maximize the likelihood of the observed data, but just as in the case of least-squares, mathematics can provide short cuts to computing the ideal coefficients. There are two fortunate short cuts for finding a maximum likelihood estimator:

- Because observations are assumed to be independent, the joint probability across the observations is the product of the probability functions for each observation.

- Because addition is easier than multiplication, instead of multiplying the probabilities to get the joint probability, you add the logarithms of the probabilities, which gives the *log-likelihood*.

This makes for easy computations. Remember that an individual response has a multinomial distribution, so the probability is $\rho_i$ for the $i=1$ to $r$ probabilities over $r$ response categories.

Consider the first five responses of the cola example: Kool Cola, Lemonitz, Sparkle Cola, Sparkle Cola, and Lemonitz. For Kool Cola, Lemonitz, and Sparkle Cola, denote the probabilities as $\rho1$, $\rho2$, and $\rho3$ respectively. The joint log-likelihood is:

$$\log(\rho1) + \log(\rho2) + \log(\rho3) + \log(\rho3) + \log(\rho2)$$

It turns out that this likelihood is maximized by setting the probability parameter estimates to the category count divided by the total count, giving

$p1 = n1/n = 1/5$

$p2 = n2/n = 2/5$

$p3 = n3/n = 2/5$

where $p1$, $p2$, and $p3$ estimate $\rho1$, $\rho2$, and $\rho3$. Substituting this into the log-likelihood gives the maximized log-likelihood of

$$\log(1/5) + \log(2/5) + \log(2/5) + \log(2/5) + \log(2/5)$$

At first it may seem that taking logarithms of probabilities is a mysterious and obscure thing to do, but it is actually very natural. You can think of the negative logarithm of $p$ as the number of binary questions you need to ask to determine which of $1/p$ equally likely outcomes happens. The negative logarithm converts units of probability into units of information. You can think of the negative loglikelihood as the *surprise* value of the data because surprise is a good word for unlikeliness.

## Likelihood Ratio Tests

One way to measure the credibility for a hypothesis is to compare how much surprise (–log-likelihood) there would be in the actual data with the hypothesized values compared with the surprise at the maximum likelihood estimates. If there is too much surprise, then you have reason to throw out the hypothesis.

It turns out that the distribution of twice the difference in these two surprise (-log-likelihood) values approximately follows a chi-square distribution.

Here is the setup: Fit a model twice. The first time you fit by maximum likelihood with no constraints on the parameters. The second time you fit by maximum likelihood, but constrain the parameters by the null hypothesis that the outcomes are equally likely. It happens that twice the difference in log-likelihoods has an approximate chi-square distribution (under the null hypothesis). These chi-square tests are called *likelihood ratio chi-squares*, or *LR chi-squares*.

Twice the difference in the log-likelihood is a likelihood ratio chi-square test.

The likelihood ratio tests are very general. They occur not only in categorical responses, but also in a wide variety of situations.

## The $G^2$ Likelihood Ratio Chi-Square Test

Let's focus on Bernoulli probabilities for categorical responses. The log-likelihood for a whole sample is the sum of natural logarithms of the probabilities attributed to the events that actually occurred.

$$\text{log-likelihood} = \sum \ln(\text{probability the model gives to event that occurred in data})$$

The likelihood ratio chi-square is twice the difference in the two likelihoods, when one is constrained by the hypothesis and the other is unconstrained.

$$G^2 = 2 \text{ (log-likelihood(unconstrained)} - \text{log-likelihood(constrained))}$$

Using the words observed and hypothesized, this is formed by the sum over all observations

$$G^2 = \sum [\log(\rho_{y_i}) - \log(p_{y_i})]$$

where $\rho_{y_i}$ is the hypothesized probability and $p_{y_i}$ is the estimated probability for the events $y_i$ that actually occurred.

If you have already collected counts for each of the responses, and bring the subtraction into the log as a division, the formula becomes

$$G^2 = 2 \sum n_i \log \frac{\rho_{y_i}}{p_{y_i}}$$

To compare with the Pearson chi-square, which is written schematically in terms of counts, the LR chi-square statistic can be written

$$G^2 = 2 \sum \text{observed} \left( \log \frac{\text{expected}}{\text{observed}} \right)$$

# Univariate Categorical Chi-Square Tests

A company gave 998 of its employees the Myers-Briggs Type Inventory (MBTI) questionnaire. The test is scored to result in a 4-character personality type for each person. There are 16 possible outcomes, represented by 16 combinations of letters (see **Figure 11.5**). The company wanted to know if its employee force was statistically different in personality types from the general population.

## Comparing Univariate Distributions

The data table Mb-dist.jmp has a column called TYPE to use as a Y response, and a Count column to use as a frequency. To see the company test results:

🖱  Open the sample table called Mb-dist.jmp.

⫐  Choose **Analyze > Distribution** and use Type as the Y variable and Count as the frequency variable to see the report in **Figure 11.5**.

**Figure 11.5**   Histogram, Mosaic Plot, and Frequencies for Myers-Briggs Data

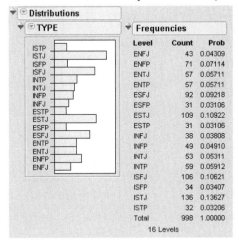

To test the hypothesis that the personalities test results at this company occur at the same rates as the general population:

⫐  Select the **Test Probabilities** command in the popup menu on the histogram title bar at the top of the report.

A dialog then appears at the end of the report.

⫐  Edit the Hypoth Prob (hypothesized probability) values by clicking and then entering the values as shown on the left in **Figure 11.6**.

These are the general population rates for each personality type.

⫐  Click the lower radio button in the dialog, then click **Done**.

You now see the test results appended to the Test Probabilities table, as shown in the table on the right in **Figure 11.6**.

Note that the company does have a significantly different profile than the general population. Both chi-square tests are highly significant. The company appears to have more ISTJ's (introvert sensing thinking judging) and fewer ESFP's (extrovert sensing feeling perceiving) than the general population.

**Figure 11.6**   Test Probabilities report for the Myers-Briggs data

| Test Probabilities | | |
| --- | --- | --- |
| Level | Estim Prob | Hypoth Prob |
| ENFJ | 0.04309 | 0.04950 |
| ENFP | 0.07114 | 0.04950 |
| ENTJ | 0.05711 | 0.04950 |
| ENTP | 0.05711 | 0.04950 |
| ESFJ | 0.09218 | 0.12871 |
| ESFP | 0.03106 | 0.12871 |
| ESTJ | 0.10922 | 0.12871 |
| ESTP | 0.03106 | 0.12871 |
| INFJ | 0.03808 | 0.00990 |
| INFP | 0.04910 | 0.00990 |
| INTJ | 0.05311 | 0.00990 |
| INTP | 0.05912 | 0.00990 |
| ISFJ | 0.10621 | 0.05941 |
| ISFP | 0.03407 | 0.04950 |
| ISTJ | 0.13627 | 0.05941 |
| ISTP | 0.03206 | 0.05941 |

Click then Enter Hypothesized Probabilities.

Choose rescaling method to sum probabilities to 1.
○ Fix omitted at estimated values, rescale hypothesis
◉ Fix hypothesized values, rescale omitted

[ Done ]  [ Help ]

| Test Probabilities | | |
| --- | --- | --- |
| Level | Estim Prob | Hypoth Prob |
| ENFJ | 0.04309 | 0.04950 |
| ENFP | 0.07114 | 0.04950 |
| ENTJ | 0.05711 | 0.04950 |
| ENTP | 0.05711 | 0.04950 |
| ESFJ | 0.09218 | 0.12871 |
| ESFP | 0.03106 | 0.14851 |
| ESTJ | 0.10922 | 0.12871 |
| ESTP | 0.03106 | 0.12871 |
| INFJ | 0.03808 | 0.00990 |
| INFP | 0.04910 | 0.00990 |
| INTJ | 0.05311 | 0.00990 |
| INTP | 0.05912 | 0.00990 |
| ISFJ | 0.10621 | 0.05941 |
| ISFP | 0.03407 | 0.04950 |
| ISTJ | 0.13627 | 0.05941 |
| ISTP | 0.03206 | 0.05941 |

| Test | ChiSquare | DF | Prob>Chisq |
| --- | --- | --- | --- |
| Likelihood Ratio | 722.1019 | 15 | <.0001 |
| Pearson | 1013.214 | 15 | <.0001 |

Method: Fix hypothesized values, rescale omitted
Note: Hypothesized probabilities did not sum to 1. Probabilities have been rescaled.

By the way, some people find it upsetting that different statistical methods get different results. Actually, the $G^2$ (likelihood ratio) and $\chi^2$ (Pearson) chi-square statistics are usually close.

## Charting to Compare Results

🖑   The report in **Figure 11.7** was done by using **Graph > Chart** with Type as X, and the actual data from General and Company as Y.

🖑   When the chart appears, select bar chart from the use the **Horizontal** command in the Chart title bar popup menu to rearrange the chart.

**Figure 11.7**    Mean sample personality scores and scores for general population

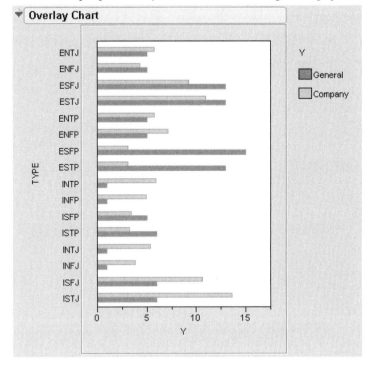

# Exercises

1. P&D Candies produces a bag of assorted sugar candies, called "Moons", in several colors. Based on extensive market research, they have decided on the following mix of Moons in each bag: Red, 20%, Yellow 10%, Brown 30%, Blue 20%, and Green, 20%. A consumer advocate suspects that the mix is not what the company claims, so he gets a bag containing 100 Moons. The 100 pieces of candy are represented in the file Candy.jmp (fictional data).

    (a)    Can the consumer advocate reasonably claim that the company's mix is not as they say?

    (b)    Do you think a single-bag sample is representative of all the candies produced?

2. There is an urban myth that people tend to get "wilder" during phases of a full moon. To study one facet of this phenomenon, two researchers in Germany studied police records for three different conditions of the moon. The number of drunk drivers arrested for days with a full moon, two days prior to the full moon, and "other" days were recorded. Their results showed 175 drunk drivers per day for the two days prior to the full moon, 161 drunk drivers per day during the full moon, and 120 per day at other times.

(a)   Test these data against the null hypothesis that the moon has no effect on the number of drunk-driving arrests.

(b)   Policemen, firemen, and emergency room workers often believe strongly in the lunar association. Does this study confirm or deny their suspicions?

3.  One of the ways that public schools can make extra money is to install vending machines for students to access between classes. Suppose a high school installed three drink machines for different manufacturers in a common area of the school. After one week, they collected information on the number of visits to each machine, as shown in the following table:

| Machine A | Machine B | Machine C |
|-----------|-----------|-----------|
| 1546      | 1982      | 1221      |

Is there evidence of the students preferring one machine over another?

# Categorical Models

## Overview

Chapter 11, "Categorical Distributions," introduced the distribution of a single categorical response. You were introduced to the Pearson and the likelihood ratio chi-square tests and saw how to compare univariate categorical distributions.

This chapter covers multivariate categorical distributions. In the simplest case, the data can be presented as a two-way contingency table of frequency counts, with the expected cell probabilities and counts formed from products of marginal probabilities and counts. The chi-square test again is used for the contingency table and is the same as testing multiple categorical responses for independence.

Correspondence analysis is shown as a graphical technique useful when the response and factors have many levels or values.

Also, a more general categorical response model is used to introduce nominal and ordinal logistic regression, which allows multiple continuous or categorical factors.

# Fitting Categorical Responses to Categorical Factors: Contingency Tables

When a categorical response is examined in relationship to a categorical factor (in other words, both X and Y are categorical), the question is: do the response probabilities vary across factor-defined subgroups of the population? Comparing a continuous variable and a categorical variable in this way was covered in "Comparing Many Means: One-Way Analysis of Variance" on page 199. In that chapter, means were fit for each level of a categorical variable and tested using an ANOVA. When the continuous response is replaced with a categorical response, the equivalent technique is to estimate response probabilities for each subgroup and test that they are the same across the subgroups.

The subgroups are defined by the levels of a categorical X factor. For each subgroup, the set of response probabilities must add up to 1. For example, consider the following:

- The probability of whether a patient lives or dies (response probabilities) depending on whether the treatment (the X factor) was drug or placebo.

- The probability that type of car purchased (response probabilities) depending on marital status (the X factor).

To estimate response probabilities for each subgroup, you take the count in a given response level and divide it by the total count from that subgroup.

## Testing with $G^2$ and $X^2$

You want to test whether the factor affects the response. The null hypothesis is that the response probabilities are the same across subgroups. The model comparison is to compare the fitted probabilities over the subgroups to the fitted probabilities combining all the groups into one population (a constant response model)

As a measure of fit for the models you want to compare, you can use the negative log-likelihood to compute a likelihood ratio chi-square test. To do this, subtract the log-likelihoods for the two models and multiply by 2. For each observation, the log-likelihood is the log of the probability attributed to the response level of the observation.

**Warning**: When the table is *sparse*, neither the Pearson or likelihood ratio chi-square is a very good approximation to the true distribution. The Cochran criterion, used to determine if the tests are appropriate, defines sparse as when more than 20% of the cells have expected counts less than 5.

The Pearson chi-square tends to be better behaved in sparse situations than the likelihood ratio chi-square. However, $G^2$ is often preferred over $X^2$ for other reasons, specifically because it is generalizable to general categorical models where $X^2$ is not.

"Categorical Distributions" on page 257 discussed the $G^2$ and $X^2$ test statistics in more detail.

## Looking at Survey Data

Survey data often yield categorical data suitable for contingency table analysis. For example, a company did a survey to find out what factors relate to the brand of automobile people buy—in other words, what kind of people buy what kind of cars? Cars were classified into three brands: American, European, and Japanese (which includes other Asian brands). This survey also contained demographic information (marital status and gender).

The results of the survey are in the sample table called Carpoll.jmp. The first step is to find probabilities for each brand when nothing else is known about the car buyer. Looking at the distribution of car brand gives this information. To see the report on the distribution of brand shown in **Figure 12.1**,

🖱 Open Carpoll.jmp and choose **Analyze > Distribution** with country as the variable.

Overall, the Japanese brands have a 48.8% share.

**Figure 12.1**   Histograms and Frequencies for country in Carpoll Data

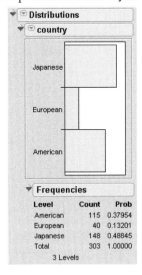

The next step is to look at the demographic information as it relates to brand of auto.

🖑  Choose **Analyze > Fit Y by X** with country as Y, and sex, marital status, and size as X variables.

🖑  Click **OK**.

The Fit Y by X platform displays Mosaic plots and Crosstabs tables for the combination of country with each of the X variables. By default, JMP displays Count, Total%, Col%, and Row% (listed in the upper-left corner of the table) for each cell in the contingency table.

🖑  Right-click in the icon in the Contingency table to see the menu of the optional items. Uncheck all items except **Count** and **Col%** to see the table shown to the right.

### Contingency Table: Country by Sex

Is the distribution of the response levels

| Count Col % | American | European | Japanese | |
|---|---|---|---|---|
| **Contingency Table** | | | | |
| | country | | | |
| Female | 54 | 19 | 65 | 138 |
| | 46.96 | 47.50 | 43.92 | |
| Male | 61 | 21 | 83 | 165 |
| | 53.04 | 52.50 | 56.08 | |
| | 115 | 40 | 148 | 303 |

different over different levels of other categorical variables? In principle, this is like a one-way analysis of variance, estimating separate means for each sample, but this time they are rates over response categories rather than means.

In the contingency table, you see the response probabilities as the Col% values in the bottom of each cell. The column percents are not much different between "Female" and "Male."

### Mosaic Plot

The Fit Y by X platform for a categorical variable displays information graphically with mosaic plots like the one shown to the right.

A mosaic plot is a set of side-by-side divided bar plots to compare the subdivision of response probabilities for each sample. The mosaic is formed by first dividing up the horizontal axis according to the sample proportions. Then each of these cells is subdivided

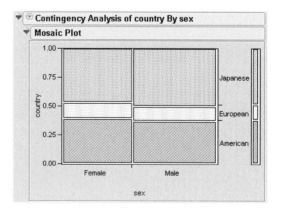

vertically by the estimated response probabilities. The area of each rectangle is proportional to the frequency count for that cell.

## Testing Marginal Homogeneity

Now ask the question, "Are the response probabilities significantly different across the samples (in this example, male and female)?" Specifically, is the proportion of sales by country the same for males and females? The hypothesis that the distributions are the same across the sample's subgroup is sometimes referred to as the hypothesis of *marginal homogeneity.*

Instead of regarding the categorical X variable as fixed, you can consider it as another Y response variable and look at the relationship between two Y response variables. The test would be the same, but it would be known by a different name, as the *test for independence.*

When the response was continuous, there were two ways to get a test statistic that turned out to be equivalent:

- Look at the distribution of the estimates, usually leading to a *t*-test.

- Compare the fit of a model with a submodel, leading to an *F*-test.

The same two approaches work for categorical models. The two approaches to getting a test statistic for a contingency table both result in chi-square tests.

- If the test is derived in terms of the distribution of the estimates, then you are led to the Pearson $X^2$ form of the $\chi^2$ test.

- If the test is derived by comparing the fit of a model with a submodel, then you are led to the likelihood-ratio $G^2$ form of the $\chi^2$ test.

For the likelihood ratio chi-square ($G^2$), two models are fit by maximum likelihood. One model is constrained by the hypothesis that assumes a single response population and the other is not constrained. Twice the difference of the log-likelihoods from the two models is a chi-square statistic for testing the hypothesis. The table here has the chi-square tests that test whether country of car purchased is a function of sex.

| Source | DF | -LogLike | RSquare (U) |
|--------|-----|----------|-------------|
| Model | 2 | 0.15594 | 0.0005 |
| Error | 299 | 298.29531 | |
| C. Total | 301 | 298.45125 | |
| N | 303 | | |

| Test | ChiSquare | Prob>ChiSq |
|------|-----------|------------|
| Likelihood Ratio | 0.312 | 0.8556 |
| Pearson | 0.312 | 0.8556 |

The top portion of the table shows the comparison of models. The line C Total shows that the model constrained by the hypothesis (fitting only one set of response probabilities) has a negative log-likelihood of 298.45. After you partition the sample by the gender factor, the negative log-likelihood is reduced to 298.30 as reported in the Error line. The difference in log-likelihoods is 0.1559, reported in the Model line. This isn't accounting for much of the variation. The likelihood ratio (LR) chi-square is twice this difference, that is, $G^2 = 0.312$, and has a nonsignificant *p*-value of 0.8556.

These statistics don't support the conclusion that the car country purchase response depends on the gender of the driver.

If you want to think about the distribution of the estimates, then in each cell you can compare the actual proportion to the proportion expected under the hypothesis, square it, and divide by something close to its variance, giving a cell chi-square. The sum of these cell chi-square values is the Pearson chi-square statistic $X^2$, which here is also 0.312, with a $p$-value of 0.8556. In this example, the Pearson chi-square happens to be the same as the likelihood ratio chi-square.

## Car Brand by Marital Status

Let's look at the relationships of country to other categorical variables. In the case of marital status (**Figure 12.2**), there is a more significant result, with the $p$-value of 0.076. Married people are more likely to buy the American brands. Why? Perhaps because the American brands are generally larger vehicles, which make them more comfortable for families.

**Figure 12.2**   Mosaic Plot, Crosstabs, and Tests Table for country by marital status

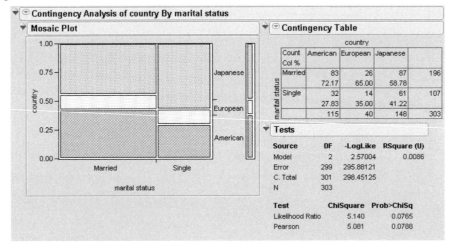

## Car Brand by Size of Vehicle

If marital status is a proxy for size of vehicle, looking at country by size will give more direct information.

The Tests table for country by size (**Figure 12.3**) shows a strong relationship with a very significant chi-square. The Japanese dominate the market for small cars, the Americans dominate the market for large cars, and the European share is about the same in all three markets. The relationship is highly significant, with $p$-values less than 0.0001.

**Figure 12.3**   Mosaic Plot, Crosstabs, and Tests Table for country by size

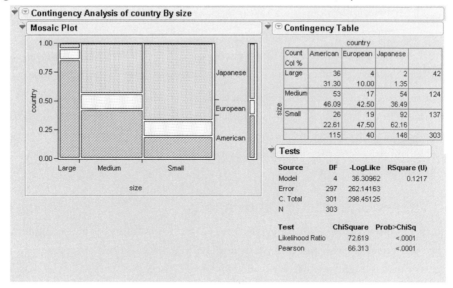

# Two-Way Tables: Entering Count Data

Often, raw categorical data is presented in a *two-way table* like the one shown below. The levels of one variable are the rows, the levels of the other variable are the columns, with cells containing frequency counts. For example, data for a study of alcohol and smoking (based on Schiffman, 1982) is arranged in a two-way table, like this:

|  |  | Relapsed | |
|---|---|---|---|
|  |  | **Yes** | **No** |
| **Alcohol Consumption** | **Consumed** | 20 | 13 |
|  | **Did Not Consume** | 48 | 96 |

This arrangement shows the two levels of alcohol consumption ("Consumed" or "Did Not Consume") and levels of whether the subject relapsed and returned to smoking (reflected in the "Yes" column) or managed to stay smoke-free (reflected in the "No" column).

In the following discussion, keep the following things in mind:

- The two variables in this table do not fit neatly into independent and dependent classifications. The subjects in the study were not separated into two groups, with one group given alcohol, and the other not. The interpretation of the data, then, needs to be limited to association, and not cause-and-effect. The tests are regarded as tests of independence of two responses, rather than the marginal homogeneity of probabilities across samples.

- Because this is a $2 \times 2$ table, JMP produces Fisher's Exact Test in its results. This test, in essence, computes exact probabilities for the data rather than relying on approximations.

Does it appear that alcohol consumption is related to the subject's relapse status? Phrased more statistically, if you assume these variables are independent, are there surprising entries in the two-way table? To answer this question, we must know what values would be expected in this table, and then determine if there are observed results that are different than these expected values.

## Expected Values Under Independence

To further examine the data, the following table shows the totals for the rows and columns of the two-way table. The row and column totals have been placed along the right and bottom margins of the table, and are therefore called *marginal totals*.

| | | Relapsed | | |
| --- | --- | --- | --- | --- |
| | | Yes | No | Total |
| Alcohol Consumption | Consumed | 20 | 13 | 33 |
| | Did Not Consume | 48 | 96 | 144 |
| | Total | 68 | 109 | 177 |

These totals aid in determining what values would be expected if alcohol consumption and relapse to smoking were not related.

As is usual in statistics, assume at first that there is no relationship between these variables. If this assumption is true, then the proportion of people in the "Yes" and "No" columns should be equal for each level of the alcohol consumption variable. If there was no effect for consumption of alcohol, then we expect these values to be the same except for random variation. To determine the *expected value* for each cell, compute

$$\frac{\text{Row total} \times \text{Column Total}}{\text{Table Total}}$$

for each cell. Instead of computing it by hand, let's enter the data into JMP to perform the calculations.

## Entering Two-Way Data into JMP

Before two-way table data can be analyzed, it needs to be *flattened* or *stacked* so that it is arranged in two data columns for the variables, and one data column for frequency counts. These steps can be completed as follows:

- Select **File > New > Data Table** to create a new data table.

- Right-click on the title of Column 1 and select **Column Info.**

- Make the name of the column Alcohol Consumption and its data type **Character.** JMP automatically changes the modeling type to **Nominal.**

- Select **Columns > New Column** to create a second column in the data table. Repeat the above process to make this a character column named Relapsed.

- Create a third column named Count to hold the cell counts from the two-way table. Since this column will hold numbers, make sure its data type is **Numeric** and its modeling type is **Continuous.**

- Select **Rows > Add Rows** and add four rows to the table—one for each cell in the two-way table.

- Enter the data so that the data table looks like the one shown to the right.

These steps have been completed, and the resulting table is included in the sample data as Alcohol.jmp

| | Alcohol Consumption | Relapsed | Count |
|---|---|---|---|
| 1 | Consumed | Yes | 20 |
| 2 | Consumed | No | 13 |
| 3 | Didn't Consume | Yes | 48 |
| 4 | Didn't Consume | No | 96 |

## Testing for Independence

One explanatory note is in order at this point. Although the computations in this situation use the counts of the data, the statistical test deals with proportions. The *independence* we are concerned with is the independence of the probabilities associated with each cell in the table. Specifically, let

$$\rho_i = \frac{n_i}{n} \text{ and } \rho_j = \frac{n_j}{n}.$$

where $\rho_i$ and $\rho_j$ are, respectively, the probabilities associated with each of the $i$ rows and $j$ columns. Now, let $\rho_{ij}$ be the probability associated with the *cell* located at the $i$th row and $j$th column. The null hypothesis of independence is that $\rho_{ij}=\rho_i\rho_j$. Although the computations we present use counts, do not forget that the essence of the null hypothesis is about probabilities.

The test for independence is the $X^2$ statistic, whose formula is

$$\sum \frac{(\text{Observed} - \text{Expected})^2}{\text{Expected}}$$

To compute this statistic in JMP:

🖱 Select **Analyze > Fit Y By X.** Set Alcohol Consumption as the X, Relapsed as the Y, and Count as the Freq.

This produces a report that contains the contingency table of counts, which should agree with the two-way table used as the source of the data. To see the information relevant to the computation of the $X^2$ statistic,

🖱 Right-click inside the contingency table and uncheck **Row%**, **Col%** and **Total%**.

🖱 Again, right-click inside the contingency table and make sure that **Count, Expected, Deviation**, and **Cell Chi-Square** are checked.

The Tests table (**Figure 12.4**) shows the Likelihood Ratio and Pearson Chi-square statistics. This represents the chi square for the table (the sum of all the cell chi-square values). The *p*-values for the chi-square tests are less than 0.05, so the null hypothesis is rejected. Alcohol consumption seems to be associated with whether the patient relapsed into smoking.

**Figure 12.4**   Contingency Report

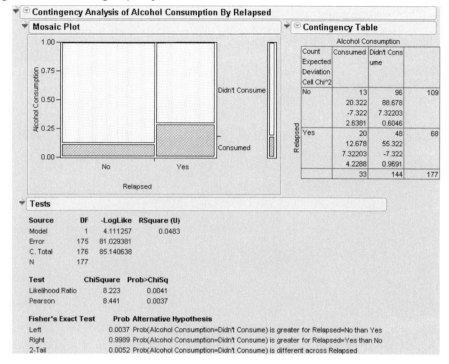

The composition of the Pearson $X^2$ statistic can be seen cell by cell. The cell for "Yes" and "Consumed" in the upper right has an actual count of 20 and an expected count of 12.678. Their difference (deviation) is 7.322. This cell's contribution to the overall chi-square is

$$\frac{(20 - 12.678)^2}{12.678}$$

which is 4.22. Repeating this procedure for each cell shows the chi-square as 2.63 + 0.60 + 4.22 + 0.97, which is 8.441.

# If You Have a Perfect Fit

If a fit is perfect, every response's category is predicted with probability 1. The response is completely determined by which sample it is in. In the other extreme, if the fit contributes nothing, then each distribution of the response in each sample subgroup is the same.

As an example, consider collecting information for 156 people on what city and state they live in. It's likely that one would think that there is a perfect fit between the city and the state of a person's residence. If the city is known, then the state is almost surely known. **Figure 12.5** shows what this perfect fit looks like.

**Figure 12.5**   Mosaic Plot, Crosstabs, and Tests Table for City by State

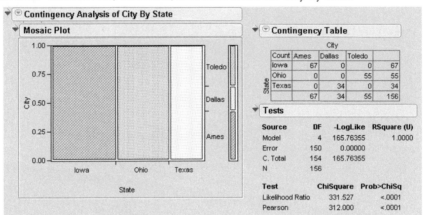

Now suppose the analysis includes people from Austin, a second city in Texas. City still predicts state perfectly, but not the other way around (state does not predict city). Conducting these two analyses shows that the chi-squares are the same. They are invariant if you switch the Y and X variables. However, the mosaic plot, the attribution of the log-likelihood, and $R^2$ are different (**Figure 12.6**).

What happens if the response rates are the same in each cell as shown in **Figure 12.5**? Examine the artificial data for this situation and notice the mosaic levels line up perfectly and the chi-squares are zero.

**Figure 12.6**  Comparison of Plots, Tables, and Tests When X and Y are Switched

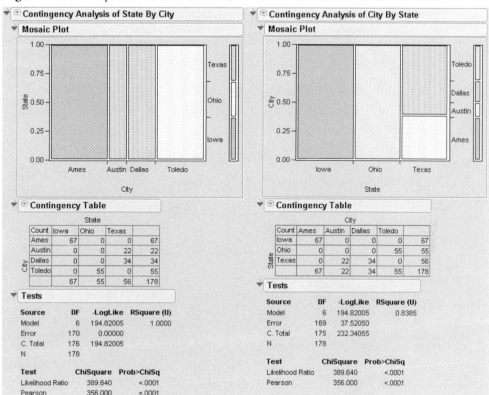

# Special Topic: Correspondence Analysis— Looking at Data with Many Levels

Correspondence analysis is a graphical technique that shows which rows or columns of a frequency table have similar patterns of counts. Correspondence analysis is particularly valuable when you have many levels, because it is difficult to find patterns in tables or mosaic plots with many levels.

The data table **Mbtied.jmp** has counts of personality types by educational level (**Educ**) and gender (Myers and McCaulley). The values of educational level are D for dropout, HS for high school graduate, and C for college graduate. **Gender** and **Educ** are concatenated to form the variable **GenderEd**. The goal is to determine the relationships between **GenderEd** and personality type. Remember, there is no implication of any cause-and-effect relationship because there is no way to tell whether personality affects education or education affects personality. The data can, however, show trends. The following example shows how correspondence analysis can help identify trends in categorical data:

🖰 Open the data table called Mbtied.jmp and choose **Analyze > Fit Y by X** with MBTI as X and GenderEd as Y, and Count as the Freq variable.

Now try to make sense out of the resulting mosaic plot and contingency table shown in **Figure 12.7**. It has with 96 cells—too big to understand at a glance. A correspondence analysis will clarify some patterns.

**Figure 12.7**   Mosaic Plot and Table for MBTI by GenderEd

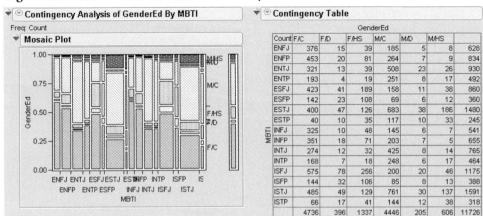

🖰 Select **Correspondence Analysis** from the popup menu next to the Contingency Table title to see the plot in **Figure 12.8**.

The Correspondence Analysis plot organizes the row and column profiles in a two-dimensional space so that the X values that have similar Y profiles tend to cluster together, and the Y values that have similar X profiles tend to cluster together. In this case, you want to see how the GenderEd groups are associated with the personality groups.

This plot quickly shows patterns. Gender and the Feeling(F)/Thinking(T) component form a cluster, and education clusters with the Intuition(N)/Sensing(S) personality indicator. The Extrovert(E)/Introvert(I) and Judging(J)/Perceiving(P) types do not separate much. The most separation among these is the Judging(J)/Perceiving(P) separation among the Sensing(S)/Thinking (T) types (mostly non-college men).

**Figure 12.8** Correspondence Analysis Plot

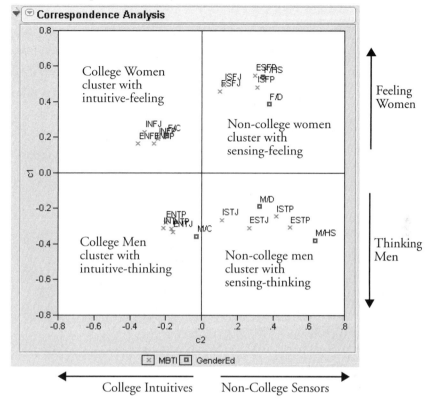

The correspondence analysis indicates that the Extrovert/Introvert and Judging/Perceiving do not separate well for education and gender.

# Continuous Factors with Categorical Responses: Logistic Regression

Suppose that a response is categorical, but the probabilities for the response change as a function of a continuous predictor. In other words, you are presented with a problem with a continuous X and a categorical Y. Some situations like this are the following:

*   Whether you bought a car this year (categorical) as a function of your disposable income (continuous).

*   The kind of car you bought (categorical) as a function of your age (continuous).

- The probability of whether a patient lived or died (categorical) as a function of blood pressure (continuous).

Problems like these call for *logistic regression*. Logistic regression provides a method to estimate the probability of choosing one of the response levels as a smooth function of the factor. It is called logistic regression because the S-shaped curve it uses to fit the probabilities is called the logistic function.

## Fitting a Logistic Model

The Spring.jmp sample is a weather record for the month of April. The variable Precip measures rainfall and the variable Rained categorizes rainfall using the formula shown to the right.

$$\text{If} \left\{ \begin{array}{ll} Precip > 0.02 \Rightarrow \text{"Rainy"} \\ else \quad\quad \Rightarrow \text{"Dry"} \end{array} \right\}$$

🖰 Open Spring.jmp and choose **Analyze > Distribution** to generate a histogram and frequency table of the Rained variable as shown here.

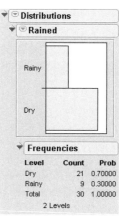

Out of the 30 days in April, there were 9 rainy days. Therefore, with no other information, you predict a 9/30 = 30% chance of rain for every day.

Now suppose you look at the temperature in the morning and then try to give a precipitation probability. Knowing the barometric pressure also might help in finding a more informative prediction.

In each case, the thing being modeled is the probability of getting one of several responses. The probabilities are constrained to add to 1. In the simplest situation, like this rain example, the response has two levels (a *binary* response). Remember that statisticians like to take logs of probabilities. In this case, what they fit is the difference in logs of the two probabilities as a linear function of the factor variable.

If $p$ denotes the probability for the first response level, then $1-p$ is the probability of the second, and the linear model is written

$$\log(p) - \log(1-p) = b0 + b1 {}^{*}\text{X} \text{ or } \log(p/(1-p)) = b0 + b1 {}^{*}\text{X}$$

where $\log(p/(1-p))$ is called the *logit* of $p$ or the *log odds-ratio*.

There is no error term here because the predicted value is not a response level; it is a probability distribution for a response level. For example, if the weatherman predicts a 90% chance of rain, you don't say he erred if it didn't rain.

The accounting is done by summing the negative logarithms of the probabilities attributed by the model to the events that actually did occur. So if $p$ is the precipitation probability from the weather model, then the score is $-\log(p)$ if it rains, and $-\log(1-p)$ if it doesn't. A weather reporter that is a perfect predictor comes up with a $p$ of 1 when it rains ($-\log(p)$ is zero if $p$ is 1) and a $p$ of zero when it doesn't rain ($-\log(1-p)=0$ if $p=0$). The perfect score is zero. No surprise $-\log(p) = 0$ means perfect predictions. If you attributed a probability of zero to an event that occurred, then the $-\log$-likelihood would be infinity, a pretty bad score for a forecaster.

So the inverse logit of the model $b0+b1*X$ expresses the probability for each response level, and the estimates are found so as to maximize the likelihood. That is the same as minimizing the negative sum of logs of the probabilities attributed to the response levels that actually occurred for each observation.

You can graph the probability function as shown in **Figure 12.9**. The curve is just solving for $p$ in the expression

$$\log(p/(1-p)) = b0+b1*X$$

which is

$$p = 1/(1+\exp(-(b0+b1*X))).$$

For a given value of $X$, this expression evaluates the probability of getting the first response. The probability for the second response is the remaining probability, $1-p$, so that they sum to 1.

**Figure 12.9**   Logistic Regression Fits Probabilities of a Response Level

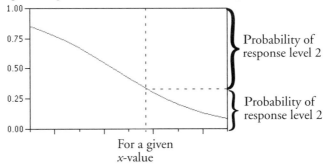

To fit the rain column by temperature and barometric pressure for the spring rain data:

- Choose **Analyze > Fit Y by X** specifying the nominal column Rained as Y, and the continuous columns Temp and Pressure as the X variables.

The Fit Y by X platform produces a separate logistic regression for each predictor variable.

The cumulative probability plot on the left in **Figure 12.10** shows that the relationship with temperature is very weak. As the temperature ranges from 35 to 75, the probability of dry weather only changes from 0.73 to 0.66. The line of fit partitions the whole probability into the response categories. In this case, you read the probability of Dry directly on the vertical axis. The probability of Rained is the distance from the line to the top of the graph, which is 1 minus the axis reading. The weak relationship is evidenced by the very flat line of fit; the precipitation probability doesn't change much over the temperature range.

The plot on the right in **Figure 12.10** indicates a much stronger relationship with barometric pressure. When the pressure is 29.0 inches, the fitted probability of rain is near 100% (0 probability for Dry at the left of the graph). The curve crosses the 50% level at 29.32. (You can use the crosshair tool to see this.) At 29.8, the probability of rain drops near zero (near 1.0 for Dry).

You can also add reference lines at the known X and Y values.

- Double-click on the Rained (Y) axis to bring up an axis modification dialog. Enter 0.5 as a reference line.

- Double-click on the Pressure (X) axis and enter 29.32 in the axis modification dialog.

When both reference lines appear, they intersect on the logistic curve as shown in the plot on the right in **Figure 12.10**.

**Figure 12.10**   Cumulative Probability Plot for Discrete Rain Data

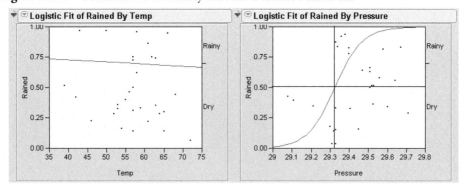

The Whole-Model Test table and the Parameter Estimates table support the plot. The $R^2$ measure of fit, which can be interpreted on a scale of 0 to 100%, is only 0.07% (see **Figure 12.11**). A 100% $R^2$ would indicate a model that predicted outcomes with certainty. The

likelihood ratio chi-square is not at all significant. The coefficient on temperature is a very small -0.008. The parameter estimates can be unstable because they have high standard errors with respect to the estimates.

**Figure 12.11**   Logistic Regression for Discrete Rain Data

| Whole Model Test | | | | | |
|---|---|---|---|---|---|
| Model | -LogLikelihood | DF | ChiSquare | Prob>ChiSq | |
| Difference | 0.013334 | 1 | 0.026668 | 0.8703 | |
| Full | 18.312595 | | | | |
| Reduced | 18.325929 | | | | |

| RSquare (U) | 0.0007 |
|---|---|
| Observations (or Sum Wgts) | 30 |

Converged by Gradient

| Parameter Estimates | | | | |
|---|---|---|---|---|
| Term | Estimate | Std Error | ChiSquare | Prob>ChiSq |
| Intercept | 1.34073823 | 3.0620868 | 0.19 | 0.6615 |
| Temp | -0.0086266 | 0.0529847 | 0.03 | 0.8707 |

For log odds of Dry/Rainy

| Whole Model Test | | | | | |
|---|---|---|---|---|---|
| Model | -LogLikelihood | DF | ChiSquare | Prob>ChiSq | |
| Difference | 6.250851 | 1 | 12.5017 | 0.0004 | |
| Full | 12.075078 | | | | |
| Reduced | 18.325929 | | | | |

| RSquare (U) | 0.3411 |
|---|---|
| Observations (or Sum Wgts) | 30 |

Converged by Gradient

| Parameter Estimates | | | | |
|---|---|---|---|---|
| Term | Estimate | Std Error | ChiSquare | Prob>ChiSq |
| Intercept | -405.36267 | 169.29517 | 5.73 | 0.0166 |
| Pressure | 13.8233881 | 5.7651316 | 5.75 | 0.0165 |

For log odds of Dry/Rainy

In contrast, the overall $R^2$ measure of fit with barometric pressure is 34%. The likelihood ratio chi-square is highly significant and the parameter coefficient for Pressure increased to 13.8 (**Figure 12.11**).

The conclusion is that if you want to predict whether the weather will be rainy, it doesn't help to know the temperature, but it does help to know the barometric pressure.

## Degrees of Fit

The illustrations in **Figure 12.12** summarize the degree of fit as shown by the cumulative logistic probability plot.

When the fit is weak, the parameter for the slope term (X factor) in the model is small, which gives a small slope to the line in the range of the data. A perfect fit means that before a certain value of X, all the responses are one level, and after that value of X, all the responses are another level. A strong model can bet almost all of its probability on one event happening. A weak model has to bet conservatively with the background probability, less affected by the X factor's values.

**Figure 12.12**  Strength of Fit in Logistic Regression

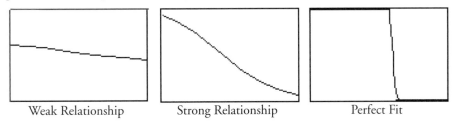

| Weak Relationship | Strong Relationship | Perfect Fit |

Note that when the fit is perfect, as shown on the rightmost graph of **Figure 12.12**, the slope of the logistic line approaches infinity. This means that the parameter estimates are also infinite. In practice, the estimates are allowed to inflate only until the likelihood converges and are marked as unstable by the computer program. You can still test hypotheses, because they are handled through the likelihood, rather than using the estimate's (theoretically infinite) values.

## A Discriminant Alternative

There is another way to think of the situation where the response is categorical and factor is continuous. You can reverse the roles of the Y and X and treat this problem as one of finding the distribution of temperature and pressure on rainy and dry days. Then, work backwards to obtain prediction probabilities. This technique is called *discriminant analysis*.

- ⌐ For this example, open (or make active) the Spring.jmp data table. (Note: If you open it from scratch, you need to add the Rained variable as detailed on page 290.)

- ⌐ Choose **Analyze > Fit Y by X** specifying Temp and Pressure as the Y variables and Rained as X.

- ⌐ Select **Means/Anova/Pooled t** from the Display popup menu showing beneath the plots to see the results in **Figure 12.13**.

You can quickly see that the difference between the relationships of temperature and pressure to raininess. However, the discriminant approach is a somewhat strange way to go about this example and has some problems:

- The standard analysis of variance assumes that the factor distributions are Normal.

- Discriminant analysis works backwards: First, in the weather example you are trying to predict rain. But the ANOVA approach designates Rained as the independent variable, from which you can say something about the predictability of temperature and pressure.

Then, you have to reverse-engineer your thinking to infer raininess from temperature and pressure.

**Figure 12.13**   Temperature and Pressure as a Function of (Discrete) Rain Variable

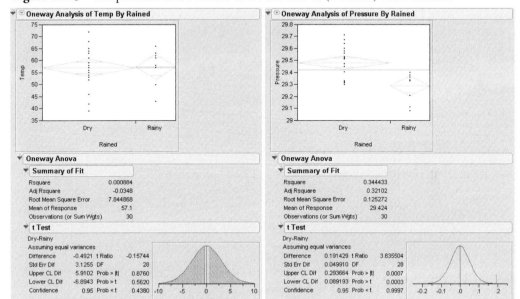

# Special Topics

## Inverse Prediction

If you want to know what value of the $X$ regressor yields a certain probability, you can solve the equation, $\log(p/(1-p)) = b0 + b1*X$, for $X$, given $p$. This is often done for toxicology situations, where the $X$-value for $p=50\%$ is called an *LD50* (Lethal Dose for 50%). Confidence intervals for these inverse predictions (called *fiducial confidence intervals*) can be obtained.

The Fit Model platform has an inverse prediction facility. Let's find the LD50 for pressure in the rain data—that is, the value of pressure that gives a 50% chance of rain.

   🖱 Choose **Analyze > Fit Model**.

When the Fit Model dialog appears,

   🖱 Select **Rained** in the variable selection list and assign it as Y. Select **Pressure** and assign it as a Model Effect.

🖱  Click **Run Model**.

🖱  When the platform appears, select **Inverse Prediction** from the popup menu on the analysis title bar.

This displays the dialog at the left in **Figure 12.14**.

The Probability and 1–Alpha fields are editable. You can fill the dialog with any values of interest. The result is an inverse probability for each probability request value you entered, at the alpha level specified.

🖱  For this example, enter 0.5 as the first entry in the Probability column.

🖱  Click **Run**.

The Inverse Prediction table shown in **Figure 12.14** appears appended to the platform tables.

The inverse prediction computations say that there is a 50% chance of rain when the barometric pressure is 29.32.

**Figure 12.14**  Inverse Prediction Dialog

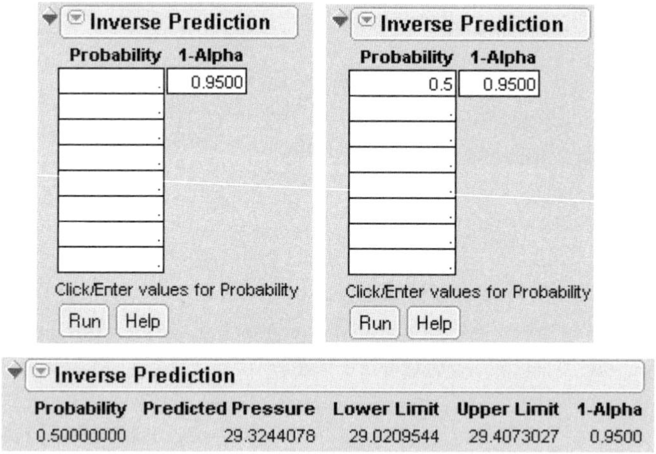

| Inverse Prediction | | | | |
|---|---|---|---|---|
| Probability | Predicted Pressure | Lower Limit | Upper Limit | 1-Alpha |
| 0.50000000 | 29.3244078 | 29.0209544 | 29.4073027 | 0.9500 |

To confirm this on the graph,

- Get the crosshair tool from the Tools menu or toolbar. Click and drag the crosshair tool on the logistic plot until the horizontal line is at the 0.50 value of Rained.

- Hold the crosshair at that value and drag to the logistic curve. You can then read the Pressure value, slightly more than 29.3, on the *x*-axis.

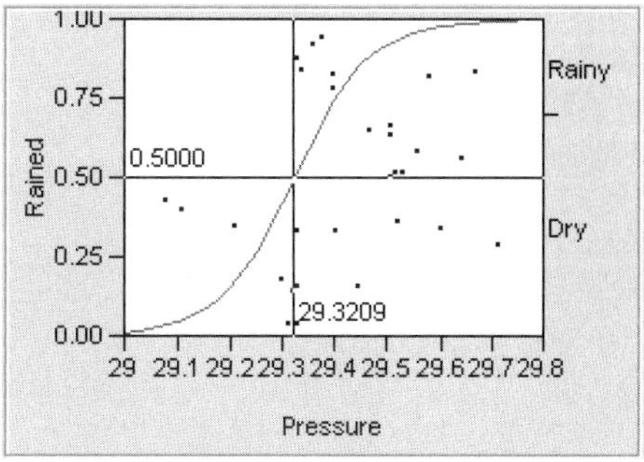


## Polytomous: More Than Two Response Levels

If there are more than two response categories, the response is said to be *polytomous* and a generalized logistic model is used. For the curves to be very flexible, you have to fit a set of linear model parameters for each of $r - 1$ response levels. The logistic curves are accumulated in such a way as to form a smooth partition of the probability space as a function of the regression model. In **Figure 12.15**, the probabilities are the distances between the curves, which add up to 1.

Where the curves come close together, the model is saying the probability of a response level is very low. Where the curves separate widely, the fitted probabilities are large.

**Figure 12.15** Polytomous Logistic Regression with 5 Response Levels

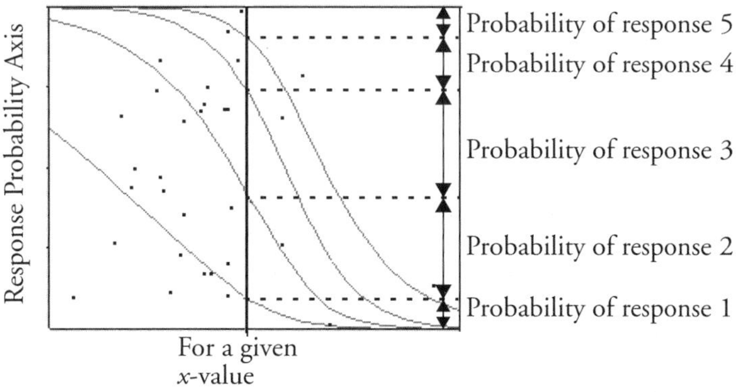


For example, consider fitting the probabilities for brand with the carpoll.jmp data as a smooth function of age. The result (**Figure 12.16**) shows the relationship where younger individuals tend to buy more Japanese cars and older individuals tend to buy more American cars. Note the double set of estimates (two curves) needed to describe three responses.

**Figure 12.16** Cumulative Probability Plot and Logistic Regression for Country by Age

**Whole Model Test**

| Model | -LogLikelihood | DF | ChiSquare | Prob>ChiSq |
|---|---|---|---|---|
| Difference | 3.47924 | 2 | 6.958482 | 0.0308 |
| Full | 294.97201 | | | |
| Reduced | 298.45125 | | | |

| | |
|---|---|
| RSquare (U) | 0.0117 |
| Observations (or Sum Wgts) | 303 |

Converged by Gradient

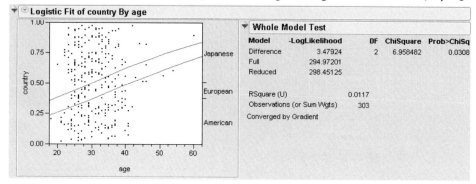

## Ordinal Responses: Cumulative Ordinal Logistic Regression

In some cases, you don't need the full generality of multiple linear model parameter fits for the $r - 1$ cases, but can assume that the logistic curves are the same, only shifted by a different amount over the response levels. This means that there is only one set of regression parameters on the factor, but $r - 1$ intercepts for the $r$ responses.

**Figure 12.17** Ordinal Logistic Regression Cumulative Probability Plot

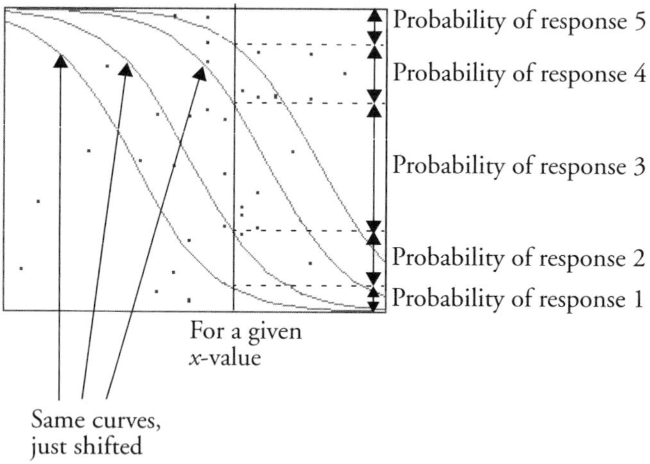

The logistic curve is actually fitting the sum of the probabilities for the responses at or below it, so it is called a *cumulative* ordinal logistic regression. In the Spring data table, there is a column called SkyCover with values 1 to 10.

First, note that you don't need to treat the response as nominal because the data have a natural order. Also, in this example, there is not enough data to support the large number of parameters needed by a 10-response level nominal model. Instead, use a logistic model that fits SkyCover as an ordinal variable with the continuous variables Temp and Humid1: PM.

   🖱️  Change the modeling type of the SkyCover column to Ordinal by clicking the icon next to the column name in the Columns Panel, located to the left of the data grid.

   🖱️  Chose **Analyze > Fit Y by X** and specify SkyCover as Y, and columns Temp and Humid1:PM as the X variables.

**Figure 12.18** indicates that the relationship of SkyCover to Temp is very weak, with an $R^2$ of 0.09%, fairly flat lines, and a nonsignificant chi-square. The direction of the relation is that the higher sky covers are more likely with the higher temperatures.

**Figure 12.18**   Ordinal Logistic Regression for Ordinal Sky Cover with Temperature

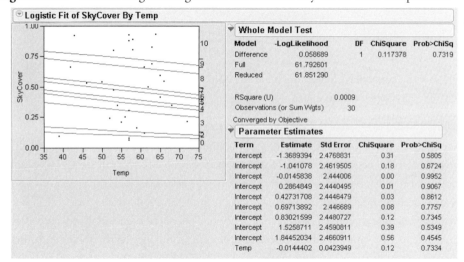

**Figure 12.19** indicates that the relationship with humidity is quite strong. As the humidity rises to 70%, it predicts a 50% probability of a sky cover of 10. At 100% humidity, the sky cover will be almost certainly 10. The $R^2$ is 29% and the likelihood ratio chi-square is highly significant.

**Figure 12.19**  Logistic Regression for Ordinal Sky Cover with Humidity

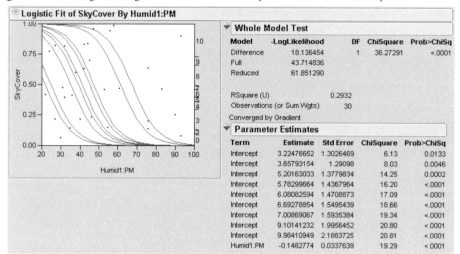

Note that the curves have bigger shifts in the popular 30% and 80% categories. Also, no data occurs for SkyCover = 2, so that is not even in the model.

There is a useful alternative interpretation to this ordinal model. Suppose you assume there is some continuous response with a random error component that the linear model is really fitting. But, for some reason, you can't observe the response directly. You are given a number that indicates which of $r$ ordered intervals contains the actual response, but you don't know how the intervals are defined. You assume that the error term follows a logistic distribution, which has a similar shape to a Normal distribution. This case is identical to the ordinal cumulative logistic model, and the intercept terms are estimating the threshold points that define the intervals corresponding to the response categories.

Unlike the nominal logistic model, the ordinal cumulative logistic model is efficient to fit for even hundreds of response levels. It can even be used effectively for continuous responses when there are $n$ unique response levels for $n$ observations. In such a situation, there are $n - 1$ intercept parameters constrained to be in order, and there is one parameter for each regressor.

# Surprise: Simpson's Paradox: Aggregate Data versus Grouped Data

Several statisticians have studied the "hot hand" phenomenon in basketball. The idea is that basketball players seem to have hot streaks, when they make most of their shots, alternating with cold streaks when they shoot poorly. The Hothand.jmp table contains the free throw shooting records for two Boston Celtics players (Larry Bird and Rick Robey) over the 1980-81 and 1981-82 seasons (Tversky and Gilovich, 1989).

The null hypothesis is that two sequential free throw shots are independent. There are two directions in which they could be non-independent, the positive relationship (hot hand) and a negative relationship (cold hand).

The Hothand.jmp sample data have the columns First and Second (first shot and second shot) for the two players and a count variable. There are 4 possible shooting combinations: hit-hit, hit-miss, miss-hit, and miss-miss.

- 🖰 Open Hothand.jmp.

- 🖰 Choose **Analyze > Fit Y by X**. Select Second as Y, First as X, and Count as the Freq variable, then click **OK**.

- 🖰 When the report appears, right-click in the contingency table and deselect all displayed numbers except **Col%**.

The results in **Figure 12.20** show that if the first shot is made, then the probability of making the second is 0.758; if the first shot is missed the probability of making the second is 0.241. This tends to support the hot hand hypothesis. The two chi-square statistics are on the border of 0.05 significance.

**Figure 12.20**  Crosstabs and Tests for Hothand Basketball Data

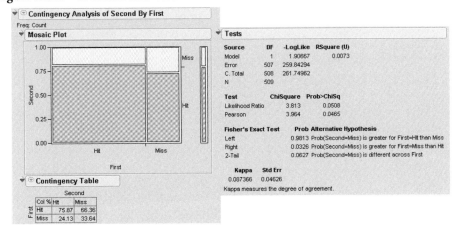

Does this really work? A researcher (Wardrop 1995), looked at contingency tables for each player. You can do this using the By-groups.

🖑   Again, Choose **Analyze > Fit Y by X** with Second as Y, First as X, and Count as the
Freq variable.

🖑   This time, assign Player as the **By** variable in the Launch dialog.

The results for the two players are shown in **Figure 12.21**.

**Figure 12.21**   Crosstabs and Tests for grouped High-end Basketball Data

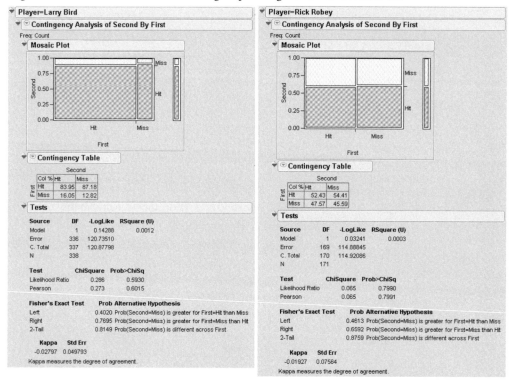

Contrary to the first result, both players shot better the second time after a miss than after a
hit. So how can this be when the aggregate table gives the opposite results from both
individual tables? This is an example of a phenomenon called *Simpson's paradox* (Simpson,
1951; Yule, 1903).

In this example, it is not hard to understand what happens if you think how the aggregated
table works. If you see a hit on the first throw, the player is probably Larry Bird, and since he
is usually more accurate he will likely hit the second basket. If you see a miss on the first
throw, the player is likely by Rick Robey, so the second throw will be less likely to score. The

hot hand relationship is an artifact that the players are much different in scoring percentages generally and populate the aggregate unequally.

A better way to summarize the aggregate data, taking into account these background relationships, is to use a blocking technique called the *Cochran-Mantel-Haenszel* test.

- Click the aggregate Fit Y by X platform to make it active and choose **Cochran Mantel Haenszel** from the popup menu on the report's title bar.

A grouping dialog appears that lists the variables in the data table.

- Select Player as the grouping variable in this dialog and click **OK** to see the table in **Figure 12.22**.

These results are more accurate because they are based on the grouping variable instead of the ungrouped data.

**Figure 12.22** Crosstabs and Tests for Grouped Hothand Basketball Data

**Cochran-Mantel-Haenszel Tests**

Stratified by Player

| CMH Test | ChiSquare | DF | Prob>Chisq |
|---|---|---|---|
| Correlation of Scores | 0.2507 | 1 | 0.6166 |
| Row Score by Col Categories | 0.2507 | 1 | 0.6166 |
| Col Score by Row Categories | 0.2507 | 1 | 0.6166 |
| General Assoc. of Categories | 0.2507 | 1 | 0.6166 |

**Frequency Counts**

Player=Larry Bird First

| Second | Hit | Miss |
|---|---|---|
| Hit | 251 | 48 |
| Miss | 34 | 5 |

Player=Rick Robey First

| Second | Hit | Miss |
|---|---|---|
| Hit | 54 | 49 |
| Miss | 37 | 31 |

# Exercises

1. M.A. Chase and G.M. Dummer conducted a study in 1992 to determine what traits children regarded as important to popularity. Their data is represented in the file Children's Popularity.jmp. Demographic information was recorded, as well as the rating given to four traits assessing their importance to popularity: Grades, Sports, Looks, and Money.

   (a) Is there a difference based on gender on the importance given to making good grades?

(b)  Is there a difference based on gender on the importance of excelling in sports?

(c)  Is there a difference based on gender on the importance of good looks or on having money?

(d)  Is there a difference between Rural, Suburban, and Urban students on rating these four traits?

(e)  Investigate a model that predicts the importance placed on grades, based on the student's Grade, Age, Race, and Rural/Urban status. Are all the variables necessary?

2.  One of the concerns of textile manufacturers is the absorbency of materials that clothes are made out of. Clothes that can quickly absorb sweat (such as cotton) are often thought of as more comfortable than those that cannot (such as polyester). To increase absorbency, material is often treated with chemicals. In this fictional experiment, several swatches of denim were treated with two acids to increase their absorbency. They were then assessed to determine if their absorbency had increased or not. The investigator wanted to determine if there is a difference in absorbency change for the two acids under consideration. The results are presented in the following table:

|  |  | Acid | |
|---|---|---|---|
|  |  | A | B |
| Absorbency | Increased | 54 | 40 |
|  | Did Not Increase | 25 | 40 |

Does the researcher have evidence to say that there is a difference in absorbency between the two acids?

3.  The taste of cheese can be affected by the additives that it contains. McCullagh and Nelder (1983) report a study (conducted by Dr. Graeme Newell) to determine the effects of four different additives on the taste of a cheese. The tasters responded by rating the taste of each cheese on a scale of 1 to 9. The results are in the data table Cheese.jmp.

(a)  Produce a mosaic plot to examine the difference of taste among the four cheese additives.

(b)  Do the statistical tests say that the difference amongst the additives is significant?

(c)  Conduct a contingency analysis to determine which of the four additives results in the best tasting cheese.

4. The file Titanic.jmp contains information on the Passengers of the RMS Titanic. The four variables represent the class (first, second, third, and crew), age, sex, and survival status (yes or no) for each passenger. Use JMP to answer the following questions:

   (a) How many passengers were in each class?

   (b) How many passengers were children?

   (c) How many passengers were on the boat? How many survived?

   (d) Test the hypothesis that there is no difference in the survival rate among classes.

   (e) Test the hypothesis that there is no difference in the survival rate between males and females.

5. Do dolphins alter their behavior based on the time of day? To study this phenomenon, a marine biologist in Denmark gathered the data presented in Dolphins.jmp (Rasmussen, 1998). The variables represent different activities observed in groups of dolphins, with the Groups variable showing the number of groups observed.

   (a) Does this data show evidence that dolphins exhibit different behaviors during different times of day?

   (b) There is a caution displayed with the chi-square statistic. Should you reject the results of this analysis based on the warning?

# Multiple Regression

# 13

## Overview

Multiple regression is the technique of fitting or predicting a response variable from a linear combination of several other variables. The fitting principle is least squares, the same as with simple linear regression.

Many regression concepts were introduced in previous chapters, so this chapter concentrates on showing some new concepts not encountered in simple regression: the point-by-point picture of a hypothesis test with the leverage plot, collinearity, (the situation in which one regressor variable is closely related to another), and the case of exact linear dependencies.

# Parts of a Regression Model

Linear regression models are the sum of the products of coefficient parameters and factors. In addition, linear models for continuous responses are usually specified with a Normally-distributed error term. The parameters are chosen such that their values minimize the sum of squared residuals. This technique is called estimation by *least squares*.

**Figure 13.1**   Parts of a Linear Model

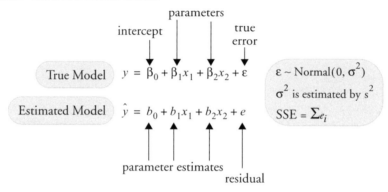

Note in **Figure 13.1** the differences in notation between the assumed true model with unknown parameters and the estimated model.

### response, Y

The *response* (or *dependent*) *variable* is the one you want to predict. Its estimates are the dependent variable, *y*, in the regression model.

### regressors, X's

The *regressors* (X) in the regression model are also called *independent variables*, *factors*, *explanatory variables*, and other discipline-specific terms. The regression model uses a linear combination of these effects to fit the response value.

### coefficients, parameters

The fitting technique produces estimates of the parameters, which are the coefficients for the linear combination that defines the regression model.

### intercept term

Most models have intercept terms to fit a constant in a linear equation. This is equivalent to having a regressor variable that always has the value 1. The intercept is meaningful by itself only if it is meaningful to know the predicted value where all the regressors are zero. However, the intercept plays a strong role in testing the rest of the model, because it represents the mean if all the other terms are zero.

## error, residual

If the fit isn't perfect, then there is error left over. *Error* is the difference between an actual value and its predicted value. When speaking of true parameters, this difference is called *error*. When using estimated parameters, this difference is called a *residual*.

# A Multiple Regression Example

Aerobic fitness can be evaluated using a special test that measures the oxygen uptake of a person running on a treadmill for a prescribed distance. However, it would be more economical to evaluate fitness with a formula that predicts oxygen uptake using simple measurements, such as running time and pulse measurements.

To develop such a formula, run time and pulse measurements were taken for 31 participants who each ran 1.5 miles. Their oxygen uptake, pulses, times, and other descriptive information was recorded. (Rawlings 1988, data courtesy of A.C. Linnerud). **Figure 13.2** shows a partial listing of the data, with variables Age, Weight, O2 Uptake (the response measure), Run Time, Rest Pulse, Run Pulse, and Max Pulse.

**Figure 13.2**   The Oxygen Uptake Data Table

| | Age | Weight | O2 Uptake | Run Time | Rest Pulse | Run Pulse | Max Pulse |
|---|---|---|---|---|---|---|---|
| 1 | 38 | 81.87 | 60.055 | 8.63 | 48 | 170 | 186 |
| 2 | 38 | 89.02 | 49.874 | 9.22 | 55 | 178 | 180 |
| 3 | 40 | 75.07 | 45.313 | 10.07 | 62 | 185 | 185 |
| 4 | 40 | 75.98 | 45.681 | 11.95 | 70 | 176 | 180 |
| 5 | 42 | 68.15 | 59.571 | 8.17 | 40 | 166 | 172 |
| 6 | 44 | 85.84 | 54.297 | 8.65 | 45 | 156 | 184 |
| 7 | 43 | 81.19 | 49.091 | 10.85 | 64 | 162 | 170 |
| 8 | 44 | 73.03 | 50.541 | 10.13 | 45 | 168 | 168 |
| 9 | 44 | 89.47 | 44.609 | 11.37 | 62 | 178 | 182 |
| 10 | 44 | 81.42 | 39.442 | 13.08 | 63 | 174 | 176 |
| 11 | 45 | 66.45 | 44.754 | 11.12 | 51 | 176 | 176 |
| 12 | 45 | 87.66 | 37.388 | 14.03 | 56 | 186 | 192 |

(Linnerud / Notes Linneruds / Columns (7/0): Age, Weight, O2 Uptake, Run Time, Rest Pulse, Run Pulse, Max Pulse / Rows / All Rows  31)

Now, investigate Run Time and Run Pulse as predictors of oxygen uptake (O2 Uptake):

🖰  Open the Linnerud.jmp sample data table. Choose **Analyze > Fit Model** to see the Fit Model dialog.

🖱 Select the O2 Uptake column and click **Y** to make it the response (Y) variable. Select Run Time and Run Pulse, and click **Add** to make them the Effects in Model.

Your dialog should look like the one in **Figure 13.3**.

🖱 Click **Run Model** to launch the platform.

**Figure 13.3**  Fit Model Dialog for Multiple Regression

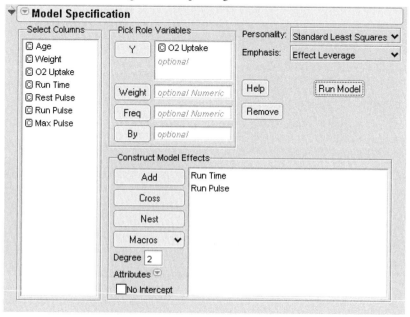

Now you have tables, shown in **Figure 13.4**, that report on the regression fit:

- The Summary of Fit table shows that the model accounted for 76% of the variation around the mean (RSquare). The remaining residual error is estimated to have a standard deviation of 2.69 (Root Mean Square Error).

- The Parameter Estimates table shows Run Time to be highly significant ($p < 0.0001$), but Run Pulse is not significant ($p = 0.15$). Using these parameter estimates, the prediction equation is

  O2 Uptake = 93.089 – 3.14 Run Time – 0.0735 Run Pulse

- The Effect Test table shows details of how each regressor contributes to the fit.

**Figure 13.4**  Statistical Tables for Multiple Regression Example

**Response O2 Uptake**

**Whole Model**

**Summary of Fit**

| | |
|---|---|
| RSquare | 0.761424 |
| RSquare Adj | 0.744383 |
| Root Mean Square Error | 2.693374 |
| Mean of Response | 47.37581 |
| Observations (or Sum Wgts) | 31 |

**Analysis of Variance**

| Source | DF | Sum of Squares | Mean Square | F Ratio |
|---|---|---|---|---|
| Model | 2 | 648.26218 | 324.131 | 44.6815 |
| Error | 28 | 203.11936 | 7.254 | **Prob > F** |
| C. Total | 30 | 851.38154 | | <.0001 |

**Parameter Estimates**

| Term | Estimate | Std Error | t Ratio | Prob>|t| |
|---|---|---|---|---|
| Intercept | 93.088766 | 8.248823 | 11.29 | <.0001 |
| Run Time | -3.140188 | 0.373265 | -8.41 | <.0001 |
| Run Pulse | -0.073509 | 0.050514 | -1.46 | 0.1567 |

**Effect Tests**

| Source | Nparm | DF | Sum of Squares | F Ratio | Prob > F |
|---|---|---|---|---|---|
| Run Time | 1 | 1 | 513.41745 | 70.7746 | <.0001 |
| Run Pulse | 1 | 1 | 15.36208 | 2.1177 | 0.1567 |

# Residuals and Predicted Values

The residual is the difference between the actual response and the response predicted by the model. The residuals represent the error in the model. Points that don't fit very well have large residuals. It is helpful to look at a plot of the residuals and the predicted values, so JMP automatically appends a residual plot to the bottom of the Whole Model report, as shown in **Figure 13.5**.

> **Note:** The points in **Figure 13.5** show as a medium-sized circle marker instead of the default dot. To change the marker or marker size,
>
> 🖱 Select all of the rows in the data table (Ctrl+A or ⌘+A)
>
> 🖱 Right-click anywhere in the plot frame and choose **Row Markers** from the menu that appears.
>
> 🖱 Select the marker you want to use from the Markers palette.

**Figure 13.5**  Residual Plot for Multiple Regression Example

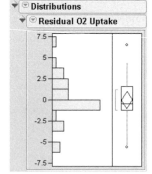

You can save these residuals as a column in the data table:

🖑  Select **Save Columns > Residuals** from the triangle popup menu found on the title of the report.

The result is a new column in the data table called Residual O2 Uptake, which contains the residual for each response point. To examine them in more detail:

🖑  Choose **Analyze > Distribution** and select the column of residuals to view the distribution of the residuals, as shown to the right.

Many researchers do this routinely to verify that the residuals are not too non-Normal to warrant concern about violating Normality assumptions.

You might also want to store the prediction formula from the multiple regression:

🖑  Select **Save Columns > Prediction Formula** from the popup menu to create a new column in the data table called Predicted O2 Uptake. Its values are the calculated predicted values for the model.

🖑  To see the formula used to generate the values in the column, right-click at the top of the Predicted O2 Uptake column and choose **Formula**. The Formula Editor window opens and displays the formula

93.0087761+–3.1410876•Run Time+–0.0735095•Run Pulse

This formula defines a plane for O2 Uptake as a function of Run Time and Run Pulse. The formula stays in the column and is evaluated whenever new rows are added, or when variables used in the expression change their values. You can cut-and-paste or drag this formula into other JMP data tables.

## The Analysis of Variance Table

The Whole-Model report consists of several tables that compare the full model fit to the simple mean model fit.

The Analysis of Variance table (shown to the right) lists the sums of squares and degrees of freedom used to form the whole model test:

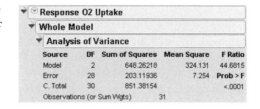

- The **Error Sum of Squares** (SSE) is 203.1. It is the sum of squared residuals after fitting the full model.

- The **C. Total Sum of Squares** is 851.4. It is the sum of squared residuals if you removed all the regression effects except for the intercept and, therefore, fit only the mean.

- The **Model Sum of Squares** is 648.3. It is the sum of squares caused by the regression effects, which measures how much variation is accounted for by the regressors. It is the difference between the Total Sum of Squares and the Error Sum of Squares.

The Error, C Total, and Model sums of squares are the ingredients needed to test the whole-model hypothesis that all the parameters in the model are zero except for the intercept (the simple mean model).

## The Whole Model *F*-Test

To form an *F*-test,

1. Divide the Model Sum of Squares (648.3 in this example) by the number of terms (effects) in the model excluding the intercept. That divisor (2 in this case) is found in the column labeled DF (Degrees of Freedom). The result is the *Mean Square for the Model*.

2. Divide the Error Sum of Squares (208.119 in this example) by the its associated degrees of freedom, 28, giving the *Mean Square for Error*.

3. Compute the *F*-ratio as the Model Mean Square divided by the Mean Square for Error.

   The significance level, or *p*-value, for this ratio is then calculated for the proper degrees of freedom (2 used in the numerator and 28 used in the denominator). The *F*-ratio, 44.6815, in the analysis of variance table shown above, is highly significant ($p<0.0001$), which indicates that the model does fit better than simply the mean.

## Whole-Model Leverage Plot

There is a good way to view this whole-model hypothesis graphically using a scatterplot of actual response values against the predicted values. The plot below shows the actual values versus predicted values for this aerobic exercise example.

A $45°$ line from the origin shows where the actual response and predicted response are equal. The vertical distance from a point to the $45°$ line of fit is the difference of the actual and the predicted values—the *residual error*. The mean is shown by the horizontal dashed line. The distance from a point to the horizontal line at the mean is what the residual would be if you removed all the effects from the model.

The portrayal of a plot that compares residuals from the two models in this way is called a *leverage plot*. The idea is to get a feel for how much better the sloped line fits than the horizontal line.

Superimposed on the plot are the confidence curves representing the 0.05-level whole-model hypothesis. If the confidence curves do not contain the horizontal line, the whole-model $F$-test is significant.

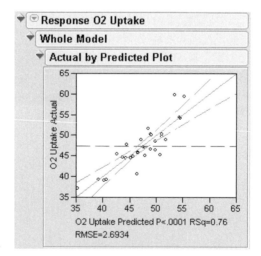

The leverage plot shown to the right is for the whole model, which includes both Run Time and Run Pulse.

## Details on Effect Tests

You can explore the significance of an effect in a model by looking at the distribution of the estimate, or by looking at the contribution of the effect to the model:

- To look at the distribution of the estimate, compute its standard error. The standard error can be used either to construct confidence intervals for the parameter or to perform a *t*-test that the parameter is equal to some value (usually zero). The *t*-tests are given in the Parameter Estimates table. Confidence intervals can also be requested.

- If you take an effect out of the model, then the error sum of squared increases. That difference in sums of squares (with the effect included and excluded) can be used to construct an *F*-test on whether the contribution of the effect to the model is significant. The *F*-tests are given in the Effect Tests table.

It turns out that *F*-tests and *t*-tests are equivalent. The square of the *t*-value in the Parameter Estimates table is the same as the *F*-statistic in the Effect Test table. For example, the square of the *t*-ratio (8.41) for Run Time is 70.77, which is the *F*-ratio for Run Time.

## Effect Leverage Plots

Scroll to the right on the regression platform to see details of how each regressor contributes to the model fit. The plots for the effect tests are *also* called leverage plots, although they are not the same as the leverage plots encountered in the whole-model test. The effect leverage plots (see **Figure 13.6**) show how each effect contributes to the fit after all the other effects have been included in the model. A leverage plot for a hypothesis test (an *effect leverage plot*) is any plot with horizontal and sloped reference lines and points laid out having the following two properties:

- The distance from each point to the sloped line measures the residual for the full model. The sums of squares of these residuals form the error sum of squares (SSE).

- The distance from each point to the horizontal line measures the residual for a restricted model without the effect. The sums of squares of these residuals form the SSE for the constrained model (the model without the effect). In this way, it is easy to see point by point how the sum of squares for the effect is formed. The difference in sums of squares of the two residual distances forms the numerator for the *F*-test for the effect.

**Figure 13.6**   Leverage Plots for Significant Effect and Nonsignificant Effect

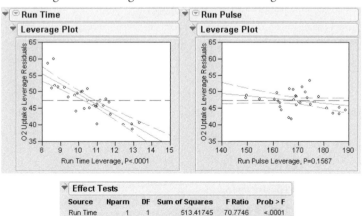

The leverage plot is interpreted for an effect in the same way as a simple regression plot. In fact, JMP superimposes a kind of 95% confidence curve on the sloped line that represents the full model. If the line is sloped significantly away from horizontal, then the confidence curves don't surround the horizontal line that represents the constrained model, and the effect is significant. Alternatively, when the confidence curves enclose the horizontal line, the effect is not significant at the 0.05 level.

The leverage plots in **Figure 13.6** show that Run Time is significant and Run Pulse is not. You can see the significance by how the points support (or don't support) the line of fit in the plot and by whether the confidence curves for the line cross the horizontal line.

There is a leverage plot for any kind of effect or set of effects in a model, or for any linear hypothesis. Leverage plots in the special case of single regressors are also known by the terms *partial plot*, *partial regression leverage plot*, and *added variable plot*.

# Collinearity

Sometimes with a regression analysis, there is a close linear relationship between two or more regressors. These two regressors are said to have a *collinearity* problem. It is a problem because the regression points do not occupy all the directions of the regression space very well. The fitting plane is not well supported in certain directions. The fit is weak in those directions, and the estimates become unstable, which means they are sensitive to small changes in the data.

In the statistical results, this phenomenon translates into high standard errors for the parameter estimates and potentially high values for the parameter estimates themselves. This occurs because a small random change in the narrow direction can have a huge effect on the slope of the corresponding fitting plane. An indication of collinearity in leverage plots is when the points tend to collapse horizontally toward the center of the plot.

To see an example of collinearity, consider the aerobic exercise example with the correlated regressors Max Pulse and Run Pulse:

- ⌐ With the Linnerud exercise table active, choose **Analyze > Fit Model** (or click on the Fit Model dialog if it is still open).

- ⌐ Complete the Fit Model dialog by adding Max Pulse as an effect in the model after Run Time and Run Pulse.

- ⌐ Click **Run Model**.

When the new analysis appears, scroll to the Run Pulse and Max Pulse leverage plots. Note in **Figure 13.7** that Run Pulse is very near the boundary of 0.05 significance, and thus the confidence curves almost line up along the horizontal axis, without actually crossing it.

**Figure 13.7**   Leverage Plots for Effects in Model

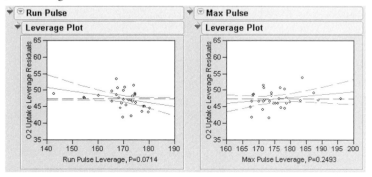

Now, as an example, let's change the relationship between the two regressors by changing a few values to cause collinearity.

🖑   Choose **Analyze > Fit Y by X**, selecting Max Pulse as the Y variable and Run Pulse as the X variable.

This produces a scatterplot showing the bivariate relationship.

🖑   Select **Density Ellipse > .90** from the popup menu on the title bar.

You should now see the scatterplot with ellipse shown to the right. The **Density Ellipse** command also generates the Correlation table beneath the plot, which shows the current correlation to be 0.55.

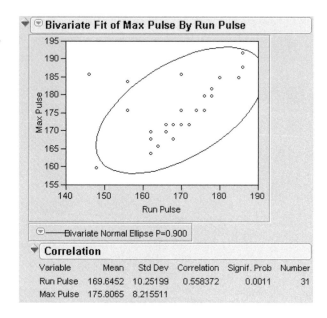

**Bivariate Fit of Max Pulse By Run Pulse**

——Bivariate Normal Ellipse P=0.900

**Correlation**

| Variable | Mean | Std Dev | Correlation | Signif. Prob | Number |
|----------|------|---------|-------------|--------------|--------|
| Run Pulse | 169.6452 | 10.25199 | 0.558372 | 0.0011 | 31 |
| Max Pulse | 175.8065 | 8.215511 | | | |

The variables don't appear to be collinear, since points are scattered in all directions. However, it appears that if four points are excluded, the correlation would increase dramatically. To do this, exclude these points and rerun the analysis:

☞ Highlight the points shown in the scatterplot below. You can do this by highlighting rows in the spreadsheet, or you can Shift-click the points in the scatterplot. With these points highlighted, choose **Rows > Label/Unlabel** to identify them.

☞ Choose **Rows > Exclude** while the rows are highlighted.

Notice in the spreadsheet that these points are now marked with a label tag and the not symbol(⊘)

☞ Again select a 0.90 **Density Ellipse** from the popup menu on the analysis title bar.

Now the ellipse and the Correlation table shows the relationship without the excluded points. The new correlation is 0.95, and the ellipse is much narrower, as shown in the plot to the right. Also, run the regression model again to see the effect of excluding these points to create collinearity:

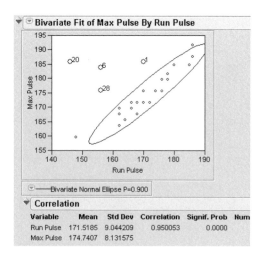

**Bivariate Fit of Max Pulse By Run Pulse**

——Bivariate Normal Ellipse P=0.900

**Correlation**

| Variable | Mean | Std Dev | Correlation | Signif. Prob | Num |
|----------|------|---------|-------------|--------------|-----|
| Run Pulse | 171.5185 | 9.044209 | 0.950053 | 0.0000 | |
| Max Pulse | 174.7407 | 8.131575 | | | |

🖱 Click **Run Model** again in the Fit Model dialog (with the same model as before).

Examine both the Parameter Estimates table and the leverage plots for Run Pulse and Max Pulse, comparing them with the previous report (see **Figure 13.8**).

The parameter estimates and standard errors for the last two regressors have more than doubled in size.

The leverage plots now have confidence curves that flare out because the points themselves collapse towards the middle. When a regressor suffers collinearity, the other variables have already absorbed much of that variable's variation, and there is less left to help predict the response. Another way of thinking about this is that there is less leverage of the points on the hypothesis. Points that are far out horizontally are said to have high leverage for the hypothesis test; points in the center have little leverage.

**Figure 13.8**   Comparison of Model Fits

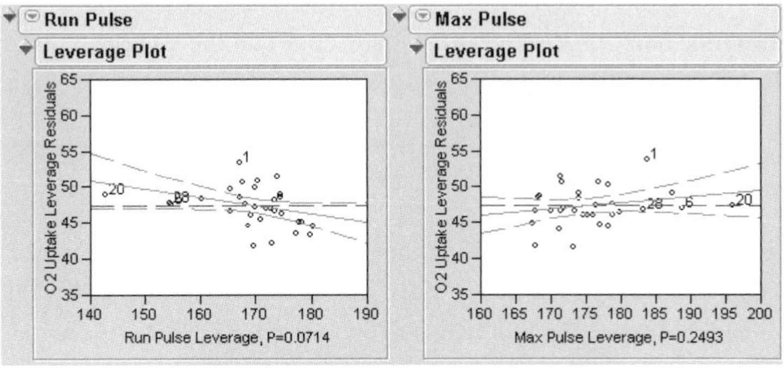

## Exact Collinearity, Singularity, Linear Dependency

Here we construct a variable to show what happens when there is an exact linear relationship, the extreme of collinearity, among the regressors:

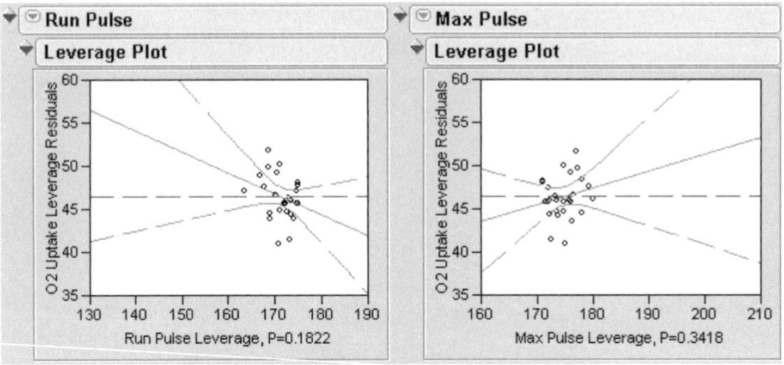

- ⫐ Click on the Linnerud data table (or re-open it if it was accidentally closed).

- ⫐ Choose **Columns > New Column**, to add a new variable (call it Run-Rest) to the data table.

- ⫐ Use the Formula Editor to create a formula that computes the difference between Run Pulse and Rest Pulse.

🖱 Now run a model of **O2 Uptake** against all the response variables, including the new variable **Run-Rest**.

The report in **Figure 13.9** shows the signs of trouble. In the parameter estimates table, there are notations on **Rest Pulse** and **Run Pulse** that the estimates are biased, and on **Run-Rest** that it is zeroed. With exact linear dependency, the least squares solution is no longer unique, so JMP chooses the solution that zeroes out every parameter estimate for variables that are linearly dependent on previous variables. The Singularity table shows what the exact relationship is, in this case expressed in terms of **Rest Pulse**. The $t$-tests for the parameter estimates must now be interpreted in a conditional sense. JMP refuses to make tests for the non-estimable hypotheses for **Rest Pulse**, **Run Pulse**, and **Run-Rest**, and shows them with no degrees of freedom.

**Figure 13.9**  Report When There is a Linear Dependency

**Response O2 Uptake**

**Singularity Details**

Run Pulse = Run - Rest + Rest Pulse

**Whole Model**

**Summary of Fit**

| | |
|---|---|
| RSquare | 0.785309 |
| RSquare Adj | 0.720901 |
| Root Mean Square Error | 2.461207 |
| Mean of Response | 46.42937 |
| Observations (or Sum Wgts) | 27 |

**Analysis of Variance**

| Source | DF | Sum of Squares | Mean Square | F Ratio |
|---|---|---|---|---|
| Model | 6 | 443.15097 | 73.8585 | 12.1928 |
| Error | 20 | 121.15077 | 6.0575 | **Prob > F** |
| C. Total | 26 | 564.30174 | | <.0001 |

**Parameter Estimates**

| Term | | Estimate | Std Error | t Ratio | Prob>|t| |
|---|---|---|---|---|---|
| Intercept | | 105.03367 | 14.34429 | 7.32 | <.0001 |
| Run Time | | -2.613413 | 0.420172 | -6.22 | <.0001 |
| Run Pulse | Biased | -0.289744 | 0.18468 | -1.57 | 0.1324 |
| Max Pulse | | 0.1845045 | 0.191153 | 0.97 | 0.3460 |
| Run - Rest | Biased | 0.0159465 | 0.07543 | 0.21 | 0.8347 |
| Age | | -0.219157 | 0.112002 | -1.96 | 0.0645 |
| Weight | | -0.05691 | 0.059584 | -0.96 | 0.3509 |
| Rest Pulse | Zeroed | 0 | 0 | . | . |

**Effect Tests**

| Source | Nparm | DF | Sum of Squares | F Ratio | Prob > F | |
|---|---|---|---|---|---|---|
| Run Time | 1 | 1 | 234.34659 | 38.6868 | <.0001 | |
| Run Pulse | 1 | 0 | 0.00000 | . | | LostDFs |
| Max Pulse | 1 | 1 | 5.64347 | 0.9316 | 0.3460 | |
| Run - Rest | 1 | 0 | 0.00000 | . | | LostDFs |
| Age | 1 | 1 | 23.19306 | 3.8288 | 0.0645 | |
| Weight | 1 | 1 | 5.52602 | 0.9123 | 0.3509 | |
| Rest Pulse | 1 | 0 | 0.00000 | . | | LostDFs |

You can see in the leverage plots for the three variables involved in the exact dependency, **Max Pulse**, **Run Pulse**, and **Run-Rest**, that the points have completely collapsed horizontally—

nothing has any leverage for these effects (see **Figure 13.10**). However, you can still test the unaffected regressors, like Max Pulse, and make good predicted values.

**Figure 13.10**    Leverage Plots When There is a Linear Dependency

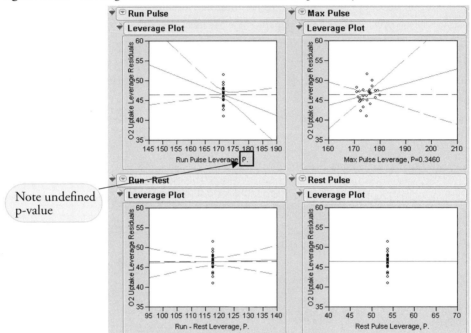

# The Longley Data: An Example of Collinearity

The Longley data is famous, and is run routinely on most statistical packages to test accuracy of calculations. Why is it a challenge? Look at the data:

🖰    Open the Longley.jmp data table.

🖰    Choose **Analyze > Fit Model.**

🖰    Enter Y as Y, and all the X columns as the model effects.

🖰    Make sure **Effect Leverage** is selected as the Fit Personality.

🖰    Click **Run Model** to see results shown in **Figure 13.11**.

**Figure 13.11**  Multiple Regression Report for Model with Collinearity

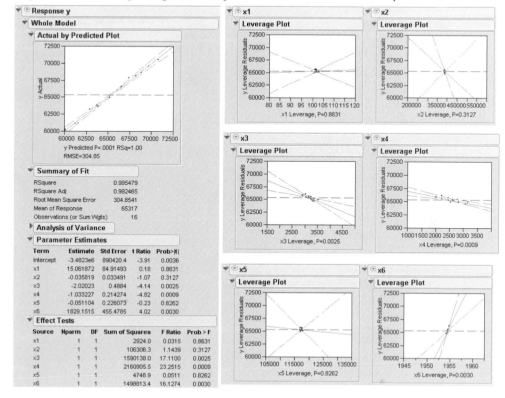

**Figure 13.11** shows the whole-model regression analysis. Looking at this overall picture doesn't give information on which (if any) of the six regressors are affected by collinearity. It's not obvious what the problems are in this regression until you look at leverage plots for the effects, which show that x1, x2, x5, and x6 have collinearity problems. Their leverage plots appear very unstable with the points clustered at the center of the plot and the confidence lines showing no confidence at all.

# The Case of the Hidden Leverage Point

Data was collected in a production setting where the yield of a process was related to three variables called Aperture, Ranging, and Cadence. Suppose you want to find out which of these effects are important, and in what direction:

🖱 Open the Ro.jmp data table and choose **Analyze > Fit Model**.

🖱  Enter Yield as Y, and Aperture, Ranging, and Cadence as the Effects in Model.

🖱  Click **Run Model**.

Everything looks fine in the tables shown in **Figure 13.12**. The Summary of Fit table shows an $R^2$ of 99.7%, which makes the regression model look like a great fit. All $t$-statistics are highly significant—but don't stop there.

JMP automatically creates the residual plot on the right in **Figure 13.12**.

**Figure 13.12**   Tables and Plots for Model with Collinearity

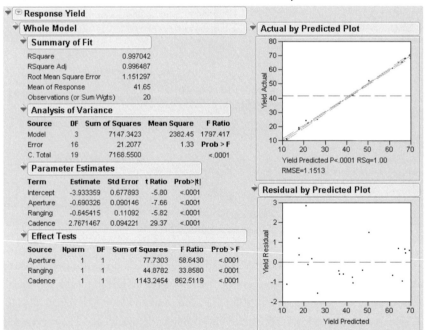

JMP produces all the standard regression results, and many more graphics. For each regression effect, there is a leverage plot showing what the residuals would be without that effect in the model. Note in **Figure 13.13** that row 20, which appeared unremarkable in the whole-model leverage and residual plots, is far out into the extremes of the effect leverage plots.

It turns out that row 20 has monopolistic control of the estimates on all the parameters. All the other points appear wimpy because they track the same part of the shrunken regression space.

**Figure 13.13**   Leverage Plots That Detect Unusual Points

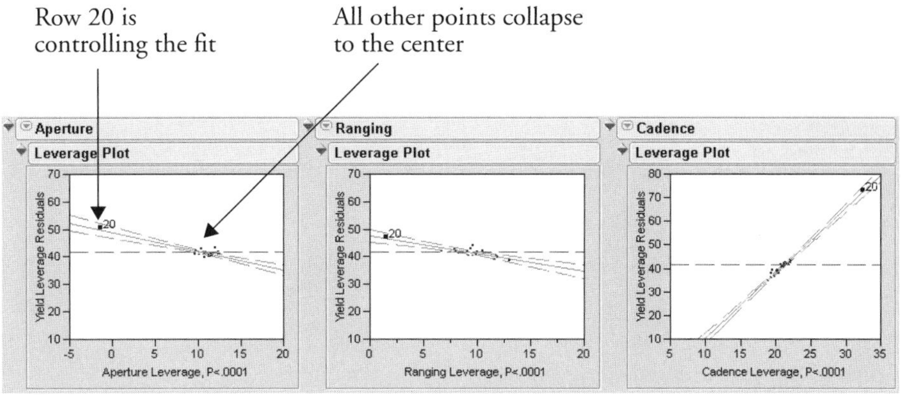

In a real analysis, row 20 would warrant special attention. Suppose, for example, that row 20 had an error, and was really 32 instead of 65:

👆 Change the value of Yield in row 20 from 65 to 32 and run the model again.

The Parameter Estimates for both the corrected table and incorrect table are shown to the right. The top table shows the parameter estimates computed from the data with an incorrect point. The bottom table has the corrected estimates. In high response ranges, the first prediction equation would give very different results than the second equation. The $R^2$ is again high and the parameter estimates are all significant—but every estimate is completely different even though only one point changed!

**Parameter Estimates**

| Term | Estimate | Std Error | t Ratio | Prob>|t| |
|------|----------|-----------|---------|----------|
| Intercept | -3.933359 | 0.677893 | -5.80 | <.0001 |
| Aperture | -0.690326 | 0.090146 | -7.66 | <.0001 |
| Ranging | -0.645415 | 0.11092 | -5.82 | <.0001 |
| Cadence | 2.7671467 | 0.094221 | 29.37 | <.0001 |

**Parameter Estimates**

| Term | Estimate | Std Error | t Ratio | Prob>|t| |
|------|----------|-----------|---------|----------|
| Intercept | -0.035267 | 0.246935 | -0.14 | 0.8882 |
| Aperture | 1.7097579 | 0.032837 | 52.07 | <.0001 |
| Ranging | 1.7318952 | 0.040405 | 42.86 | <.0001 |
| Cadence | 0.2853072 | 0.034322 | 8.31 | <.0001 |

# Mining Data with Stepwise Regression

Let's try a regression analysis on the O2 Uptake variable with a set of 30 randomly-generated columns as regressors. It seems like all results should be nonsignificant with random regressors, but that's not always the case.

👆 Open the Linnrand.jmp data table.

The data table has 30 columns named X1 to X30 that were generated by a uniform random number generator and stored as a formula in each column.

🖰 Choose **Analyze > Fit Model** and use O2 Uptake as Y. Select X1 through X30 as the Effects in Model.

🖰 Select **Stepwise** from the fitting personality popup menu, as shown in **Figure 13.14**, then click **Run Model**.

This stepwise approach launches a different regression platform, geared to playing around with different combinations of regressors.

**Figure 13.14**   Fit Model Dialog for Stepwise Regression

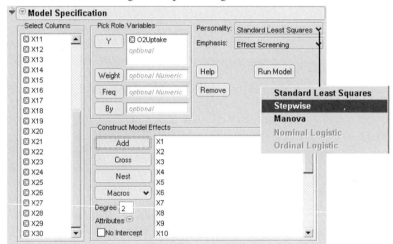

To run a stepwise regression, use the control panel that appears after you run the model from the Fit Model dialog (see **Figure 13.15**).

🖰 Click **Go** in the Stepwise Regression Control panel to begin the stepwise variable selection process.

By default, stepwise runs a forward selection process. At each step, it adds the variable to the regression that is most significant. You can also select **Backward** or **Mixed** as the stepwise direction from the control panel popup menu.

The process selects variables that are significant at the level specified in the **Prob to Enter** and **Prob to Leave** fields on the control panel. The process stops when no more variables are significant, and displays the Current Estimates table, shown in **Figure 13.16**. You will also see an Iteration table (not shown here) that lists the order in which the variables entered the model.

**Figure 13.15**   Stepwise Regression Control Panel

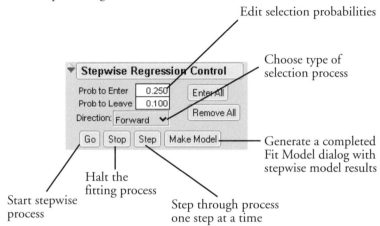

Edit selection probabilities

Choose type of selection process

Generate a completed Fit Model dialog with stepwise model results

Halt the fitting process

Step through process one step at a time

Start stepwise process

**Note:** The example in **Figure 13.16** was run with Forward direction to enter variables, 0.1 Prob to Enter, and 0.1 Prob to Leave. If you run this problem, the results may differ because the X variables are generated by random number functions.

**Figure 13.16** Current Estimates Table Showing Selected Variables

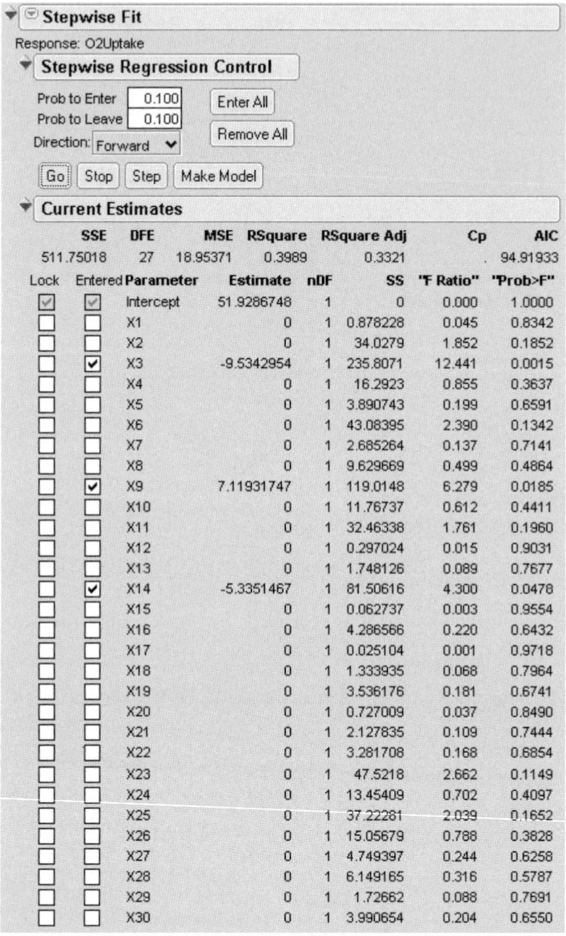


| SSE | DFE | MSE | RSquare | RSquare Adj | Cp | AIC |
|---|---|---|---|---|---|---|
| 511.75018 | 27 | 18.95371 | 0.3989 | 0.3321 | . | 94.91933 |

| Lock | Entered | Parameter | Estimate | nDF | SS | "F Ratio" | "Prob>F" |
|---|---|---|---|---|---|---|---|
| ☑ | ☑ | Intercept | 51.9286748 | 1 | 0 | 0.000 | 1.0000 |
| ☐ | ☐ | X1 | 0 | 1 | 0.878228 | 0.045 | 0.8342 |
| ☐ | ☐ | X2 | 0 | 1 | 34.0279 | 1.852 | 0.1852 |
| ☐ | ☑ | X3 | -9.5342954 | 1 | 235.8071 | 12.441 | 0.0015 |
| ☐ | ☐ | X4 | 0 | 1 | 16.2923 | 0.855 | 0.3637 |
| ☐ | ☐ | X5 | 0 | 1 | 3.890743 | 0.199 | 0.6591 |
| ☐ | ☐ | X6 | 0 | 1 | 43.08395 | 2.390 | 0.1342 |
| ☐ | ☐ | X7 | 0 | 1 | 2.685264 | 0.137 | 0.7141 |
| ☐ | ☐ | X8 | 0 | 1 | 9.629669 | 0.499 | 0.4864 |
| ☐ | ☑ | X9 | 7.11931747 | 1 | 119.0148 | 6.279 | 0.0185 |
| ☐ | ☐ | X10 | 0 | 1 | 11.76737 | 0.612 | 0.4411 |
| ☐ | ☐ | X11 | 0 | 1 | 32.46338 | 1.761 | 0.1960 |
| ☐ | ☐ | X12 | 0 | 1 | 0.297024 | 0.015 | 0.9031 |
| ☐ | ☐ | X13 | 0 | 1 | 1.748126 | 0.089 | 0.7677 |
| ☐ | ☑ | X14 | -5.3351467 | 1 | 81.50616 | 4.300 | 0.0478 |
| ☐ | ☐ | X15 | 0 | 1 | 0.062737 | 0.003 | 0.9554 |
| ☐ | ☐ | X16 | 0 | 1 | 4.286566 | 0.220 | 0.6432 |
| ☐ | ☐ | X17 | 0 | 1 | 0.025104 | 0.001 | 0.9718 |
| ☐ | ☐ | X18 | 0 | 1 | 1.333935 | 0.068 | 0.7964 |
| ☐ | ☐ | X19 | 0 | 1 | 3.536176 | 0.181 | 0.6741 |
| ☐ | ☐ | X20 | 0 | 1 | 0.727009 | 0.037 | 0.8490 |
| ☐ | ☐ | X21 | 0 | 1 | 2.127835 | 0.109 | 0.7444 |
| ☐ | ☐ | X22 | 0 | 1 | 3.281708 | 0.168 | 0.6854 |
| ☐ | ☐ | X23 | 0 | 1 | 47.5218 | 2.662 | 0.1149 |
| ☐ | ☐ | X24 | 0 | 1 | 13.45409 | 0.702 | 0.4097 |
| ☐ | ☐ | X25 | 0 | 1 | 37.22281 | 2.039 | 0.1652 |
| ☐ | ☐ | X26 | 0 | 1 | 15.05679 | 0.788 | 0.3828 |
| ☐ | ☐ | X27 | 0 | 1 | 4.749397 | 0.244 | 0.6258 |
| ☐ | ☐ | X28 | 0 | 1 | 6.149165 | 0.316 | 0.5787 |
| ☐ | ☐ | X29 | 0 | 1 | 1.72662 | 0.088 | 0.7691 |
| ☐ | ☐ | X30 | 0 | 1 | 3.990654 | 0.204 | 0.6550 |

- After the stepwise selection finishes selecting variables, click **Make Model** on the control panel.

The Fit Model dialog shown in **Figure 13.17** then appears, and you can run a standard least squares regression with the effects that were selected by the stepwise process as most active.

- Select **Run Model** on the Fit Model Dialog

**Figure 13.17**   Fit Model Dialog Generated by the Make Model Option

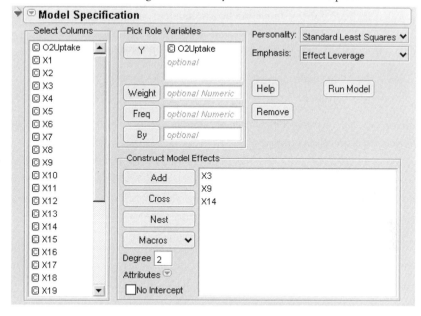

When you run the model, you get the standard regression reports, shown to the right. The Parameter Estimates table shows all effects significant at the 0.05 level.

However, what has just happened is that we created enough data to generate a number of coincidences, and then gathered those coincidences into one analysis and ignored the rest of the variables. This is like gambling all night in a casino, but exchanging money only for those hands where you win. When you mine data to the extreme, you get results that are too good to be true.

**Summary of Fit**

| | |
|---|---|
| RSquare | 0.398918 |
| RSquare Adj | 0.332131 |
| Root Mean Square Error | 4.353586 |
| Mean of Response | 47.37581 |
| Observations (or Sum Wgts) | 31 |

**Parameter Estimates**

| Term | Estimate | Std Error | t Ratio | Prob>|t| |
|---|---|---|---|---|
| Intercept | 51.928675 | 2.366656 | 21.94 | <.0001 |
| X3 | -9.534295 | 2.70307 | -3.53 | 0.0015 |
| X9 | 7.1193175 | 2.841091 | 2.51 | 0.0185 |
| X14 | -5.335147 | 2.572754 | -2.07 | 0.0478 |

# Exercises

1. The file Grandfather Clocks.jmp (Smyth, 2000) contains data on grandfather clocks sold at open auction. Included are the selling price, age of the clock, and number of bidders on the clock. You are interested in predicting price based on the other two variables.

(a) Use the Fit Model platform to construct a model using age as the only predictor of price. What is the $R^2$ for this single predictor model?

(b) Add the number of bidders to the model. Does the $R^2$ increase markedly?

2. In *Gulliver's Travels*, the Lilliputians make an entire set of clothes for the (giant) Gulliver by only taking a few measurements from his body:

> "*The seamstresses took my measure as I lay on the ground, one standing at my neck, and another at my mid-leg, with a strong cord extended, that each held by the end, while a third measured the length of the cord with a rule of an inch long. Then they measured my right thumb, and desired no more; for by a mathematical computation, that twice round the thumb is once round the wrist, and so on to the neck and the waist, and by the help of my old shirt, which I displayed on the ground before them for a pattern, they fitted me exactly.*" (Swift, 1735)

Is there a relationship among the different parts of the body? Body Measurements.jmp (Larner, 1996) contains measurements collected as part of a statistics project in Australia from 22 male subjects. In this exercise, you will construct a model to predict the mass of a person based on other characteristics.

(a) Using the Fit Model platform with personality **Standard Least Squares**, construct a model with mass as Y, and all other variables as effects in the model.

(b) Examine the resulting report and determine the effect that has the least significance to the model. In other words, find the effect with the largest Prob >F. Remove this effect from the model and re-run the analysis.

(c) Repeat part (b) until all effects have significance at the 0.05 level.

(d) Now, use the Fit Model platform with personality **Stepwise** to produce a similar model. Initially, enter all effects with Prob to Leave set at 0.05. Compare this model to the one you generated in part (c).

3. Biologists are interested in determining factors that will predict the amount of time an animal will sleep during the day. To investigate the possibilities, Allison and Ciccetti (1976) gathered information on 62 different mammals. Their data is presented in the file Sleeping Animals.jmp. The variables describe body weight, brain weight, total time spent sleeping in two different states ("dreaming" and "non-dreaming"), life span, and gestation time. The researchers also calculated indices to represent predation (1 meaning unlikely to be preyed upon, 5 meaning likely to be preyed upon), exposure (1 meaning the animal sleeps in a well-protected den, 5 meaning most exposure), and an overall danger index, based upon predation, exposure, and other factors (1 meaning least danger from other animals, 5 meaning most danger).

(a) Use the Fit Y By X platform to examine the single-variable relationships between TotalSleep and the other variables. Which two variables look like they have the highest correlation with TotalSleep?

(b) Construct a model using the two explanatory variables you found in part (a) and note its effectiveness.

(c) Construct a model using forward stepwise regression, with 0.10 as the probability to enter and leave the model. Compare this model to the one you constructed in part (b).

(d) Construct two models using forward and mixed stepwise regressions and compare it to the other models you have found. Which is the most effective at predicting total amount of sleep?

(e) Comment on the generalizability of this model. Would it be safe to use it to predict sleep times for a llama? Or a gecko? Explain your reasoning.

(f) Explore models that predict sleep in the dreaming and non-dreaming stages. Do the same predictors appear to be valid?

4. The file Cities.jmp contains a collection of pollution data for 52 cities around the country.

(a) Use the techniques of this chapter to build a model predicting Ozone for the cities listed. Use any of the continuous variables as effects.

(b) After you are satisfied with your model, determine whether there is an additional effect caused by the region of the country the city is in.

(c) In your model, interpret the coefficients of the significant effects.

(d) Comment on the generalizability of your model to other cities.

# Fitting Linear Models

14

## Overview

Several techniques, of increasing complexity, have been covered in this book. From fitting single means, to fitting multiple means, to fitting situations where the regressor is a continuous function, specific techniques have been demonstrated to address a wide variety of statistical situations. This chapter introduces a new approach involving *general linear models*, which will encompass all the models covered so far and extend to many more situations. They are all unified under the technique of least squares, fitting parameters to minimize the sum of squared residuals.

The techniques can be generalized even further to cover categorical response models, and other more specialized applications.

# The General Linear Model

Linear models are the sum of the products of coefficient parameters and factor columns. The linear model is rich enough to encompass most statistical work. By using a coding system, you can map categorical factors to regressor columns. You can also form interactions and nested effects from products of coded terms. **Table 12.1** lists many of the situations handled by the general linear model approach. To read the model notation in **Table 12.1**, suppose that factors A, B, and C are categorical factors, and that X1, X2, and so forth, are continuous factors.

**Table 12.1 Different Linear Models**

| Situation | Model Notation | Comments |
|---|---|---|
| one-way ANOVA | Y = A | add a different value for each level |
| two-way ANOVA no interaction | Y = A, B | additive model with terms for A and B |
| two-way ANOVA with interaction | Y = A, B, A*B | each combination of A and B has a unique add-factor |
| three-way factorial | Y = A, B, A*B, C, A*C, B*C, A*B*C | for $k$-way factorial, $2^k-1$ terms. The higher order terms are often dropped |
| nested model | Y = A, B[A] | The levels of B are only meaningful within the context of A levels, e.g. City[State], pronounced "city within state" |
| simple regression | Y = X1 | an intercept plus a slope coefficient times the regressor |
| multiple regression | Y = X1, X2, X3, X4,... | there can be dozens of regressors |
| polynomial regression | Y = X1, $X1^2$, $X1^3$, $X1^4$ | linear, quadratic, cubic, quartic,... |
| quadratic response surface model | Y = X1, X2, X3 $X1^2$, $X2^2$, $X3^2$, X1*X2, X1*X3, X2*X3 | all the squares and cross products of effects define a quadratic surface with a unique critical value where the slope is zero, which can be a minimum or a maximum, or a saddle point. |
| analysis of covariance | Y = A, X1 | main effect (A), adjusting for the covariate (X1) |

| Situation | Model Notation | Comments |
|---|---|---|
| analysis of covariance with different slopes | Y = A, X1, A*X1 | tests that the covariate slopes are different in different A groups |
| nested slopes | Y = A, X1[A] | separate slopes for separate groups |
| multivariate regression | Y1, Y2 = X1, X2... | the same regressors affect several responses |
| MANOVA | Y1, Y2, Y3 = A | a categorical variable affects several responses |
| multivariate repeated measures | sum and contrasts of (Y1 Y2 Y3)= A and so on. | the responses are repeated measurements over time on each subject. |

## Kinds of Effects in Linear Models

The richness of the general linear model results from the kind of effects you can include as columns in a coded model. Special sets of columns can be constructed to support various effects.

### Intercept term

Most models have an intercept term to fit the constant in the linear equation. This is equivalent to having a regressor variable that is always 1. If this is the only term in the model, it serves to estimate the mean. This part of the model is so automatic that it becomes part of the background, the submodel to which the rest of the model is compared.

### Continuous effects

These values are direct regressor terms, taken into the model without modification. If all your variables are continuous effects, then the linear model is called multiple regression (See Chapter 11 for an extended discussion of multiple regression).

### Categorical effects

The model must fit a separate constant for each level of a categorical effect. These effects lead to columns through an internal coding scheme,

The only case in which intercepts are not used is the one in which the surface of fit must go through the origin. This happens in mixture models, for example. If you suppress the intercept term, then certain statistics (such as the whole-model $F$-test and the $R^2$) do not apply because the question is no longer of just fitting a grand mean submodel against a full model.

In some cases (like mixture models) the intercept is suppressed but there is a hidden intercept in the factors. This case is detected and the $R^2$ and $F$ are reported as usual.

which is described in the next section. These are also called *main effects* when contrasted with compound effects, such as interactions.

### Interactions

These are crossings of categorical effects, in which you fit a different constant for each combination of levels of the interaction terms. Interactions are often written with an asterisk between terms, such as Age*Sex. If continuous effects are crossed, they are multiplied together.

### Nested effects

Nested effects occur when a term is only meaningful in the context of another term, and thus is kind of a combined main effect and interaction with the term within which it is nested. For example, city is nested within state, because if you know you're in Chicago, you also know you're in Illinois. If you specify a city name alone, like Trenton, then Trenton, New Jersey could be confused with Trenton, Michigan. Nested effects are written with the upper level term in parentheses or brackets, like City[State].

It is also possible to have combinations of continuous and categorical effects, and to combine interactions and nested effects.

## Coding Scheme to Fit a One-Way ANOVA as a Linear Model

When you include categorical variables in a model, JMP converts the categorical values (levels) into internal columns of numbers and analyzes the data as a linear model. The rules to make these columns are the *coding scheme*. These columns are sometimes called *dummy* or *indicator* variables. They make up an internal design matrix used to fit the linear model.

Coding determines how the parameter estimates are interpreted. However, note that the interpretation of parameters is different than the construction of the coded columns. In JMP, the categorical variables in a model are such that:

- There is an indicator column for each level of the categorical variable except the last level. An indicator variable is 1 for a row that has the value represented by that indicator, is −1 for rows that have the last categorical level (for which there is no indicator variable), and zero otherwise.

- A parameter is interpreted as the comparison of that level with the average effect across all levels. The effect of the last level is the negative of the sum of all the parameters for that effect. That is why this coding scheme is often called *sum-to-zero coding*.

Different coding schemes are the reason why different answers are reported in different software packages. The coding scheme doesn't matter in many simple cases, but it does matter in more complex cases, because it affects the hypotheses that are tested.

It's best to start learning the new approach covered in this chapter by looking at a familiar model. In Chapter 7, "Comparing Many Means: One-Way Analysis of Variance," you saw the DrugLBI.jmp data, comparing three drugs (Snedecor and Cochran, 1986). Let's return to this data table and see how the general linear model handles a one-way ANOVA.

The sample table called DrugLBI.jmp (**Figure 14.1**) contains the results of a study that measured the response of 30 subjects after treatment by one of three drugs. First, look at the one-way analysis of variance given previously by the Fit Y by X continuous-by-nominal platform.

**Figure 14.1** Drug Data Table with One-Way Analysis of Variance Report

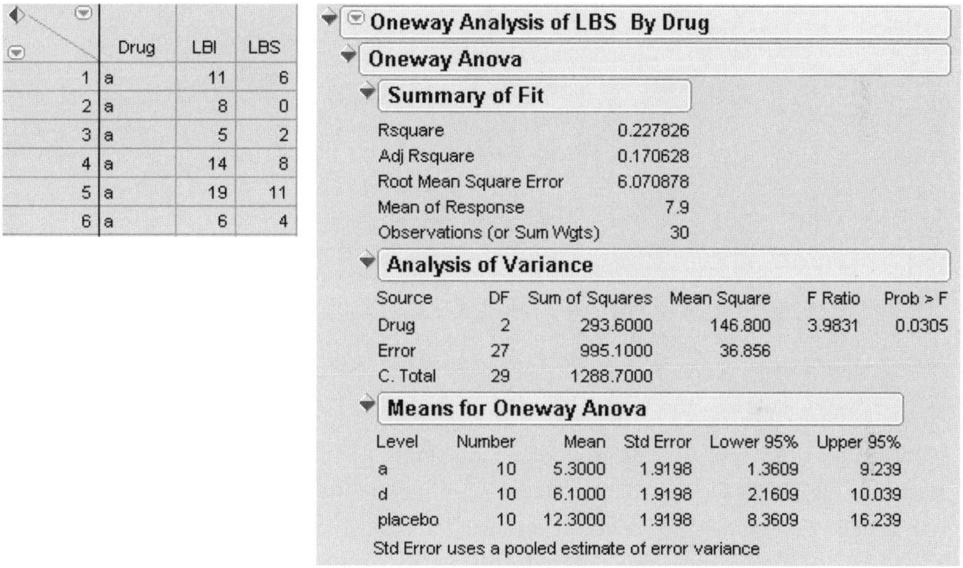

| | Drug | LBI | LBS |
|---|---|---|---|
| 1 | a | 11 | 6 |
| 2 | a | 8 | 0 |
| 3 | a | 5 | 2 |
| 4 | a | 14 | 8 |
| 5 | a | 19 | 11 |
| 6 | a | 6 | 4 |

Oneway Analysis of LBS By Drug

Oneway Anova

Summary of Fit

| | |
|---|---|
| Rsquare | 0.227826 |
| Adj Rsquare | 0.170628 |
| Root Mean Square Error | 6.070878 |
| Mean of Response | 7.9 |
| Observations (or Sum Wgts) | 30 |

Analysis of Variance

| Source | DF | Sum of Squares | Mean Square | F Ratio | Prob > F |
|---|---|---|---|---|---|
| Drug | 2 | 293.6000 | 146.800 | 3.9831 | 0.0305 |
| Error | 27 | 995.1000 | 36.856 | | |
| C. Total | 29 | 1288.7000 | | | |

Means for Oneway Anova

| Level | Number | Mean | Std Error | Lower 95% | Upper 95% |
|---|---|---|---|---|---|
| a | 10 | 5.3000 | 1.9198 | 1.3609 | 9.239 |
| d | 10 | 6.1000 | 1.9198 | 2.1609 | 10.039 |
| placebo | 10 | 12.3000 | 1.9198 | 8.3609 | 16.239 |

Std Error uses a pooled estimate of error variance

Now, do the same analysis using the Fit Model command.

- Open DrugLBI.jmp, which has variables called Drug, LBI, and, LBS.

Drug has values "a", "d", and "placebo". LBS is bacteria count after treatment, and LBI is a baseline count.

- Choose **Analyze > Fit Model**.
- When the Fit Model dialog appears, select LBS as the Y (response) variable. Select Drug and click **Add** to see it included as a Model Effect.
- Select **Minimal Report** from the Emphasis drop-down menu.

The completed dialog should look like the one in **Figure 14.2**.

🖱  Click **Run Model** to see the analysis result in **Figure 14.3**.

**Figure 14.2**  Fit Model Dialog for Simple One-Way ANOVA

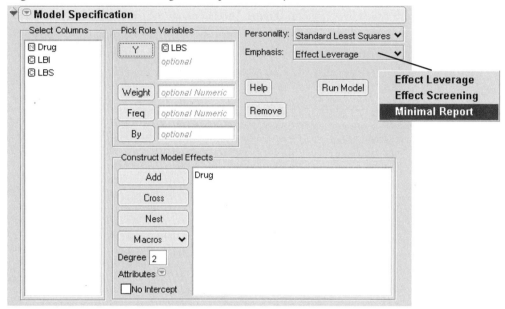

Now compare the reports from this new analysis (**Figure 14.3**) with the one-way ANOVA reports in **Figure 14.1**. Note that the statistical results are the same: the same $R^2$, ANOVA $F$-test, means, and standard errors on the means. (The ANOVA $F$-test is in both the Whole-Model Analysis of Variance table and in the Effect Test table because there is only one effect in the model.)

Although the two platforms produce the same results, the way the analyses were run internally was not the same. The Fit Model analysis ran as a regression on an intercept and two regressor variables constructed from the levels of the model main effect. The next section describes how this is done.

**Figure 14.3** ANOVA Results Given by the Fit Model Platform

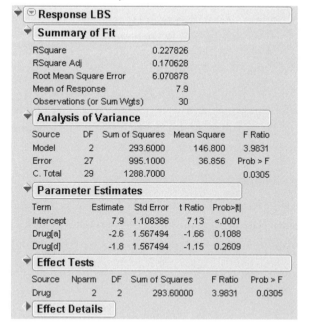

## Regressor Construction

The terms in the Parameter Estimates table are named according to what level each is associated with.

The terms are called Drug[a] and Drug[d]. Drug[a] means that the regressor variable is coded as 1 when the level is "a", −1 when the level is "placebo", and 0 otherwise. Drug[d] means that the variable is 1 when the level is

| Parameter Estimates | | | | |
|---|---|---|---|---|
| Term | Estimate | Std Error | t Ratio | Prob>|t| |
| Intercept | 7.9 | 1.108386 | 7.13 | <.0001 |
| Drug[a] | -2.6 | 1.567494 | -1.66 | 0.1088 |
| Drug[d] | -1.8 | 1.567494 | -1.15 | 0.2609 |

"d", −1 when the level is "placebo", and 0 otherwise. You can write the notation for Drug[a] as ([Drug=a]-[Drug=placebo]), where [Drug=a] is a one-or-zero indicator of whether the drug is "a" or not. The regression equation then looks like this:

$$y = b_0 + b_1*((Drug=a)-(Drug=placebo)) + b_2*((Drug=d)-(Drug=placebo)) + error$$

So far, the parameters associated with the regressor columns in the equation are represented by the names $b_0$, $b_1$ and so forth.

## Interpretation of Parameters

What is the interpretation of the parameters for the two regressors, now named in the equation as $b_1$, and $b_2$? The equation can be rewritten as

$$y = b_0 + b_1*[Drug=a] + b_2* [Drug=d] + (-b_1-b_2)*[Drug=placebo] + error$$

The sum of the coefficients ($b_1$, $b_2$, and $-b_1-b_2$) on the three indicators is always zero (again, sum-to-zero coding). The advantage of this coding is that the regression parameter tells you immediately how its level differs from the average response across all the levels.

## Predictions Are the Means

To verify that the coding system works, calculate the means, which are the predicted values for the levels "a", "d", and "placebo", by substituting the parameter estimates shown previously into the regression equation

$$Pred \ y = b_0 + b_1*([Drug=a]-[Drug=placebo]) + b_2*([Drug=d]-[Drug=placebo)$$

For the "a" level,

$$Pred \ y = 7.9 + -2.6*(1-0) + -1.8*(0-0) = 5.3, \text{ which is the mean } y \text{ for "a".}$$

For the "d" level,

$$Pred \ y = 7.9 + -2.6*(0-0) + -1.8*(1-0) = 6.1, \text{ which is the mean } y \text{ for "d".}$$

For the "placebo" level,

$$Pred \ y = 7.9 + -2.6*(0-1) + -1.8*(0-1) = 12.3, \text{ which is the mean } y \text{ for "placebo".}$$

## Parameters and Means

Now, substitute the means symbolically and solve for the parameters as functions of these means. First, write the equations for the predicted values for the three levels, called A for "a", D for "d" and P for "placebo".

$$MeanA = b_0 + b_1*1 + b_2*0$$

$$MeanD = b_0 + b_1*0 + b_2*1$$

$$\text{MeanP} = b_0 + b_1{}^*(-1) + b_2{}^*(-1)$$

After solving for the $b$'s, the following coefficients result:

$$b_1 = \text{MeanA} - (\text{MeanA} + \text{MeanD} + \text{MeanP})/3$$

$$b_2 = \text{MeanD} - (\text{MeanA} + \text{MeanD} + \text{MeanP})/3$$

$$(-b_1 - b_2) = \text{MeanP} - (\text{MeanA} + \text{MeanD} + \text{MeanP})/3$$

> Each level's parameter is interpreted as how different the mean for that group is from the mean of the means for each level.

In the next sections you will meet the generalization of this and other coding schemes, with each coding scheme having a different interpretation of the parameters.

Keep in mind that the coding of the regressors does not necessarily follow the same rule as the interpretation of the parameters. (This is a result from linear algebra, resting on the fact that the inverse of a matrix is its transpose only if the matrix is orthogonal).

Overall, analysis using this coding technique is a way to convert estimating group means into an equivalent regression model. It's all the same least-squares results using a different approach.

## Analysis of Covariance: Putting Continuous and Classification Terms into the Same Model

Now take the previous drug example that had one main effect (Drug), and now add the other term (LBI) to the model. LBI is a regular regressor, meaning it is a continuous effect, and is called a *covariate*.

⤴ In the Fit Model dialog window used earlier, click LBI and then **Add**. Now both Drug and LBI are effects, as shown in **Figure 14.4**.

⤴ Click **Run Model** to see the results in **Figure 14.5**.

**Figure 14.4**   Fit Model Dialog for Analysis of Covariance

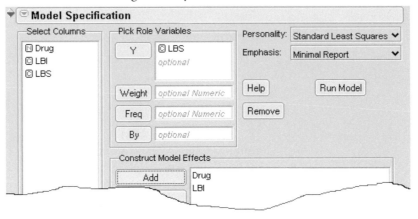

**Figure 14.5**   Analysis of Covariance Results Given by the Fit Model Platform

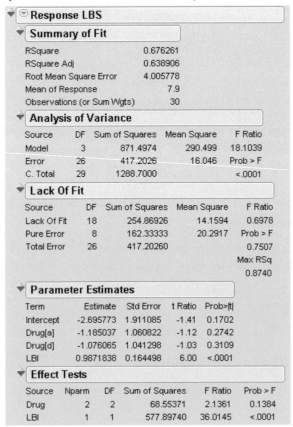

This new model is a hybrid between the ANOVA models with nominal effects and the regression models with continuous effects. Because the analysis method uses a coding scheme, the categorical term can be put into the model with the regressor.

The new results show that adding the covariate LBI to the model raises the $R^2$ from 22.78% (from **Figure 14.3**) to 67.62%. The parameter estimate for LBI is 0.987, which is not unexpected because the response is the bacteria count, and LBI is the baseline count before treatment. With a coefficient of nearly 1 for LBI, the model is really fitting the difference in bacteria counts. The difference in counts has a smaller variation than the absolute counts.

The $t$-test for or LBI is highly significant. Because the Drug effect uses two parameters, refer to the $F$-tests to see if Drug is significant. The $F$-test is testing that both parameters are zero. The surprising $p$-value for Drug is now 0.1384.

Drug, which was significant in the previous model, is no longer significant! How could this be? The error in the model has been reduced, so it should be easier for differences to be detected.

One possible explanation is that there might be a relationship between LBI and Drug.

    🖰   Choose **Analyze > Fit Y by X**, selecting LBI as Y and Drug as X.

    🖰   When the one-way platform appears, select the **Means/Anova** command from the popup menu on the title of the scatterplot.

Look at the results, shown to the right, to examine the relationship of the covariate LBI to Drug.

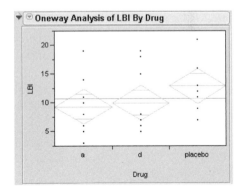

It appears that the drugs have not been randomly assigned—or, if they were, they drew an unlikely unbalanced distribution. The toughest cases (with the most bacteria) tended to be given the inert drug "placebo." This gave the "a" and "d" drugs a head start at reducing the bacteria count until LBI was brought into the model.

When fitting models where you don't control all the factors, you may find that the factors are interrelated, and the significance of one depends on what else is in the model.

## The Prediction Equation

The prediction equation generated by JMP can be stored as a formula in its own column.

⌐ Close the Fit Y by X window and return to the Fit Model results.

⌐ Select **Save Columns > Prediction Formula** command from the red triangle popup menu on the uppermost title bar of the report.

This command creates a new column in the data table called Pred Formula Y. To see this formula (the prediction formula),

⌐ Right-click in the column heading area and select **Formula**.

This opens a calculator window with the following formula for the prediction equation:

$$-2.695772906127$$
$$+\text{Match}(Drug)\begin{bmatrix}"a" & \Rightarrow -1.1850365373806\\ "d" & \Rightarrow -1.0760652051714\\ "placebo" & \Rightarrow 2.261101742552\\ else & \Rightarrow .\end{bmatrix}$$
$$+0.98718381112985*LBI$$

## The Whole-Model Test and Leverage Plot

The whole-model test shows how the model fits as a whole compared with fitting only a mean. This is equivalent to testing that all the parameters in the linear model are zero except for the intercept. This fit has three degrees of freedom, 2 from Drug and 1 from the covariate LBI. The $F$ of 18.1 is highly significant (see **Figure 14.6**).

The whole-model leverage plot is a plot of the actual value versus its predicted value. The residual is the distance from each point to the $45°$ line of fit (where the actual is equal to the predicted).

**Figure 14.6**  Whole-Model Test and Leverage Plot for Analysis of Covariance

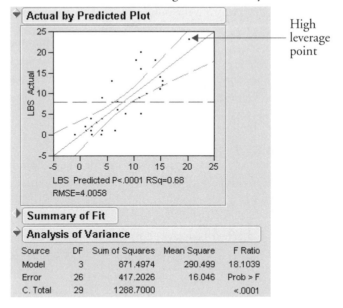

The leverage plot in **Figure 14.6** shows the hypothesis test point by point. The points that are far out horizontally (like point 25) tend to contribute more to the test because the predicted values from the two models differ more there. Points like this are called *high leverage points*.

## Effect Tests and Leverage Plots

Now look at the effect leverage plots in **Figure 14.7** to examine the details for testing each effect in the model. Each effect test is computed from a difference in the residual sums of squares that compare the fitted model to the model without that effect.

For example, the sum of squares (SS) for LBI can be calculated by noting that the SS(error) for the full model is 417.2, but the SS(error) was 995.1 for the model that had only the Drug main effect (see **Figure 14.3**). The SS(error) for the model that includes the covariate is 417.2. The reduction in sum of squares is the difference, 995.1 − 417.2 = 577.9, as you can see in the Effect Test table (**Figure 14.7**). Similarly, if you remove Drug from the model, the SS(Error) grows from 417.2 to 485.8, a difference of 68.6 from the full model.

The leverage plot shows the composition of these sums of squares point by point. The Drug leverage plot in **Figure 14.7** shows the effect on the residuals that would result from removing Drug from the model. The distance from each point to the sloped line is its residual. The distance from each point to the horizontal line is what its residual would be if Drug were removed from the model. The difference in the sum of squares for these two sets of residuals is the sum of squares for the effect, which is the numerator of the *F*-test for the **Drug** effect.

You can evaluate leverage plots in a way that is similar to evaluating a plot for a simple regression. The effect is significant if the points are able to support the sloped line significantly away from the horizontal. The confidence curves are placed around the sloped line to show the 0.05-level test. The curves cross the horizontal line if the effect is significant (as on the right in **Figure 14.7**), but they encompass the horizontal line if the effect is not significant (as on the left).

Click on points to see which ones are high leverage—away from the middle on the horizontal axis. Note whether they seem to support the test or not. If the points support the test, they are on the side trying to pull the line to have a higher slope.

**Figure 14.7**   Effect Tests and Effect Leverage Plots for Analysis of Covariance

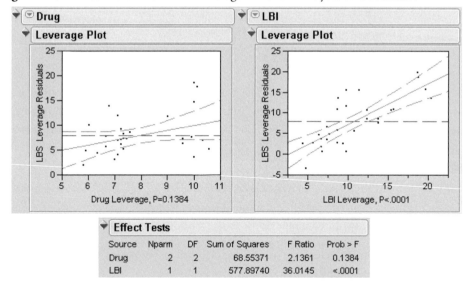

**Figure 14.8** summarizes the elements of a leverage plot. A leverage plot for a specific hypothesis is any plot with the following properties:

- There is a sloped line representing the full model and a horizontal line representing a model constrained by an hypothesis.
- The distance from each point to the sloped line is the residual from the full model.
- The distance from each point to the horizontal line is the residual from the constrained model.

**Figure 14.8** Schematic Defining a Leverage Plot

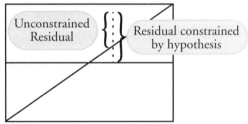

## Least Squares Means

It might not be fair to make comparisons between raw cell means in data that you fit to a linear model. Raw cell means do not compensate for different covariate values and other factors in the model. Instead, construct predicted values that are the expected value of a typical observation from some level of a categorical factor when all the other factors have been set to neutral values. These predicted values are called *least squares means*. There are other terms used for this idea: *marginal means*, *adjusted means*, and *marginal predicted values*.

The role of adjusted or least-squares means is that they allow comparisons of levels with the control of other factors being held fixed (*ceteris paribus*).

In the drug example, the least squares means are the predicted values expected for each of the three values of Drug, given that the covariate LBI is held at some constant value. The constant value is chosen for convenience to be the mean of the covariate, which is 10.7333. The prediction equation gives the least squares means as follows:

fit equation:

$$-2.695 - 1.185 \text{ drug[a-placebo]} - 1.0760 \text{ drug[d-placebo]} + 0.98718 \text{ LBI}$$

for a:

$$-2.695 - 1.185 \ (1) - 1.0760 \ (0) + 0.98718 \ (10.7333) = 6.71$$

for d:

$$-2.695 - 1.185 \ (0) - 1.0760 \ (1) + 0.98718 \ (10.7333) = 6.82$$

for placebo:

$$-2.695 - 1.185 \ (-1) - 1.0760 \ (-1) + 0.98718 \ (10.7333) = 10.16$$

To verify these results, return to the Fit Model platform and open the Least Squares Means table for the **Drug** effect (shown to the right).

**Least Squares Means Table**

| Level | Least Sq Mean | Std Error | Mean |
|-------|---------------|-----------|---------|
| a | 6.714963 | 1.2884943 | 5.3000 |
| d | 6.823935 | 1.2724690 | 6.1000 |
| placebo | 10.161102 | 1.3159234 | 12.3000 |

In the diagram shown at the top in **Figure 14.9**, The ordinary means are taken with different values of the covariate, so it is not fair to compare them. In the diagram at the bottom in **Figure 14.9**, the least squares means for this model are the intersections of the lines of fit for each level with the **LBI** value of 10.733. With this data, the least squares means are less separated than the raw means.

**Figure 14.9** Diagram of Ordinary Means (top) and Least Squares Means (bottom)

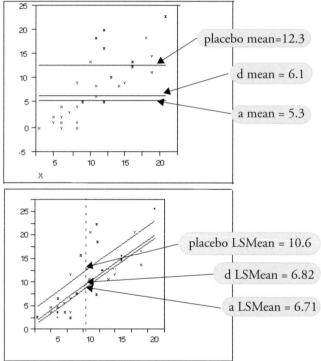

## Lack of Fit

The lack-of-fit test is the opposite of the whole-model test. Where the whole-model tests whether anything in the model is significant, the lack-of-fit tests whether anything left out of the model is significant. Unlike all other tests, it is usually desirable for the lack-of-fit test to be nonsignificant. A significant lack-of-fit test is advice to add more effects to the model using higher orders of terms already in the model.

But how is it possible to test effects that haven't been put in the model? All tests in linear models compare a model with a constrained or reduced version of that model. To test all the terms that could be in the model but are not—now *that* would be amazing!

Lack-of-fit compares the fitted model with a saturated model using the same terms. A *saturated* model is one that has a parameter for each combination of factor values that exist in the data. For example, a one-way analysis of variance is already saturated because it has a parameter for each level of the single factor. A complete factorial with all higher order interactions is completely saturated. For simple regression, saturation would be like having a separate coefficient to estimate for each value of the regressor.

If the lack-of-fit test is significant, there is some significant effect that has been left out of the model, and that effect is a function of the factors already in the model. It could be a higher-order power of a regressor variable, or some form of interaction among classification variables. If a model is already saturated, there is no lack-of-fit test possible.

The other requirement for a lack-of-fit test in continuous responses is that there be some exact replications of factor combinations in the data table. These exact duplicate rows (except for responses) allow the test to estimate the variation to use as a denominator in the lack-of-fit *F*-test. The error variance estimate from exact replicates is called *pure error* because it is independent of whether the model is right or wrong (assuming that it includes all the right factors).

In the drug model with covariate, the observations shown in **Table 12.2** form exact replications of data for Drug and LBI. The sum of squares around the mean in each replicate group reveals the contributions to pure error.

This pure error represents the best that can be done in fitting these terms to the model for this data. Whatever is done to the model involving Drug and LBI, these replicates and this error always exists. Pure error exists in the model regardless of the exact form of the model.

**Table 12.2 Lack-of-Fit Analysis**

| Replicate Rows | Drug | LBI | LBS | Pure Error DF | Contribution to Pure Error |
|---|---|---|---|---|---|
| 6 | a | 6 | 4 | 1 | $4.5=(4-2.5)^2+(1-2.5)^2$ |
| 8 | a | 6 | 1 | | |
| 1 | a | 11 | 6 | 1 | $2.0 = (6-7)^2+(6-8)^2$ |
| 9 | a | 11 | 8 | | |
| 11 | d | 6 | 0 | 1 | 2.0 |
| 12 | d | 6 | 2 | | |
| 14 | d | 8 | 1 | 2 | 32.667 |
| 16 | d | 8 | 4 | | |
| 18 | d | 8 | 9 | | |
| 27 | placebo | 12 | 5 | 2 | 120.667 |
| 28 | placebo | 12 | 16 | | |
| 30 | placebo | 12 | 20 | | |
| 21 | placebo | 16 | 13 | 1 | 0.5 |
| 26 | placebo | 16 | 12 | | |
| Total | | | | 8 | 162.333 |

Pure error can reveal how complete the model is. If the error variance estimate from the model is much greater than the pure error, then adding higher order effects of terms already in the model improves the fit.

The Lack-of-Fit table for this example is shown to the right. The difference between the total error from the fitted model and pure error is called Lack-of-Fit error. It represents all the terms that might have been added to the model,

**Lack Of Fit**

| Source | DF | Sum of Squares | Mean Square | F Ratio |
|---|---|---|---|---|
| Lack Of Fit | 18 | 254.86926 | 14.1594 | 0.6978 |
| Pure Error | 8 | 162.33333 | 20.2917 | Prob > F |
| Total Error | 26 | 417.20260 | | 0.7507 |
| | | | | Max RSq |
| | | | | 0.8740 |

but were not. The ratio of the lack-of-fit mean square to the pure error mean square is the *F*-test for lack-of-fit. For the covariate model, the lack-of-fit error is not significant, which is good because it is an indication that the model is adequate with respect to the terms included in the model.

# Separate Slopes: When the Covariate Interacts with the Classification Effect

When a covariate model includes a main effect and a covariate regressor, the analysis uses a separate intercept for the covariate regressor for each level of the main effect.

If the intercepts are different, could the slopes of the lines also be different? To find out, a method to capture the interaction of the regression slope with the main effect is needed. This is accomplished by introducing a crossed term, the interaction of Drug and LBI, into the model:

🖰 Return to the Fit Model dialog, which already has Drug and LBI as effects in the model.

🖰 Shift-click on Drug and LBI in the column selector list so that both columns are highlighted as shown in **Figure 14.10**.

🖰 Click the **Cross** button. This gives an effect in the model called Drug*LBI.

🖰 Click **Run Model** to see the results (**Figure 14.11**).

**Figure 14.10** Fit Model Dialog for Analysis of Covariance with Separate Slopes

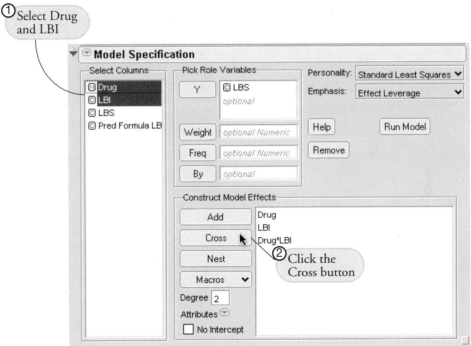

This specification adds two parameters to the linear model that allow the slopes for the covariate to be different for each Drug level. The new variable is the product of the dummy variables for Drug and the covariate values.

The Summary of Fit tables in **Figure 14.11** compare this separate slopes fit to the same slopes fit, showing an increase in $R^2$ from 67.63% to 69.15%.

**Figure 14.11**   Analysis of Covariance with Same Slopes (left) and Separate Slopes (right)

**Response LBS** (Same Slopes)

**Summary of Fit**

| | |
|---|---|
| RSquare | 0.676261 |
| RSquare Adj | 0.638906 |
| Root Mean Square Error | 4.005778 |
| Mean of Response | 7.9 |
| Observations (or Sum Wgts) | 30 |

**Analysis of Variance**

| Source | DF | Sum of Squares | Mean Square | F Ratio |
|---|---|---|---|---|
| Model | 3 | 871.4974 | 290.499 | 18.1039 |
| Error | 26 | 417.2026 | 16.046 | Prob > F |
| C. Total | 29 | 1288.7000 | | <.0001 |

**Lack Of Fit**

| Source | DF | Sum of Squares | Mean Square | F Ratio |
|---|---|---|---|---|
| Lack Of Fit | 18 | 254.86926 | 14.1594 | 0.6978 |
| Pure Error | 8 | 162.33333 | 20.2917 | Prob > F |
| Total Error | 26 | 417.20260 | | 0.7507 |
| | | | | Max RSq |
| | | | | 0.8740 |

**Parameter Estimates**

| Term | Estimate | Std Error | t Ratio | Prob>|t| |
|---|---|---|---|---|
| Intercept | -2.695773 | 1.911085 | -1.41 | 0.1702 |
| Drug[a] | -1.185037 | 1.060822 | -1.12 | 0.2742 |
| Drug[d] | -1.076065 | 1.041298 | -1.03 | 0.3109 |
| LBI | 0.9871838 | 0.164498 | 6.00 | <.0001 |

**Effect Tests**

| Source | Nparm | DF | Sum of Squares | F Ratio | Prob > F |
|---|---|---|---|---|---|
| Drug | 2 | 2 | 68.55371 | 2.1361 | 0.1384 |
| LBI | 1 | 1 | 577.89740 | 36.0145 | <.0001 |

**Response LBS** (Separate Slopes)

**Summary of Fit**

| | |
|---|---|
| RSquare | 0.691505 |
| RSquare Adj | 0.627235 |
| Root Mean Square Error | 4.070002 |
| Mean of Response | 7.9 |
| Observations (or Sum Wgts) | 30 |

**Analysis of Variance**

| Source | DF | Sum of Squares | Mean Square | F Ratio |
|---|---|---|---|---|
| Model | 5 | 891.1420 | 178.228 | 10.7594 |
| Error | 24 | 397.5580 | 16.565 | Prob > F |
| C. Total | 29 | 1288.7000 | | <.0001 |

**Lack Of Fit**

| Source | DF | Sum of Squares | Mean Square | F Ratio |
|---|---|---|---|---|
| Lack Of Fit | 16 | 235.22462 | 14.7015 | 0.7245 |
| Pure Error | 8 | 162.33333 | 20.2917 | Prob > F |
| Total Error | 24 | 397.55795 | | 0.7231 |
| | | | | Max RSq |
| | | | | 0.8740 |

**Parameter Estimates**

| Term | Estimate | Std Error | t Ratio | Prob>|t| |
|---|---|---|---|---|
| Intercept | -3.108215 | 2.06108 | -1.51 | 0.1446 |
| Drug[a] | -1.28643 | 1.115235 | -1.15 | 0.2601 |
| Drug[d] | -0.770981 | 1.09551 | -0.70 | 0.4884 |
| LBI | 1.0027452 | 0.171762 | 5.84 | <.0001 |
| Drug[a]*(LBI-10.7333) | -0.257522 | 0.237815 | -1.08 | 0.2896 |
| Drug[d]*(LBI-10.7333) | 0.0658032 | 0.227523 | 0.29 | 0.7749 |

**Effect Tests**

| Source | Nparm | DF | Sum of Squares | F Ratio | Prob > F |
|---|---|---|---|---|---|
| Drug | 2 | 2 | 52.05637 | 1.5713 | 0.2284 |
| LBI | 1 | 1 | 564.56753 | 34.0821 | <.0001 |
| Drug*LBI | 2 | 2 | 19.64465 | 0.5930 | 0.5606 |

The separate slopes model shifts two degrees of freedom from the lack-of-fit error to the model, increasing the model degrees of freedom from 3 to 5. The pure error seen in both Lack-of-Fit tables is the same because there are no new variables in the separate slopes covariance model. The new effect in the separate slopes model is constructed from terms already in the original analysis of covariance model.

The Effect Test table in the report on the right in **Figure 14.11** shows that the test for the new term Drug*LBI for separate slopes is not significant; the *p*-value is 0.56 (shown below the plot on the right). The confidence curves on the leverage plot for the Effect Test do not cross the horizontal mean line, showing that the interaction term doesn't significantly contribute to the model. The least squares means for the separate slopes model have a more dubious value now.

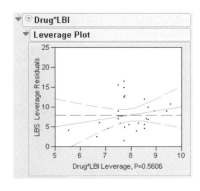

Previously, with the same slopes on LBI as shown in **Figure 14.11**, the least squares means changed with whatever value of LBI was used, but the separation between them did not. Now, with separate slopes as shown in **Figure 14.12**, the separation of the least squares means is also a function of LBI. The least squares means are more or less significantly different depending on whatever value of LBI is used. JMP uses the overall mean, but this does not represent any magic standard base. Notice that a and d cross near the mean LBI, so their least squares means happen to be the same only because of where the LBI covariate was set.

**Figure 14.12**   Illustration of Covariance with Separate Slopes

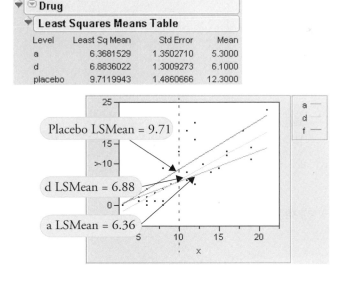

Interaction effects always have the potential to cloud the main effect, as will be seen again with the two-way model in the next section.

# Two-Way Analysis of Variance and Interactions

This section shows how to analyze a model in which there are two nominal or ordinal classification variables—a two-way model instead of a one-way model.

For example, a popcorn experiment was run, varying three factors and measuring the popped volume yield per volume of kernels. The goal was to see which factors gave the greatest volume of popped corn. **Figure 14.13** shows a listing of the popcorn data.

⌐🖰   To see the data, open the sample table Popcorn.jmp.

It has variables type with values "plain" and "gourmet"; batch, which designates whether the popcorn was popped in a large or small batch; and oil amt with values "lots" or "little." Start with two of the three factors, type and batch.

**Figure 14.13**    Listing of the Popcorn Data Table

| | type | oil amt | batch | yield | trial |
|---|---|---|---|---|---|
| 1 | plain | little | large | 8.2 | 1 |
| 2 | gourmet | little | large | 8.6 | 1 |
| 3 | plain | lots | large | 10.4 | 1 |
| 4 | gourmet | lots | large | 9.2 | 1 |
| 5 | plain | little | small | 9.9 | 1 |
| 6 | gourmet | little | small | 12.1 | 1 |
| 7 | plain | lots | small | 10.6 | 1 |
| 8 | gourmet | lots | small | 18.0 | 1 |
| 9 | plain | little | large | 8.8 | 2 |
| 10 | gourmet | little | large | 8.2 | 2 |
| 11 | plain | lots | large | 8.8 | 2 |
| 12 | gourmet | lots | large | 9.8 | 2 |
| 13 | plain | little | small | 10.1 | 2 |
| 14 | gourmet | little | small | 15.9 | 2 |
| 15 | plain | lots | small | 7.4 | 2 |
| 16 | gourmet | lots | small | 16.0 | 2 |

Popcorn
Notes Artificial dat
Fit Model

Columns (5/0)
⬚ type
⬚ oil amt
⬚ batch
⬚ yield
⬚ trial

Rows
All Rows    16
Selected     0
Excluded     0

⌐🖰   Choose **Analyze > Fit Model**.

🖑  When the Fit Model dialog appears, select
yield as the Y (response) variable. Select type
and batch and click **Add** to use them as the
Model Effects.

The Fit Model dialog should look like the one
shown to the right.

🖑  Click **Run Model** to see the analysis.

**Figure 14.14** shows the analysis tables for the two-
factor analysis of variance:

- The model explains 56% of the variation in yield (the $R^2$).

- The remaining variation has a standard error of 2.248 (Root Mean Square Error).

- The significant lack-of-fit test ($p$-value of 0.0019) says that there is something in the two factors that is not being captured by the model. The factors are affecting the response in a more complex way than is shown by main effects alone. The model needs an interaction term.

- Each of the two effects has two levels, so they each have a single parameter. Thus the $t$-test results are identical to the $F$-test results. Both factors are significant.

**Figure 14.14**    Two-Factor Analysis of Variance for Popcorn Experiment

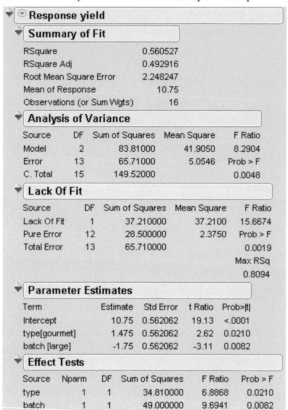

The leverage plots in **Figure 14.15** show the point-by-point detail for the fit as a whole and the fit as it is carried by each factor partially. Because this is a balanced design, all the points have the same leverage. This means they are spaced out horizontally the same in the leverage plot for each effect.

**Figure 14.15**  Leverage Plots for Two-Factor Popcorn Experiment with Interaction

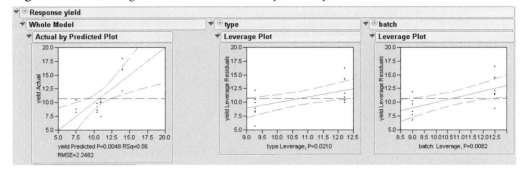

The lack-of-fit test shown in **Figure 14.14** suggests adding a higher order effect, such as an interaction term, also called a *crossed effect*. An interaction means that the response is not simply the sum of a separate function for each term. In addition, each term affects the response differently depending on the level of the other term in the model.

The type by batch interaction is added to the model as follows:

    🖱   Return to the Fit Model dialog, which already has the type and batch terms in the model.

We want to select both type and batch in the **Select Columns** list. To do so,

    🖱   Click on type and Control-click (⌘-click on the Macintosh) on batch to extend the selection.

    🖱   Click the **Cross** button to see the type*batch interaction effect in the Fit Model dialog as shown to the right.

    🖱   Click **Run Model** to see the tables in **Figure 14.16**.

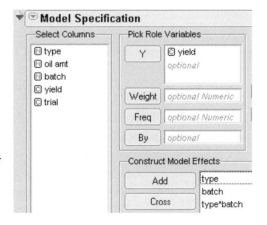

Including the interaction term increased the $R^2$ from 56% to 81%. The standard error of the residual (Root Mean Square Error) has gone down from 2.2 to 1.54.

**Figure 14.16**  Statistical Analysis of Two-Factor Experiment with Interaction

**Response yield**

**Summary of Fit**

| | |
|---|---|
| RSquare | 0.80939 |
| RSquare Adj | 0.761738 |
| Root Mean Square Error | 1.541104 |
| Mean of Response | 10.75 |
| Observations (or Sum Wgts) | 16 |

**Analysis of Variance**

**Parameter Estimates**

| Term | Estimate | Std Error | t Ratio | Prob>|t| |
|---|---|---|---|---|
| Intercept | 10.75 | 0.385276 | 27.90 | <.0001 |
| type[gourmet] | 1.475 | 0.385276 | 3.83 | 0.0024 |
| batch [large] | -1.75 | 0.385276 | -4.54 | 0.0007 |
| type[gourmet]*batch [large] | -1.525 | 0.385276 | -3.96 | 0.0019 |

**Effect Tests**

| Source | Nparm | DF | Sum of Squares | F Ratio | Prob > F |
|---|---|---|---|---|---|
| type | 1 | 1 | 34.810000 | 14.6568 | 0.0024 |
| batch | 1 | 1 | 49.000000 | 20.6316 | 0.0007 |
| type*batch | 1 | 1 | 37.210000 | 15.6674 | 0.0019 |

The Effect Test table shows that all effects are significant. The type*batch effect has a *p*-value of 0.0019, highly significant. The number of parameters (and degrees of freedom) of an interaction are the product of the number of parameters of each term in the interaction. The type*batch interaction has one parameter (and one degree of freedom) because the type and batch terms each have only one parameter.

An interesting phenomenon, which is true only in balanced designs, is that the parameter estimates and sums of squares for the main effects are the same as in the previous fit without interaction. The *F*-tests are different only because the error variance (Mean Square Error) is smaller in the interaction model. The interaction effect test is identical to the lack-of-fit test in the previous model.

Again, the leverage plots (**Figure 14.17**) show the tests in point-by-point detail. The confidence curves cross the horizontal strongly. The effects tests (shown in **Figure 14.6**) confirm that the model and all effects are highly significant.

**Figure 14.17**   Leverage Plots for Two-Factor Experiment with Interaction

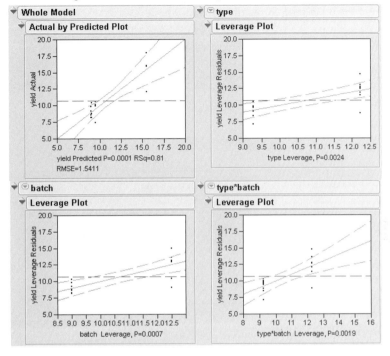

You can see some details of the means in the least squares means table, but in a balanced design (equal numbers in each level and no covariate regressors), they are equal to the raw cell means.

🖱  To see profile plots for each effect, select the **LSMeans Plot** command in the popup menu at the top of each leverage plot.

The result is a series of profile plots below each effect's report. Profile plots are a graphical form of the values in the Least Squares Means table.

- The leftmost plot in **Figure 14.18** is the profile plot for the type main effect. The "gourmet" type popcorn seems to have a higher yield.

- The middle plot is the profile plot for the batch main effect. It looks like small batches have higher yields.

- The rightmost plot is the profile plot for the type by batch interaction effect.

Looking at the effects together in an interaction plot shows that the popcorn type matters for small batches but not for big ones. Said another way, the batch size matters for gourmet popcorn, but not for plain popcorn. In an interaction profile plot, one interaction term is on the *x*-axis and the other term forms the different lines.

**Figure 14.18**   Interaction Plots and Least Squares Means

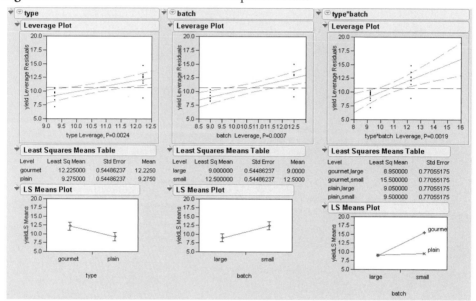

# Optional Topic: Random Effects and Nested Effects

This section talks about nested effects, repeated measures, and random effects mixed models. This is a large collection of topics to cover in a few pages, so hopefully this overview will be an inspiration to look to other textbooks and study these topics more completely.

As an example, consider the following situation. Six animals from two species were tracked, and the diameter of the area that each animal wandered was recorded. Each animal was measured four times, once per season.

**Figure 14.19** shows a listing of the Animals data.

**Figure 14.19** Listing of the Animals Data Table

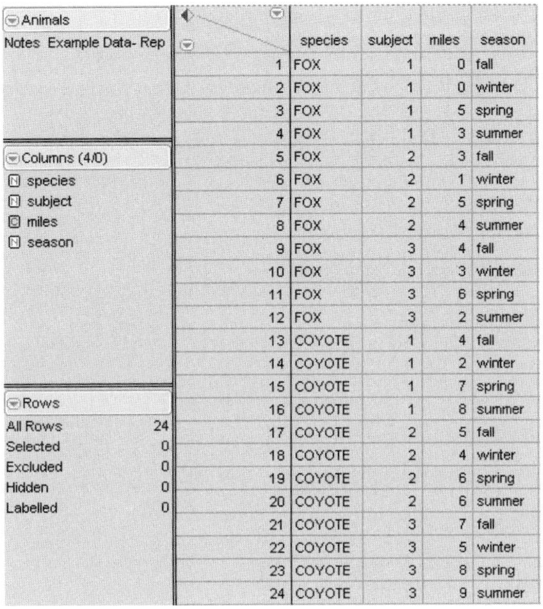
Animals
Notes Example Data- Rep
Columns (4/0)
species
subject
miles
season
Rows
All Rows 24
Selected 0
Excluded 0
Hidden 0
Labelled 0

| | species | subject | miles | season |
|---|---|---|---|---|
| 1 | FOX | 1 | 0 | fall |
| 2 | FOX | 1 | 0 | winter |
| 3 | FOX | 1 | 5 | spring |
| 4 | FOX | 1 | 3 | summer |
| 5 | FOX | 2 | 3 | fall |
| 6 | FOX | 2 | 1 | winter |
| 7 | FOX | 2 | 5 | spring |
| 8 | FOX | 2 | 4 | summer |
| 9 | FOX | 3 | 4 | fall |
| 10 | FOX | 3 | 3 | winter |
| 11 | FOX | 3 | 6 | spring |
| 12 | FOX | 3 | 2 | summer |
| 13 | COYOTE | 1 | 4 | fall |
| 14 | COYOTE | 1 | 2 | winter |
| 15 | COYOTE | 1 | 7 | spring |
| 16 | COYOTE | 1 | 8 | summer |
| 17 | COYOTE | 2 | 5 | fall |
| 18 | COYOTE | 2 | 4 | winter |
| 19 | COYOTE | 2 | 6 | spring |
| 20 | COYOTE | 2 | 6 | summer |
| 21 | COYOTE | 3 | 7 | fall |
| 22 | COYOTE | 3 | 5 | winter |
| 23 | COYOTE | 3 | 8 | spring |
| 24 | COYOTE | 3 | 9 | summer |

## Nesting

One feature of the data is that the labeling for each subject animal is nested within species. The observations for subject 1 for species Fox are not for the same animal as subject 1 for Coyote. The way to express this in a model is to always write the subject effect as subject[species], which is read as "subject nested within species" or "subject within species." The rule about nesting is that whenever you refer to a subject with a given level of factor, if that implies what another factor's level is, then the factor should only appear in nested form.

When the linear model machinery in JMP sees a nested effect such as "B within A", denoted B[A], it computes a new set of A parameters for each level of B. The Fit Model dialog allows for nested effects to be specified as in the following example.

- Open the Animals.jmp table
- Choose **Analyze > Fit Model**.
- Use miles as the Y variable.
- Add species to the Model Effects list.
- Add subject to the Model Effects list.

🖰  Select **species** in the Select Columns list and select **subject** in the Effects in Model list.

🖰  Click the **Nest** button.

This adds the nested effect subject[species] shown to the right.

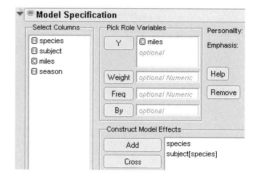

🖰  Add **season** to the Model Effects and click **Run Model** to see the results in **Figure 14.20**.

This model runs fine, but it has something wrong with it. The *F*-tests for all the effects in the model use the residual error in the denominator. The reason that this is an error, and the solution to this problem, are presented below.

**Figure 14.20**   Results for Animal Data Analysis

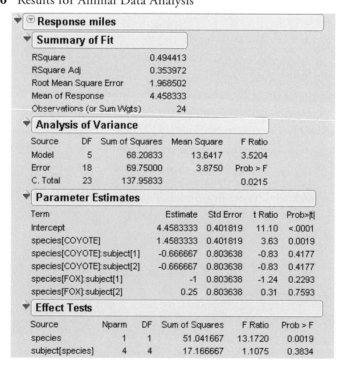

**Response miles**

**Summary of Fit**

| | |
|---|---|
| RSquare | 0.494413 |
| RSquare Adj | 0.353972 |
| Root Mean Square Error | 1.968502 |
| Mean of Response | 4.458333 |
| Observations (or Sum Wgts) | 24 |

**Analysis of Variance**

| Source | DF | Sum of Squares | Mean Square | F Ratio |
|---|---|---|---|---|
| Model | 5 | 68.20833 | 13.6417 | 3.5204 |
| Error | 18 | 69.75000 | 3.8750 | Prob > F |
| C. Total | 23 | 137.95833 | | 0.0215 |

**Parameter Estimates**

| Term | Estimate | Std Error | t Ratio | Prob>|t| |
|---|---|---|---|---|
| Intercept | 4.4583333 | 0.401819 | 11.10 | <.0001 |
| species[COYOTE] | 1.4583333 | 0.401819 | 3.63 | 0.0019 |
| species[COYOTE]:subject[1] | -0.666667 | 0.803638 | -0.83 | 0.4177 |
| species[COYOTE]:subject[2] | -0.666667 | 0.803638 | -0.83 | 0.4177 |
| species[FOX]:subject[1] | -1 | 0.803638 | -1.24 | 0.2293 |
| species[FOX]:subject[2] | 0.25 | 0.803638 | 0.31 | 0.7593 |

**Effect Tests**

| Source | Nparm | DF | Sum of Squares | F Ratio | Prob > F |
|---|---|---|---|---|---|
| species | 1 | 1 | 51.041667 | 13.1720 | 0.0019 |
| subject[species] | 4 | 4 | 17.166667 | 1.1075 | 0.3834 |

Note in **Figure 14.20** the treatment of nested effects in the model. There is one parameter for the two levels of species ("Fox" and "Coyote"). Subject is nested in species, so there is a separate set of two parameters (for three levels of subject) for subject within each level of species, giving a total of four parameters for subject. Season, with four levels, has three parameters. The total parameters for the model (not including the intercept) is 1 for species + 4 for subject + 3 for season = 8.

Use the **Save Prediction Formula** command and look at the Formula Editor window for the saved formula to see the following prediction equation using the parameter estimates:

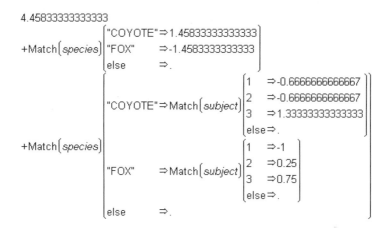

## Repeated Measures

As mentioned above, the previous analysis has a problem—the *F*-test used to test the species effect is constructed using the model residual in the denominator, which isn't appropriate for this situation. The following sections explain this problem and outline solutions.

There are three ways to understand this problem, which correspond to three different (but equivalent) resolutions:

- The effects can be declared as random, causing JMP to synthesize special *F*-tests.
- The observations can be made to correspond to the experimental unit.
- The analysis can be viewed as a multivariate problem.

The key in each method is to focus on only one of the effects in the model. In the animals example, the effect is species—how does the wandering radius differ between "Fox" and "Coyote"? The species effect is incorrectly tested in the previous example so species is the effect that needs attention.

## Method 1: Random Effects-Mixed Model

The subject effect is what is called a *random effect*. The animals were selected randomly from a large population, and the variability from animal to animal is from some unknown distribution. To generalize to the whole population, study the species effect with respect to the variability.

It turns out that if the design is balanced, it is possible to use an appropriate random term in the model as an error term instead of using the residual error to get an appropriate test. In this case subject[species], the nested effect acts as an error term for the species main effect.

To construct the appropriate *F*-test, do a hand calculation using the results from Effect Test table shown previously in **Figure 14.20**. Divide the mean square for species by the mean square for subject[species] as shown by the following formula.

$$F = \frac{\dfrac{51.041667}{1}}{\dfrac{17.166667}{4}}$$

This *F*-test has 1 numerator degree of freedom and 4 denominator degrees of freedom, and evaluates to 11.89.

Random effects in general statistics texts, often described in connection with *split plots* or *repeated measures designs*, describes which mean squares need to be used to test each model effect.

Now, let's have JMP do this calculation instead of doing it by hand. JMP will give the correct tests even if the design is not balanced.

First, specify subject[species] as a random effect.

- ✎ Click the Fit Model dialog window for the Animals data to make it active.

- ✎ Click to highlight subject[species] showing in the Model Effects list.

- ✎ Select the **Random Effect** attribute found in the **Attributes** popup menu.

The subject[species] effect then appears with &Random appended to it as shown in **Figure 14.21**.

- ✎ Select **EMS (Traditional)** from the Method popup menu as shown.

- ✎ Click **Run Model** to see the results.

**Figure 14.21**   Fit Model Dialog Using a Random Effect

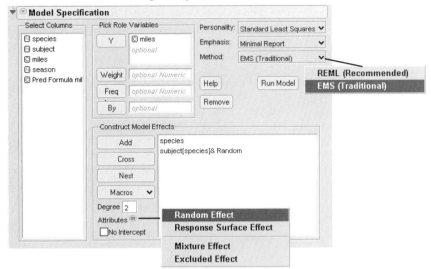

JMP constructs tests for random effects by the following steps:

1.  First, the expected mean squares are found. These are coefficients that relate the mean square to the variances for the random effects.

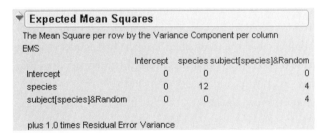

2.  Next, the variance component estimates are found using the mean square values for the random effects and their coefficients. It is possible (but rare) for a variance component estimate to be negative.

### Variance Component Estimates

| Component | Var Comp Est | Percent of Total |
|---|---|---|
| subject[species]&Random | 0.104167 | 2.618 |
| Residual | 3.875 | 97.382 |
| Total | 3.979167 | 100.000 |

These estimates based on equating Mean Squares to Expected Value.

3. For each effect in the model, JMP then determines what linear combination of other mean squares would make an appropriate denominator for an *F*-test. This denominator is the linear combination of mean squares that has the same expectation as the mean square of the effect (numerator) under the null hypothesis.

**Test Denominator Synthesis**

| Source | MS Den | DF Den | Denom MS Synthesis |
|---|---|---|---|
| species | 4.29167 | 4 | subject[species]&Random |
| subject[species]&Random | 3.875 | 18 | Residual |

4. *F*-tests are now constructed using the denominators synthesized from other mean squares. If an effect is prominent, then it will have a much larger mean square than expected under the null hypothesis for that effect.

**Tests wrt Random Effects**

| Source | SS | MS Num | DF Num | F Ratio | Prob > F |
|---|---|---|---|---|---|
| species | 51.0417 | 51.0417 | 1 | 11.8932 | 0.0261 |
| subject[species]&Random | 17.1667 | 4.29167 | 4 | 1.1075 | 0.3834 |

Again, the *F*-statistic for species is 11.89 with a *p*-value of 0.026. The tests for the other factors use the residual error mean square, which are the same as the tests done in the first model.

What about the test for **season**? Because the experimental unit for **season** corresponds to each row of the original table, the residual error in the first model is appropriate. The *F* of 10.64 with a *p*-value 0.0005 means that the **miles** (the response) does vary across **season**. If an interaction between **species** and **season** is included in the model, it is also be correctly tested using the residual mean square.

**Note 1:** There is an alternate way to define the random effects that produces slightly different expected mean squares, variance component estimates, and *F*-tests. The argument over which method of parameterization is more informative has been ongoing for 40 years.

**Note 2:** With random effects, it is not only the *F*-tests that need to be refigured. Standard deviations and contrasts on least squares means may need to be adjusted, depending on details of the situation. Consult an expert if you need to delve into the details of an analysis.

**Note 3:** Another method of estimating models with random effects, called REML, is available. Though it is the recommended method, most textbooks describe the EMS approach, which is why it is described here. Eventually, textbooks will shift to REML.

## Method 2: Reduction to the Experimental Unit

There are only 6 animals, but are there 24 rows in the data table because each animal is measured 4 times. However, taking 4 measurements on each animal doesn't make it legal to count each measurement as an observation. Measuring each animal millions of times and throwing all the data into a computer would yield an extremely powerful—and incorrect—test.

The experimental unit that is relevant to species is the individual animal, not each measurement of that animal. When testing effects that only vary subject to subject, the experimental unit should be a single response per subject, instead of the repeated measurements.

One way to handle this situation is to group the data to find the means of the repeated measures, and analyze these means instead of the individual values.

- ☝ With the Animals table active, choose **Tables > Summary**, which displays the dialog shown in **Figure 14.22**.

- ☝ Pick both species and subject as grouping variables.

- ☝ Highlight the miles variable as shown and select **Mean** from the **Stats** popup menu on the Summary dialog to see Mean(Miles) in the dialog.

This notation indicates you want to see the mean (average) miles for each subject within each species.

- ☝ Click **OK** to see the summary table shown at the bottom of **Figure 14.22**.

**Figure 14.22**   Summary Dialog and Summary Table for Animals data

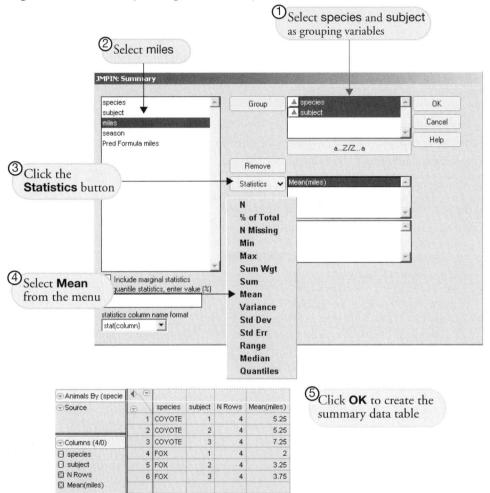

Now fit a model to the summarized data.

- ☞ With the Animals.jmp by species subject table (the Summary table) active, choose **Analyze > Fit Model**.

- ☞ Use Mean(miles) as Y, and species as the Model Effects variable.

- ☞ Click **Run Model** to see the proper *F*-test of 11.89 for species, with a *p*-value of 0.0261.

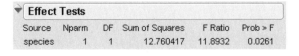

| Effect Tests | | | | | |
|---|---|---|---|---|---|
| Source | Nparm | DF | Sum of Squares | F Ratio | Prob > F |
| species | 1 | 1 | 12.760417 | 11.8932 | 0.0261 |

Note that this is the same result as the calculation shown in the previous section.

## Method 3: Correlated Measurements-Multivariate Model

In the animal example there were multiple (four) measurements for the same animal. These measurements are likely to be correlated in the same sense as two measurements are correlated in a paired *t*-test. This situation of multiple measurements of the same subject is called *repeated measures*, or a *longitudinal* situation. This kind of experimental situation can be looked at as a multivariate problem.

To use a multivariate approach, the data table must be rearranged so that there is only one row for each individual animal, with the four measurements on each animal in four columns:

⌐↑   To rearrange the Animals table, choose **Tables > Split**, which splits columns.

⌐↑   Complete the Split Columns dialog by assigning miles as the Split variable, season as the **Split Label Col** variable (its values become the new column names), and species and subject as **Group** variables

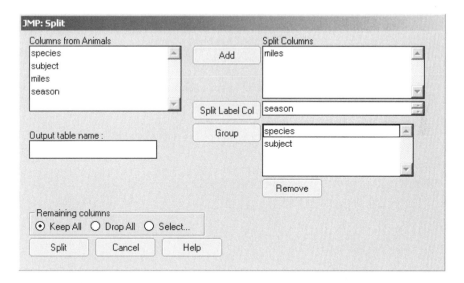

⌐↑   Click **Split** to see a new untitled table like the one shown in **Figure 14.23**.

**Figure 14.23** Rearrangement of the Animals data

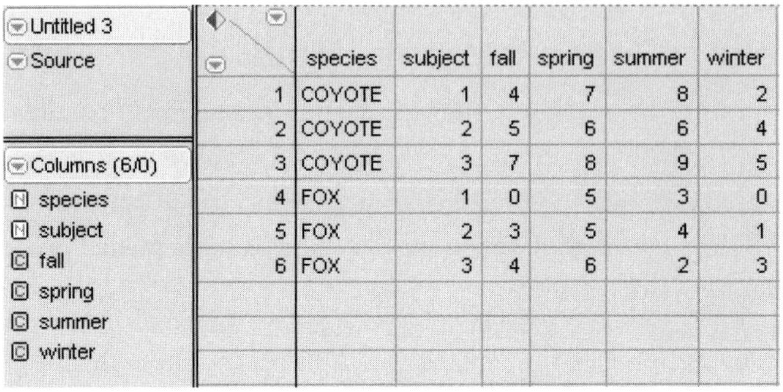


| | species | subject | fall | spring | summer | winter |
|---|---|---|---|---|---|---|
| 1 | COYOTE | 1 | 4 | 7 | 8 | 2 |
| 2 | COYOTE | 2 | 5 | 6 | 6 | 4 |
| 3 | COYOTE | 3 | 7 | 8 | 9 | 5 |
| 4 | FOX | 1 | 0 | 5 | 3 | 0 |
| 5 | FOX | 2 | 3 | 5 | 4 | 1 |
| 6 | FOX | 3 | 4 | 6 | 2 | 3 |

Then, fit a multivariate model with four Y variables and a single response:

- Choose **Analyze > Fit Model** and select fall, spring, summer, and winter as Y variables.
- Select species as the Model Effect.
- Select **Manova** from the fitting personality popup menu as shown in **Figure 14.24**.
- Click **Run Model** to see the analysis results.
- When the Multivariate control panel appears, choose **Repeated Measures** from the Response popup menu.
- When the repeated measures appears, accept the default name Time and click **OK**.

The following report is the result.

**Figure 14.24**   Model for Manova

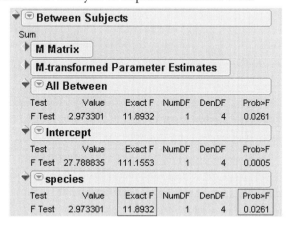

The resulting fit includes the Between Subjects report for species shown in **Figure 14.25**. The report for the **species** effect shows the same *F*-test of 11.89 with a *p*-value of 0.0261, just as with the other methods.

**Figure 14.25**   Multivariate Analysis of Repeated Measures Data

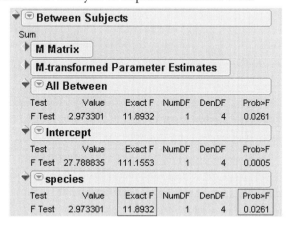

## Varieties of Analysis

In the previous cases, all the tests resulted in the same *F*-test for **species**. However, it is not generally true that different methods produce the same answer. For example, if **species** had more than two levels, the four multivariate tests (Wilk's lambda, Pillai's trace, Hotelling-

Lawley trace, and Roy's maximum root) each produce a different test result, and none of them would agree with the mixed model approach discussed previously.

Two more tests involving adjustments to the univariate method can be obtained from the multivariate fitting platform. With an unequal number of measurements per subject, the multivariate approach cannot be used. With unbalanced data, the mixed model approach also plunges into a diversity of methods that offer different answers from the Method of Moments that JMP uses.

## Summary

When using the residual error to form $F$-statistics, ask if the row in the table corresponds to the unit of experimentation. Are you measuring variation in the way appropriate for the effect you are examining?

- When the situation does not live up to the framework of the statistical method, the analysis will be incorrect, as was method 1 (treating data table observations as experimental units) in the example above for the **species** test.

- Statistics offers a diversity of methods (and a diversity of results) for the same question. In this example, the different results are not wrong, they are just different.

- Statistics is not always simple. There are many ways to go astray. Educated common sense goes a long way, but there is no substitute for expert advice when the situation warrants it.

# Exercises

1. Denim manufacturers are concerned with maximizing the comfort of the jeans they make. In the manufacturing process, starch is often built up in the fabric, creating a stiff jean that must be "broken in" by the wearer before they are comfortable. To minimize the break-in time for each pair of pants, the manufacturers often wash the jeans to remove as much starch as possible. This example concerns three methods of this washing. The data table Denim.jmp (Creighton, 2000) contains four columns: Method (describing the method used in washing), Size of Load (in pounds), Sand Blasted (recording whether or not the fabric was sand blasted prior to washing), and Starch Content (the starch content of the fabric after washing).

   (a) The three methods of washing consist of using an enzyme (alpha amalyze) to destroy the starch, chemically destroying the starch (caustic soda), and washing with an abrasive compound (pumice stone, hence the term "stone-washed jeans"). Determine if there is a significant difference due to washing method in two ways: Using the Fit Y by X platform and the Fit Model platform. Compare the Summary of Fit and Analysis of Variance displays for both analyses.

(b) Produce an LSMeans Plot for the Method factor in the above analysis. What information does this tell you? Compare the results with the means diamonds produced in the Fit Y By X plot.

(c) Fit a model of Starch Content using all three factors Size of Load, Sand Blasted, and Method. Which factors are significant?

(d) Produce LSMeans Plots for Method and Sand Blasted and interpret them.

(e) Use LSMeans Contrasts to determine if alpha amalyze is significantly different from caustic soda in its effect on Starch Content.

(f) Examine all first-level interactions of the three factors by running Fit Model and requesting **Factorial to Degree** from the Macros menu with the three factor columns selected. Which interactions are significant?

2. The file Titanic.jmp contains information on the passengers of the RMS Titanic. The four variables represent the class (first, second, third, and crew), age, sex, and survival status (yes or no) for each passenger. Use JMP to answer the following questions:

(a) Using the Fit Model platform, find a nominal logistic model to predict survival based on class, age, and sex.

(b) To evaluate the effectiveness of your model, save the probability formula from the model to the data table. Then, using Fit Y By X, prepare a contingency table to discover the percentage of times the model made the correct prediction.

(c) How do you know that the predictions in (b) are the most accurate for the given data?

3. The file Decathlon.jmp (Perkiömäki, 1995) contains decathlon scores from several winners in various competitions. Although there are ten events, they fall into groups of running, jumping, and throwing.

(a) Suppose you were going to predict scores on the 100m running event. Which of the other nine events do you think would be the best indicators of the 100m results?

(b) Run a Stepwise regression (from the Fit Model Platform) with 100m as the Y and the other nine events as X. Use the default value of 0.25 probability to enter and the Forward method. Which events are included in the resulting model?

(c)    Complete a similar analysis on Pole Vault. Examine the events included in the model. Are you surprised?

# 15

# Bivariate and Multivariate Relationships

## Overview

This chapter explores the relationship between two variables, their correlation, and the relationships between more than two variables. You look for patterns and you look for points that don't fit the patterns. You see where the data points are located, where the distribution is dense, and which way it is oriented.

Detective skills are built with the experience of looking at a variety of data, and learning to look at them in a variety of different ways. As you become a better detective, you also develop better intuition for understanding more advanced techniques.

It is not easy to look at lots of variables, but the increased range of the exploration will help you make more interesting and valuable discoveries.

# Bivariate Distributions

Previous chapters covered how the distribution of a response can vary depending on factors and groupings. This chapter returns to distributions as a simple unstructured batch of data. However, instead of a single variable, the focus is on the joint distribution of two or more responses.

# Density Estimation

As with univariate distributions, the question is where are the data? What regions of the space are dense with data, and what areas are relatively vacant? The histogram forms a simple estimate of the density of a univariate distribution. If you want a smoother estimate of the density, JMP has an option that takes a weighted count of a sliding neighborhood of points to produce the smooth curve. This idea can be extended to several variables.

One of the most classic multivariate data sets in statistics contains the measurements of iris flowers that R. A. Fisher analyzed. Fisher's iris data are in the data table called Iris.jmp, with variables Sepal length, Sepal width, Petal length, and Petal width. First, look at the variables one at a time.

- Open Iris.jmp and choose **Analyze > Distribution** on Sepal Length and Petal Length.

- When the report appears, select **Fit Distribution > Smooth Curve** from the red triangle menu on the report title bar.

- When the smooth curve appears, drag the density slider beneath the histogram to see the effect of using a wider or narrower smoothing distribution (**Figure 15.1**).

**Figure 15.1**  Univariate Distribution with Smoothing Curve

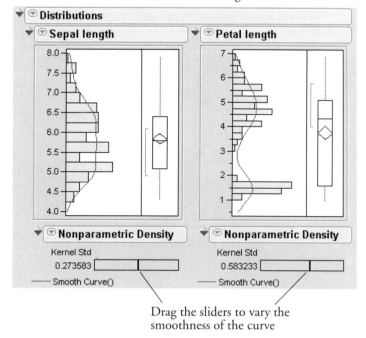

Drag the sliders to vary the
smoothness of the curve

Notice in **Figure 15.1** that Petal length has an unusual distribution with two modes and a vacant area in the middle of the range. There are no petals with a length in the range from 2 to 3.

## Bivariate Density Estimation

JMP has an implementation of a smoother that works for two variables to show bivariate densities. The goal is to draw lines around areas that are dense with points. Continue with the iris data and look at Petal length and Sepal length together:

🖰  Choose **Analyze > Fit Y by X** with Petal length as Y and Sepal Length as X.

🖰  When the scatterplot appears, select **Nonpar Density** from the menu on the title of the plot.

**Figure 15.2** Bivariate Density Estimation Curves

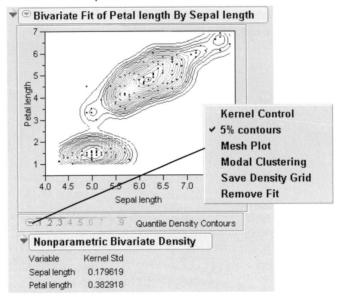

The result (**Figure 15.2**) is a contour graph, where the various contour lines show paths of equal density. The density is estimated for each point on a grid by taking a weighted average of the points in the neighborhood, where the weights decline with distance. Estimates done in this way are called *kernel smoothers*.

The Nonparametric Bivariate Density table beneath the plot has slider controls available for control of the vertical and horizontal width of the smoothing distribution.

⌐⊝ Select **Kernel Control** from the red triangle popup menu beside the legend of the contour graph.

Because it can take a while to calculate densities, they are not re-estimated until the **Apply** button is clicked.

The density contours form a map showing where the data are most dense. The contours are calculated according to the quantiles, where a certain percent of the data lie outside each contour curve. These quantile density contours show where each 5% and 10% of the data are. The innermost narrow contour line encloses the densest 5% of the data. The heavy line just outside shows the densest 10% of the data. It is labeled the 0.9 contour because 90% of the data lie outside it. Half the data distribution is inside the solid green lines, the 50% contours. Only about 5% of the data is outside the outermost 5% contour.

One of the features of the iris data is that there seem to be several local peaks in the density. There are two "islands" of data, one in the lower-left and one in the upper-right of the scatterplot.

These groups of locally dense data are called *clusters*, and the peaks of the density are called *modes*.

🖰   Select **Mesh Plot** from the popup menu on the legend of the contour plot.

This produces a 3-D surface of the density, as shown to the right.

🖰   Click and drag on the mesh plot to rotate it.

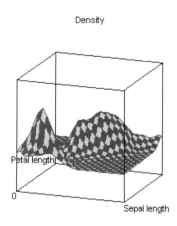

Density

## Mixtures, Modes, and Clusters

Multimodal data often comes from a mixture of several groups. Examining the data closely reveals that it is actually a collection of three species of iris: Virginica, Versicolor, and Setosa.

Conducting a bivariate density for each group results in the bivariate density plots in **Figure 15.3**. These plots have their axes adjusted to show the same scales.

**Figure 15.3**   Bivariate Density Curves

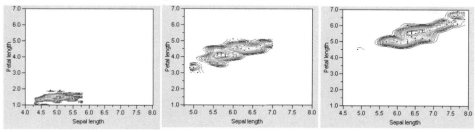

To classify an observation (an iris) into one of these three groups, a natural procedure would be to compare the density estimates corresponding to the petal and sepal length of a specimen over the three groups, and assign it to the group where its point is enclosed by the highest density curve. That kind of statistical method is called *discriminant analysis* and is shown in detail in Chapter 18, "Discriminant and Cluster Analysis" on page 467.

# The Elliptical Contours of the Normal Distribution

Notice that the contours of the distributions on each species are elliptical in shape. It turns out that ellipses are the characteristic shape for a bivariate Normal distribution. The Fit Y by X platform can show you a graph of these Normal contours.

🖰    Choose **Analyze > Fit Y by X** with Petal length as Y and Sepal length as X.

When the scatterplot appears,

🖰    Select **Group By** from the Fitting popup menu beneath the scatterplot and use Species as a grouping variable.

🖰    Select **Density Ellipse** from the Fitting popup menu with 0.50 as the level for the ellipse.

The result of these steps is shown in **Figure 15.4**. When there is a grouping variable in effect, there is a separate estimate of the bivariate Normal density (or any fit you select) for each group. The Normal density ellipse for each group encloses the densest 50% of the estimated distribution.

Notice that the two ellipses toward the top of the plot are fairly diagonally oriented, while the one at the bottom is not. The reports beneath the plot show the means, standard deviations, and correlation of Sepal length and Petal length for the distribution of each species. Note that the correlation is low for Setosa, and high for Versicolor and Virginica. The diagonal flattening of the elliptical contours is a sign of strong correlation. If variables are uncorrelated, then their Normal density contours appear to have a nondiagonal shape.

One of the main uses of a correlation is to see if variables are related. You want to know if the distribution of one variable is a function of the other. When the variables are Normally distributed and uncorrelated, then the univariate distribution of one variable is the same no matter what the value of the other variable is. When the density contours have no diagonal aspect, then the density across any slice is the same no matter where you take that slice (after you Normalize the slice to have an area of one so it becomes a univariate density).

The **Density Ellipse** command in the Fit Y by X platform also gives a significance test on the correlation, which shows the $p$-value for the hypothesis that the correlation is 0.

The bivariate Normal is quite common and is very basic for analyzing data, so let's cover it in more detail with simulations.

**Figure 15.4**   Density Ellipses for Species Grouping Variable

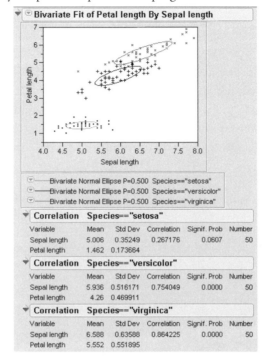

# Correlations and the Bivariate Normal

Describing Normally distributed bivariate data is easy because you need only the means, standard deviations, and the correlation of the two variables to completely characterize the distribution. If the distribution is not Normal, you might need a good deal more to summarize it.

Correlation is a measure, on a scale of -1 to 1, of how close two variables are to being linearly related. If you can draw a straight line through all the points of a scatterplot, then the correlation is one. The sign of the correlation reflects the slope of the regression line—a perfect negative correlation has value -1.

## Simulation Exercise

As in earlier chapters, it is useful to examine simulated data created with formulas. This simulated data provides a reference point when you move on to analyze real data.

🖐   Open Corrsim.jmp.

This table has no rows, but contains formulas to generate correlated data.

⊕   Choose **Rows > Add Rows** and enter 1000 when prompted for the number of rows wanted.

The formulas evaluate to give simulated correlations (**Figure 15.5**). There are two independent standard Normal random columns, labeled X and Y. The remaining columns (y.50, y.90, y.99, and y.1.00) have formulas constructed to produce the level of correlation indicated in the column names (0.5, 0.9, 0.99, and 1) The formula for generating a correlation $r$ with variable X is to make the linear mix with coefficient $r$ for X and Y computed as $\sqrt{1 - r^2}$

**Figure 15.5**   Partial Listing of Simulated Values

| Corrsim | | | X | y | y.50 | y.90 | y.99 | y1.00 |
|---|---|---|---|---|---|---|---|---|
| | | 1 | 0.570582 | 0.848661 | 1.020253 | 0.883446 | 0.684594 | 0.570582 |
| Columns (6/0) | | 2 | -0.72229 | -1.2039 | -1.40376 | -1.17483 | -0.8849 | -0.72229 |
| X ⊞ | | 3 | 1.347282 | -0.48584 | 0.25289 | 1.000781 | 1.265273 | 1.347282 |
| y ⊞ | | 4 | 1.138948 | -2.27826 | -1.40355 | 0.031984 | 0.806171 | 1.138948 |
| y.50 ⊞ | | 5 | 0.527828 | 2.061771 | 2.04946 | 1.37375 | 0.813398 | 0.527828 |
| y.90 ⊞ | | 6 | 0.916034 | -0.08285 | 0.386263 | 0.788315 | 0.895186 | 0.916034 |
| y.99 ⊞ | | 7 | -0.67314 | 0.18928 | -0.17265 | -0.52332 | -0.6397 | -0.67314 |
| y1.00 ⊞ | | 8 | -2.87152 | 3.287546 | 1.411341 | -1.15136 | -2.37903 | -2.87152 |
| Rows | | 9 | 1.990803 | 0.094391 | 1.077146 | 1.832867 | 1.98421 | 1.990803 |
| All Rows | 1000 | 10 | 1.690886 | 1.72539 | 2.339674 | 2.273877 | 1.917373 | 1.690886 |

You can use the Fit Y by X platform to examine the correlations:

⊕   Choose **Analyze > Fit Y by X** with X as the X variable, and all the Y columns as the Y's.

⊕   Hold down the Control (or ⌘) key and select **Density Ellipse** from the Fitting popup menu on the title at the top of the report. Choose **0.9** as the density level.

Holding down the Control (or ⌘) key causes the command to apply to all the open plots in the Fit Y by X window simultaneously.

⊕   Do the previous step twice more with **0.95** and **0.99** as density parameters.

These steps make Normal density ellipses (**Figure 15.6**) containing 90%, 95%, and 99% of the bivariate Normal density, using the means, standard deviations, and correlation from the data.

As an exercise, create the same plot for generated data with a correlation of -1, which is the last plot shown in **Figure 15.6**. To do this:

🖱 First create a new column.

🖱 Select **Formula** from the New Property menu on the New Column dialog or Column Info dialog.

🖱 Enter the following formula.

$$-\left[\begin{array}{l}r=1\\\sqrt{r}*X+\sqrt{1-r}*y_i\end{array}\right]$$

**Hint**: Open the Formula Editor window for the variable called Y1.00. Select its formula and drag it to the Formula Editor window for the new column you are creating. With the whole formula selected, click the unary sign change button on the Formula Editor keypad.

Note in **Figure 15.6** that as the correlation grows from 0 to 1, the relationship between the variables gets stronger and stronger. The Normal density contours are circular at correlation 0 (if the axes are scaled by the standard deviations) and collapse to the line at correlation 1.

**Figure 15.6**  Density Ellipses for Various Correlation Coefficients

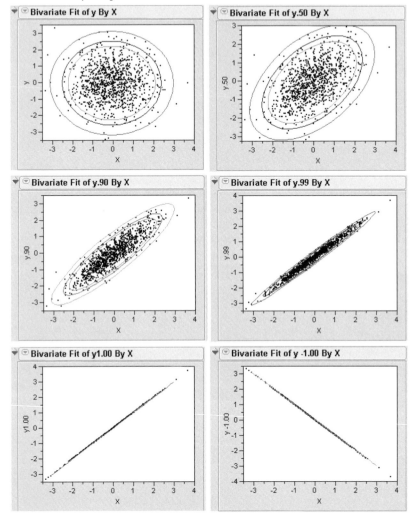

## Correlations Across Many Variables

This example uses six variables. To characterize the distribution of a six-variate Normal distribution, the means, the standard deviations, and the bivariate correlations of all the pairs of variables are needed.

In a chemistry study, the solubility of 72 chemical compounds was measured with respect to six solvents (Koehler and Dunn, 1988). One purpose of the study was to see if any of the solvents were correlated—that is, to identify any pairs of solvents that acted on the chemical compounds in a similar way.

⤒ Open Solubility.jmp to see variables Labels, 1-Octanol, Ether, Chloroform, Benzene, Carbon Tetrachloride, and Hexane.

A tag icon appears beside the Label column, which signifies that JMP will use the values in this column to label points in plots.

⤒ Choose **Analyze > Multivariate** with all the continuous solvent variables as Y's.

Initially, the scatterplot matrix looks like the one shown in **Figure 15.7**. Each small scatterplot can be identified by the name cells of its row and column.

**Figure 15.7**   Scatterplot Matrix for 6 Variables

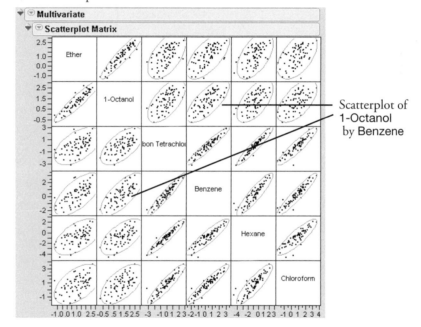

Scatterplot of
1-Octanol
by Benzene

You can resize the whole matrix by resizing one of its small scatterplots.

⤒ Move your mouse over the corner of any scatterplot until the cursor changes into a resize arrow. Click and drag to resize the plots.

Also, you can change the row and column location of a variable in the matrix by dragging its name on the diagonal with the hand tool.

Keep the correlation report for these six variables open to use again later in this chapter.

## Bivariate Outliers

Let's switch platforms to get a closer look at the relationship between ccl4 and hex using a set of density contours.

- 🖱 Choose **Analyze > Fit Y by X** with carbon tetrachloride as Y and hexane as X.

- 🖱 Select **Density Ellipse** from the Fitting popup menu four times for arguments 0.50, 0.90, 0.95, and 0.99 to add four density contours to the plot, as shown here.

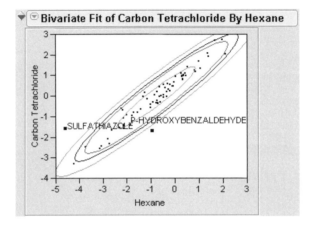

Under the assumption that the data are distributed bivariate Normal, the inside ellipse contains half the points, the next ellipse 90%, then 95%, and the outside ellipse contains 99% of the points.

Note that there are two points that are outside even the 99% ellipse.

- 🖱 To make you plot look like the one above, click and then Shift-click to highlight the two outside points. With the points highlighted, choose **Rows > Label** to label them.

The labeled points are outliers. A point can be considered an *outlier* if its bivariate Normal density contour is associated with a very low probability. Note that "P-hydroxybenzaldehyde" is not an outlier for either variable individually. In the scatterplot it is near the middle of the hex distribution, and is barely outside the 50% limit for the ccl4 distribution. However, it is a bivariate outlier because it falls outside the correlation pattern, which shows most of the points in a narrow diagonal elliptical area.

A common technique for computing outlier distance is the *Mahalanobis distance*. The Mahalanobis distance is computed with respect to the correlations as well as the means and standard deviations of both variables.

🖱 Click the Multivariate platform to make it the active window and select **Outlier Analysis** from its popup menu.

This command gives the Mahalanobis Distance outlier plot shown in **Figure 15.8**.

**Figure 15.8**   Outlier Analysis with Mahalanobis Outlier Distance Plot

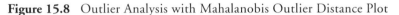
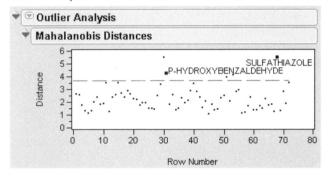

The reference line is drawn using an *F*-quantile and shows the estimated distance that contains 95% of the points. In agreement with the ellipses, "Sulfathiazole" and "P-hydroxybenzaldehyde" show as prominent outliers.

🖱 Click the scatterplot to activate it and select the brush tool from the **Tools** menu or toolbar.

🖱 Try dragging the brush tool over these two plots to confirm that the points near the central ellipse have low outlier distances and the points outside are greater distances.

🖱 Close this Multivariate platform window.

# Three and More Dimensions

To consider three variables at a time, consider the first three variables, ethane, 1-octanol, and carbon tetrachloride. You can see the distribution of points with a spinning plot:

🖱 Choose **Graph > Spinning Plot** and select ethane, 1-octanol, and carbon tetrachloride as variables to spin.

🖱 When the 3-D spinning plot appears, use the hand tool found in the **Tools** menu, or use the buttons on the spin control panel to spin the plot and look for three-variate outliers.

The spin orientation in **Figure 15.9** shows three points that appear to be outlying from the rest of the points with respect to the ellipsoid-shaped distribution.

**Figure 15.9**   Outliers as Seen in Spinning Plot

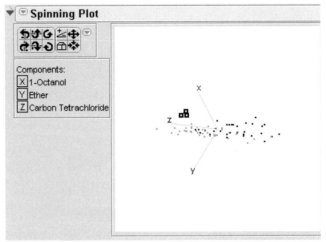

## Principal Components

As you spin the points, notice that some directions in the space show a lot more variation in the points than other directions. This was true in two dimensions when variables were highly correlated. The long axis of the Normal ellipse had the most variance; the short axis had the least. Now, in three dimensions, there are three axes showing a three-dimensional ellipsoid for trivariate Normal. The solubility data seems to have the ellipsoidal contours characteristic of Normal densities, except for a few outliers.

The directions of the axes of the Normal ellipsoids are called the *principal components*. They were mentioned in the Galton example in Chapter 8, "Fitting Curves Through Points: Regression."

The *first principal component* is defined as the direction of the linear combination of the variables that has maximum variance, subject to being scaled so the sum of squares of the coefficients is one. In a spinning plot, it is easy to rotate the plot and see which direction this is.

The *second principal component* is defined as the direction of the linear combination of the variables that has maximum variance, subject to it being at right angles (orthogonal) to the first principal component. Higher principal components are defined in the same way. There are as many principal components as there are variables. The last principal component has little or no variance if there is substantial correlation among the variables. This means that there is a direction for which the Normal density hyper-ellipsoid is very thin.

The spinning plot platform is good at showing principal components.

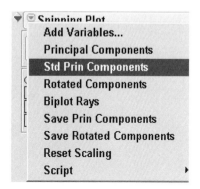

  🖱   Click the Spinning Plot platform shown in **Figure 15.9** to make it active.

  🖱   Select the **Std Prin Components** option found in the popup menu on the Spinning Plot title bar.

This adds three principal components to the variables list and creates three new rays in the spinning plot, as shown in **Figure 15.10**.

The directions of the principal components are shown as rays from the origin, labeled P1 for the first principal component, P2 for the second, and P3 for the third. As you rotate the plot, you see that the principal component rays correspond to the directions in the data in decreasing order of variance. You can also see that the principal components are at right angles in three-dimensional space.

The Principal Components report in **Figure 15.10** shows what portion of the variance among the variables is carried by each principal component. In this example, 80% of the variance is carried by the first principal component, 17% by the second, and 2% by the third. It is the correlations in the data that make the principal components interesting and useful. If the variables are not correlated, then the principal components all carry the same variance.

**Figure 15.10**  Biplot Showing Principal Components and Principal Components Report

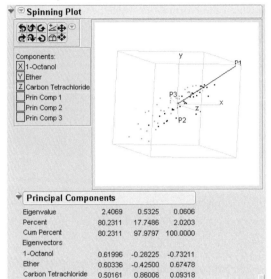

## Principal Components for Six Variables

Now let's move up to more dimensions than humans can visualize. Click the Solubility.jmp data table to make it the active window and look at principal components for all six variables in the data table.

- ☝  Proceed as before. Choose **Graph > Spinning Plot** and select the six solvent variables as spin components.

- ☝  Select **Principal Components** in the platform popup menu, which adds six principal component axes to the spinning plot and produces a principal component analysis.

**Figure 15.11**   Principal Components for Six Variables

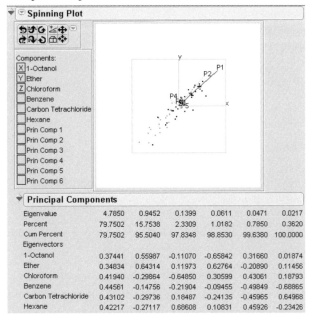

Examine the Principal Components table (**Figure 15.11**). There is a column in the table for each of the six principal components. Note in the Cum Percent row that the first three principal components work together to account for 97.8% of the variation in six dimensions.

⤵ Drag the Spinning Plot axis labels to be like those in the components panel shown to the right. (Click on X in the column of boxes beside the variable names and drag it down to the box by Prin Comp 1, drag Y to Prin Comp 2, and Z to Prin Comp 3.)

The first three principal components now become the spinning plot's $x$-, $y$-, and $z$-axes (**Figure 15.12**). As you spin the points, remember that you are seeing 97.8% of the variation in six dimensions as summarized by the first three principal components. This is the best three-dimensional view of six dimensions.

**Figure 15.12**   Principal Components Report for 6 Variables

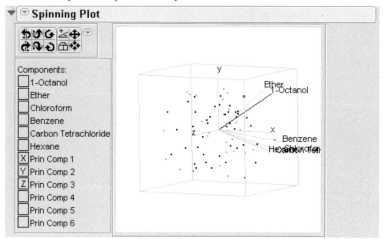

Notice that there are now rays labeled with the names of the variables. You can actually measure the coordinates of each point along these axes. The measurement is accurate to the degree that the first principal components capture the variation in the variables. This technique of showing the points on the same plot that shows directions of the variables was pioneered by Ruben Gabriel; therefore, the plot is called a *Gabriel biplot*. The rays are actually formed from the eigenvectors in the report, which are the coefficients of the principal components on the standardized variables.

## Correlation Patterns in Biplots

Note the very narrow angle between rays for ethane and 1-octanol. However, these two rays are at near right angles (in 3-D space) to the other four rays. Narrow angles between principal component rays are a sign of correlation. Principal components try to squish the data from six dimensions down to three dimensions. To represent the points most accurately in this squish, the directions for correlated variables are close together because they represent most of the same information. Thus, the Gabriel biplot shows the correlation structure of high-dimensional data.

Refer back to **Figure 15.7** to see the scatterplot matrix for all six variables. The scatterplot matrix confirms the fact that ethane and 1-octanol are highly correlated. The other four variables are also highly correlated, but there is much less correlation between these two sets of variables.

You might also consider a simple form of factor analysis, in which the components are rotated to positions so that they point in directions that correspond to clusters of variables. In JMP, this can be done in the Spinning Plot platform with the **Rotated Components** command.

## Outliers in Six Dimensions

The Fit Y by X and the Spinning plot platforms revealed extreme values in one, two, and three dimensions. These outliers also show in six dimensions. However, there could be additional outliers that violate the higher dimensional correlation pattern. In six dimensions, some of the directions of the data are quite flat, and there could be an outlier in that direction that wouldn't be revealed in an ordinary scatterplot.

   🖱 In the Spinning Plot platform with six variables, drag the X, Y, and Z axis labels to the last three principal components, as shown in **Figure 15.13**.

Now you are seeing the directions in six-dimensional space that are least prominent for the data. The data points are crowded near the center because you are looking at a small percentage of the variability.

   🖱 Use the zoom-out button (✛) on the spin control panel to expand the point cloud away from the center.

   🖱 Highlight and label the points that seem to stand out.

These are the points that are in the directions that are most unpopular. Sometimes they are called *class B outliers*.

**Figure 15.13**   Biplot Showing Six Principal Components

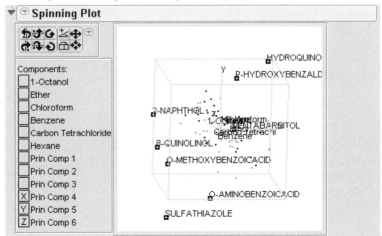

All the outliers from one dimension to six dimensions should show up on an outlier distance plot that measures distance with respect to all six variables.

  ⦿ Click (to activate) the Multivariate platform previously generated for all six variables.

  ⦿ Select **Outlier Analysis** from the platform menu to see the six-dimensional outlier distance plot in **Figure 15.14**. (If you closed the correlation window, choose **Analyze > Multivariate** with all six responses as Y variables, and then select **Outlier Analysis** from the platform menu.)

**Figure 15.14**   Outlier Distance Plot for 6 Variables

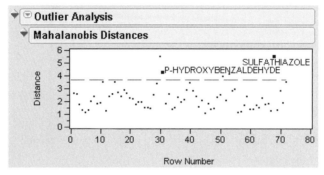

There is a refinement to the outlier distance that can help to further isolate outliers. When you estimate the means, standard deviations, and correlations, all points—including outliers—are included in the calculations and affect these estimates, causing an outlier to disguise itself.

Suppose that as you measure the outlier distance for a point, you exclude that point from all mean, standard deviation, and correlation estimates.

This technique is called *jackknifing*. The jackknifed distances often make outliers stand out better.

  ⦿ To see the Jackknifed Distance plot shown to the right, select the **Jackknife Distances** option from the popup menu at the top of the Outliers analysis.

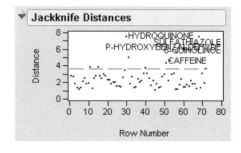

# Summary

When you have more than three variables, the relationships among them can get very complicated. Many things can be easily found in one, two, or three dimensions, but it is hard to visualize a space of more than three dimensions.

The histogram provides a good one-dimensional look at the distribution, with a smooth curve option for estimating the density. The scatterplot provides a good two-dimensional look at the distribution, with Normal ellipses or bivariate smoothers to study it. In three dimensions the spinning plot provides the third dimension. To look at more than three dimensions, you must be creative and imaginative.

One good basic strategy for high-dimensional exploration is to take advantage of correlations to reduce the number of dimensions. The technique for this is principal components, and the graph is the Gabriel biplot, which shows all the original variables as well as the points in principal component space.

You can also use highlighting tools to brush across one distribution and see how the points highlight in another view.

The hope is that you either find patterns that help you understand the data, or points that don't fit patterns. In both cases you can make valuable discoveries.

# Exercises

1.  In the data table Crime.jmp, data are given for each of the 50 U.S. states concerning crime rates per 100,000 people for seven classes of crimes.

    (a)  Use the Multivariate platform to produce a scatterplot matrix of all seven variables. What pattern do you see in the correlations?

    (b)  Conduct an outlier analysis of the seven variables, and note the states that seem to be outliers. Do the outliers seem to be states with similar crime characteristics?

    (c)  Conduct a principal components analysis on the correlations of these variables. Then, spin the principal components, assigning the $x$-, $y$-, and $z$-axes to the first three principal components. Which crimes seem to group together?

(d) It is impossible to graph all seven variables at once. Using the eigenvalues from the principal components report, how many dimensions would you retain to accurately summarize these crime statistics?

2. The data in Socioeconomic.jmp (SAS Institute 1988) consists of five socioeconomic variables for twelve census tracts in the Los Angeles Standard Metropolitan Statistical Area.

    (a) Use the Multivariate platform to produce a scatterplot matrix of all five variables.

    (b) Conduct a principal components analysis (on the correlations) of all five variables. Considering the eigenvalues produced in the report, how many factors would you use for a subsequent rotation?

    (c) Rotate the number of factors that you determined in part (b). Which variables load on each factor?

3. The file Basketball.jmp (SAS Institute, 1988) contains the pre-season ratings of collegiate basketball teams for the 1985-86 season as given by ten media outlets.

    (a) Look at a scatterplot matrix to determine which media outlets tend to agree on their ratings.

    (b) Evaluate the agreement of the raters by calculating Cronbach's $\alpha$ for this set of data. Does there seem to be a high inter-rater reliability?

# Design of Experiments

## Overview

When you have a problem in industry, you have to do a little science to fix it. When you need something to perform better, you have to do a little science to improve it. When your product is competing in a world market, you need to do a little science to make it the best. When you need evidence that your product is effective, you need to do a little science to prove it. Sometimes, you can use deductions and calculations from known science directly, but more often you need to experiment.

The use of statistical methods in industry is increasing. Arguably, the most cost-beneficial of these methods for quality and productivity improvement is statistical design of experiments. A trial-and-error search for the vital few factors that most affect quality is costly and time consuming. Fortunately, researchers in the field of experimental design have invented powerful and elegant ways of making the search process fast and effective.

The DOE platform in JMP is a tool for creating designed experiments and saving them in a JMP data table. JMP supports two ways to make a designed experiment:

- Build a new design that both matches the description of your engineering problem and remains within your budget for time and material. This method uses the **Custom Design** feature and is the recommended method for creating all designs.

- Choose a pre-formulated design from a list of designs. These lists of designs include **Screening**, **Response Surface**, **Taguchi**, and **Mixture** designs.

This chapter covers how JMP can be used to construct experimental designs and then analyze the results. See the *JMP Design of Experiments* guide in the online help for details about designs not covered in this chapter.

# Introduction

Experimentation is the fundamental tool of the scientific method. In an experiment, a response of interest is measured as some factors are changed systematically, while other factors are held as constant as possible. The goal is to determine if and how the factors affect the response. Each combination of factor settings is a *design point*. An individual response value with its factor settings is sometimes referred to as a *run*.

The challenge is to determine the fewest number of runs needed to gain adequate information for optimizing quality.

JMP IN has tools that allow you to produce almost any kind of experimental design. Several commonly used design are

- *Screening Design*s for scouting many factors: Screening designs examine many factors to see which have the greatest effect on the results of a process. To economize on the number of runs needed, each factor is usually set at only two levels, and response measurements are not taken for all possible combinations of levels. Screening designs are a prelude to further experiments.

- *Response Surface Designs* for optimization: Response surface experiments try to focus on the optimal values for a set of continuous factors. They are modeled with a curved surface so that the maximum point of the surface (optimal response) can be found mathematically.

- *Factorial Design*s: A complete factorial includes a run for all possible combinations of factor levels.

**Note:** The designs in this chapter are commonly used in industrial settings, but there are many other designs used in other settings. The DOE facility can produce additional designs such as Mixture, Taguchi, and mixed-level designs. JMP can analyze other designs, such as split plots, repeated measures, crossovers, balanced incomplete blocks, and Latin squares.

# JMP DOE

The DOE platform in JMP is an environment for describing the factors, responses, and other specifications, creating a designed experiment, and saving it in a JMP table. When you select the DOE tab on the JMP Starter window, you see the list of design command buttons shown on the tab page as in **Figure 16.1**. Alternatively, you can choose the same commands from the DOE main menu shown to the right.

**Figure 16.1**  The DOE JMP Starter Tab

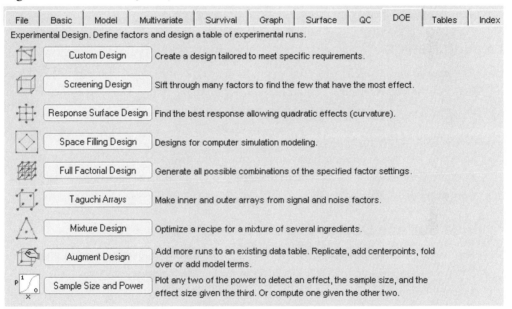

## Custom Design

Custom designs give the most flexibility of all design choices. The Custom Designer gives you the following options:

- continuous factors

- categorical factors with arbitrary numbers of levels

- mixture ingredients

- covariates (factors that already have unchangeable values and design around them)

- blocking with arbitrary numbers of runs per block

- interaction terms and polynomial terms for continuous factors

- inequality constraints on the factors

- choice of number of experimental runs to do, which can be any number greater than or equal to the number of terms in the model

- selecting factors (or combinations of factors) whose parameters are only estimated if possible.

After specifying all your requirements, the Custom Design solution generates an appropriate optimal design for those requirements. In cases where a classical design (such as a factorial) are optimal, the Custom Designer finds them. Therefore, the Custom Designer can serve any number or combination of factors.

## Screening Design

As the name suggests, screening experiments "separate the wheat from the chaff." The wheat is the group of factors having a significant influence on the response. The chaff is the rest of the factors. Typically, screening experiments involve many factors.

The Screening Designer supplies a list of popular screening designs for two or more factors. Screening factors can be continuous or categorical with two or three levels. The list of screening designs also includes designs that group the experimental runs into blocks of equal sizes where the size is a power of two.

## Response Surface Design

Response Surface Methodology (RSM) is an experimental technique invented to find the optimal response within the specified ranges of the factors. These designs are capable of fitting a second-order prediction equation for the response. The quadratic terms in these equations model the curvature in the true response function. If a maximum or minimum exists inside the factor region, RSM can find it. In industrial applications, RSM designs involve a small number of factors. This is because the required number of runs increases dramatically with the number of factors. The Response Surface designer in JMP lists well-known RSM designs for two to eight continuous factors. Some of these designs also allow blocking.

## Full Factorial Design

A full factorial design contains all possible combinations of a set of factors. This is the most conservative design approach, but it is also the most costly in experimental resources. The Full Factorial designer supports both continuous factors and categorical factors with arbitrary numbers of levels.

## Taguchi Arrays

The goal of the Taguchi Method is to find control factor settings that generate acceptable responses despite natural environmental and process variability. In each experiment, Taguchi's design approach employs two designs called the *inner* and *outer* array. The Taguchi experiment is the cross product of these two arrays. The control factors, used to tweak the process, form the inner array. The noise factors, associated with process or environmental variability, form the outer array. Taguchi's signal-to-noise ratios are functions of the observed responses over an outer array. The Taguchi designer in JMP supports all these features of the Taguchi method. The inner and outer array design lists use the traditional Taguchi orthogonal arrays such as L4, L8, L16, and so forth.

## Mixture Design

The Mixture designer lets you define a set of factors that are ingredients in a mixture. You choose among several classical mixture design approaches, such as simplex, extreme vertices, and lattice. For the extreme vertices approach you can supply a set of linear inequality constraints limiting the geometry of the mixture factor space.

## Augment Design

The Augment designer gives the following four choices for adding new runs to existing design:

- add center points

- replicate the design a specified number of times

- create a foldover design

- add runs to the design using a model, which can have more terms than the original model.

The last choice (adding runs to a design) is particularly powerful. You can use this choice to achieve the objectives of response surface methodology by changing a linear model to a full quadratic model and adding the necessary number of runs. For example, suppose you start with a two-factor, two-level, four-run design. If you add quadratic terms to the model and five new points, JMP generates the 3 by 3 full factorial as the optimal augmented design.

# Screening Design Types

Screening experiments can help identify which of many factors most affect a response. Often, screening designs are a prelude to further experiments.

## Two-Level Full Factorial

A full factorial design contains all combinations of the levels of the factors. The samples size is the product of the levels of the factors. For two-level designs, this is $2^k$ where $k$ is the number of factors. This can be expensive if the number of factors is greater than 3 or 4.

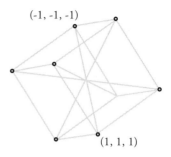

These designs are orthogonal. This means that the estimates of the effects are uncorrelated. If you remove an effect in the analysis, the values of the other estimates remain the same. Their $p$-values change slightly, because the estimate of the error variance and the degrees of freedom are different.

Full factorial designs allow the estimation of interactions of all orders up to the number of factors. Most empirical modeling involves first- or second-order approximations to the true functional relationship between the factors and the responses.

## Two-Level Fractional Factorial

A fractional factorial design also has a sample size that is a power of two. If $k$ is the number of factors, the number of runs is $2^{k-p}$ where $p < k$. Like the full factorial, fractional factorial designs are orthogonal.

The big trade-off in screening designs is between the number of runs and what is often referred to as the *resolution* of the design. If price is no object, you can run several replicates of all possible combinations of $m$ factor levels. This provides a good estimate of everything, including interaction effects to the $m$th degree. But because running experiments costs time and money, you typically only run a fraction of all possible levels. This causes some of the higher-order effects in a model to become nonestimable. An effect is *nonestimable* when it is confounded with another effect. Two effects are said to be *confounded* when the design cannot distinguish which effect affects the response. Confounding also means that an effect is a linear combination of other effects, so that if all the effects are put into the model, there is not a unique least-squares solution for the estimates.

Fractional factorials are designed by planning which interaction effects are confounded with the other interaction effects.

## Resolution Number: The Degree of Confounding

The focus of attention is usually on how the second-order interactions are confounded, because interactions above second-order are rare. The resolution number is a way to describe the degree of confounding. In practice, few experimenters worry about interactions higher than two-way interactions. These higher-order interactions are assumed to be zero. Experiments can therefore be classified by resolution number into three groups:

### Resolution=3

Main effects are not confounded with other main effects, but two-factor interactions are confounded with main effects. Only main effects are included in the model. The two-factor interactions must be assumed to be zero or negligible for the main effects to be meaningful.

### Resolution=4

Main effects are not confounded with each other or with two-factor interactions, but some two-factor interactions can be confounded with each other. Some two-factor interactions can be modeled without being confounded. Other two-factor interactions can be modeled with the understanding that they are confounded with two-factor interactions included in the model. Three-factor interactions are assumed to be negligible.

### Resolution=5

There is no confounding between main effects, between two-factor interactions, or between main effects and two-factor interactions. That is, all two-factor interactions are estimable.

All the fractional factorial designs are *minimum aberration* designs. A minimum aberration design is one in which there are a minimum number of confoundings for a given resolution.

## Plackett-Burman Designs

Plackett-Burman designs are an alternative to fractional factorials for screening. One useful characteristic is that the sample size is a multiple of 4 rather than a power of two. There are no two-level fractional factorial designs with sample sizes between 16 and 32 runs. However, there are 20-run, 24-run, and 28-run Plackett-Burman designs.

The main effects are orthogonal and two-factor interactions are only partially confounded with main effects. This is different from resolution 3 fractional factorial where two-factor interactions are indistinguishable from main effects.

In cases of effect sparsity, a stepwise regression approach can allow for removing some insignificant main effects while adding highly significant and only somewhat correlated two-factor interactions.

# Screening for Main Effects

Acme Piñata Corporation discovered that its piñatas were too easily broken. The company wants to perform experiments to discover what factors might be important for the peeling strength of flour paste. **Strength** refers to how well two pieces of paper that are glued together resist being peeled apart.

> **Note:** You can also generate all types of designs (including Screening designs) with the Custom Designer. The Custom Designer is introduced at the end of this chapter. Once you understand the fundamentals of experimental designs, the Custom Designer is the recommended approach for designing experiments.

## The Factors

Batches of flour paste were prepared to determine the effect of the following nine factors on peeling strength:

Flour: 1/8 cup of white unbleached flour or 1/8 cup of whole wheat flour

Sifted: flour was sifted or not sifted

Type: water-based paste or milk-based paste

Temp: mixed when liquid was cool or when liquid was warm

Salt: formula had a dash of salt or had no salt

Liquid: 4 teaspoons of liquid or 5 teaspoons of liquid

Clamp: pasted pieces were tightly clamped together or not clamped during drying

Sugar: formula contained 1/4 teaspoon or no sugar

Coat: whether the amount of paste applied was thin or thick

The following sections describe how to use the Screening Design platform from the **DOE** menu to generate a design for the flour paste experiment, and then analyze the results.

## Enter and Name the Factors

To begin, select a platform and define the factors of the experiment.

- ⏺ Choose **Screening Design** from the **DOE** tab on the JMP Starter or from the **DOE** main menu to see the Screening dialog in **Figure 16.2**

- ⏺ Complete the dialog as shown to define factors and factor levels.

Note that the **Responses** outline level is initially closed.

🖰   Click the disclosure diamond to open it.

You see one default response called Y.

🖰   Double click on the name and change it to **Strength**.

**Figure 16.2**   Dialog for a Screening Dialog

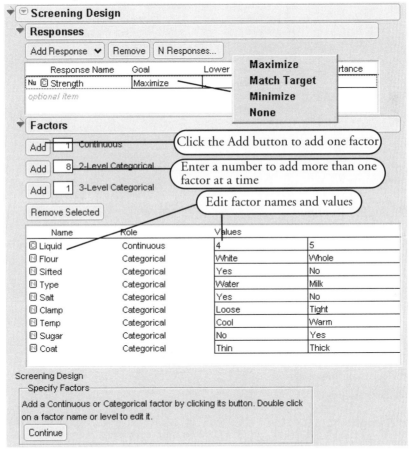

🖰   Next, click **Continue** at the bottom of the Screening Design dialog.

A list of designs appears appended to the Screening Design dialog.

🖰   For the flour paste experiment, select the **16 run Fractional Factorial**, as shown in **Figure 16.3**.

**Figure 16.3**  Screening Design Selections

| Number Of Runs | Block Size | Design Type | Resolution - what is estimable |
|---|---|---|---|
| 12 | | Plackett-Burman | 3 - Main Effects Only |
| 16 | | Fractional Factorial | 3 - Main Effects Only |
| 16 | 8 | Fractional Factorial | 3 - Main Effects Only |
| 16 | 4 | Fractional Factorial | 3 - Main Effects Only |
| 20 | | Plackett-Burman | 3 - Main Effects Only |
| 24 | | Plackett-Burman Folded | 4 - Some 2-factor interactions |
| 32 | | Fractional Factorial | 4 - Some 2-factor interactions |
| 32 | 16 | Fractional Factorial | 4 - Some 2-factor interactions |
| 32 | 8 | Fractional Factorial | 4 - Some 2-factor interactions |
| 32 | 4 | Fractional Factorial | 4 - Some 2-factor interactions |
| 64 | | Fractional Factorial | 4 - Some 2-factor interactions |
| 64 | 32 | Fractional Factorial | 4 - Some 2-factor interactions |
| 64 | 16 | Fractional Factorial | 4 - Some 2-factor interactions |
| 64 | 8 | Fractional Factorial | 4 - Some 2-factor interactions |

Screening Design
9 Factors
Choose a Design

Continue
Back

🖰   Again click **Continue**.

The list of designs is replaced by the Display and Modify Design dialog shown below.

**Change Generating Rules**

controls the choice of different fractional factorial designs for a given number of factors.

**Aliasing of Effects**

shows the confounding pattern for fractional factorial designs.

**Coded Design**

shows the pattern of high and low values for the factors in each run.

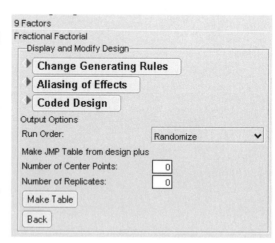

**Design Output Options**

The Design Output Options Panel supplies ways to describe and modify a design.

**Run Order**

controls sorting or randomization through the Run Order Choice popup menu.

### Number of Center Points

lets you add center points by entering the number you want in the edit box. The default is zero.

### Replicates

lets you add the desired number of replicates in an edit box. One replicate doubles the number of runs.

### Make Table

creates a JMP table of the design with columns for the factors and responses.

### Back

removes the Design Output Options Panel and re-displays the list of designs.

## Confounding Structure

Since this experiment needs to look at many factors in a very few runs, the resolution-3 design was selected. This means that main effects are confounded with two-way interactions. It is informative to look at the confounding structure of the design. To do this:

🖰   Open the **Aliasing of Effects** level showing on the Screening Design dialog.

This lists how the design you requested was generated and what factors are confounded. **Figure 16.4** shows both the Aliasing of Effects and the Coded Design.

**Figure 16.4**  Aliasing Structure and Coded Design

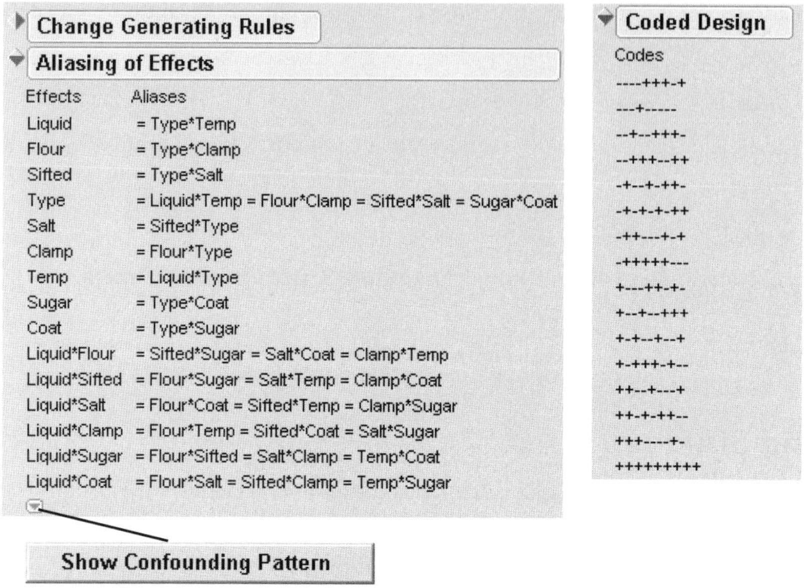

| Change Generating Rules | | Coded Design |
|---|---|---|
| **Aliasing of Effects** | | Codes |

| Effects | Aliases |
|---|---|
| Liquid | = Type*Temp |
| Flour | = Type*Clamp |
| Sifted | = Type*Salt |
| Type | = Liquid*Temp = Flour*Clamp = Sifted*Salt = Sugar*Coat |
| Salt | = Sifted*Type |
| Clamp | = Flour*Type |
| Temp | = Liquid*Type |
| Sugar | = Type*Coat |
| Coat | = Type*Sugar |
| Liquid*Flour | = Sifted*Sugar = Salt*Coat = Clamp*Temp |
| Liquid*Sifted | = Flour*Sugar = Salt*Temp = Clamp*Coat |
| Liquid*Salt | = Flour*Coat = Sifted*Temp = Clamp*Sugar |
| Liquid*Clamp | = Flour*Temp = Sifted*Coat = Salt*Sugar |
| Liquid*Sugar | = Flour*Sifted = Salt*Clamp = Temp*Coat |
| Liquid*Coat | = Flour*Salt = Sifted*Clamp = Temp*Sugar |

```
----+++-+
---+-----
--+--+++-
--+++--++
-+--+-++-
-+-+-+-++
-++---+-+
-+++++---
+---++-+-
+--+--+++
+-+--+---+
+-+++-+--
++--+---+
++-+-++--
+++----+-
+++++++++
```

**Show Confounding Pattern**

# Make a Design Data Table

🖰  The last step to generate the design is to click **Make Table** in the Screening Design dialog.

The generated design now appears in a JMP data table as shown in **Figure 16.5**. It lists 16 runs (observations) out of the $2^9 = 512$ runs possible from combinations of nine factors.

🖰  Save this data table (use **File > Save As**), naming the file MyPaste.jmp.

If you were actually conducting an experiment, you would now print a copy of the design table and use the Y column for entering the results you got from the experimental runs.

The design first included all possible combinations of the first four factors, which resulted in the 16 runs and the $2^4=16$ combinations. The runs are listed in random order. If you sort the table by the Pattern variable, you can see there are eight runs for each Liquid amount. Within each Liquid amount there are four rows for each Flour type. Within Flour type there are two runs per Sifted. Then Type is alternated every run. So for the first four factors entered into the design, there is a complete factorial design.

The last five factors are generated by multiplying internal codes for combinations of the first four factors together, the same as coding for interactions. The codes for factor levels are $-1$

and 1, so any product of the codes is also –1 or 1. For example, the Salt factor is the product of Sifted*Type; Clamp is the product Flour*Type, and so on for the remaining factors, as shown by the Aliasing of Effects table in **Figure 16.4**. If you do not enter factor names and levels into the experimental design, these codes (–1, 1) would show in the design data table (**Figure 16.5**) instead of the level names.

**Figure 16.5**   The Flour Paste Experimental Data

| | Pattern | Liquid | Flour | Sifted | Type | Salt | Clamp | Temp | Sugar | Coat | Strength |
|---|---|---|---|---|---|---|---|---|---|---|---|
| 1 | --+++--++ | 4 | White | Yes | Water | Yes | Loose | Cool | Yes | Thin | • |
| 2 | +-+++-+-- | 5 | White | Yes | Water | Yes | Loose | Warm | No | Thick | • |
| 3 | ---+----- | 4 | White | No | Water | No | Loose | Cool | No | Thick | • |
| 4 | +-+--+--+ | 5 | White | Yes | Milk | No | Tight | Cool | No | Thin | • |
| 5 | -+++++--- | 4 | Whole | Yes | Water | Yes | Tight | Cool | No | Thick | • |
| 6 | ++-+-++-- | 5 | Whole | No | Water | No | Tight | Warm | No | Thick | • |
| 7 | -+-+-+-++ | 4 | Whole | No | Water | No | Tight | Cool | Yes | Thin | • |
| 8 | ++--+---+ | 5 | Whole | No | Milk | Yes | Loose | Cool | No | Thin | • |
| 9 | -+--+-++- | 4 | Whole | No | Milk | Yes | Loose | Warm | Yes | Thick | • |
| 10 | -++---+-+ | 4 | Whole | Yes | Milk | No | Loose | Warm | No | Thin | • |
| 11 | ----+++-+ | 4 | White | No | Milk | Yes | Tight | Warm | No | Thin | • |
| 12 | +++----+- | 5 | Whole | Yes | Milk | No | Loose | Cool | Yes | Thick | • |
| 13 | +--+--+++ | 5 | White | No | Water | No | Loose | Warm | Yes | Thin | • |
| 14 | --+--+++- | 4 | White | Yes | Milk | No | Tight | Warm | Yes | Thick | • |
| 15 | +++++++++ | 5 | Whole | Yes | Water | Yes | Tight | Warm | Yes | Thin | • |
| 16 | +---++-+- | 5 | White | No | Milk | Yes | Tight | Cool | Yes | Thick | • |

The resolution 3 design assumes there are no interactions of consequence, which makes it possible to use a small number of runs to look at a large number of factors. Verify that the design is balanced.

🖱  Choose **Analyze > Distribution** for all nine factor variables.

🖱  Click in each histogram bar to see that the distribution is flat for all the other variables.

The highlighted area representing the distribution for a factor level is equal in each of the other histograms, as shown in **Figure 16.6**.

**Figure 16.6**   Histograms Verify that the Design is Balanced

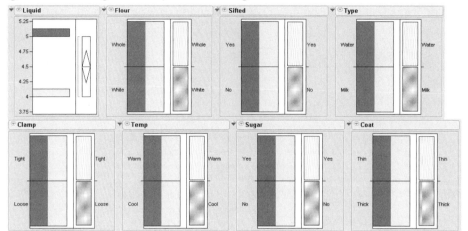

# Perform Experiment and Enter Data

Suppose you complete the experiment and collect and enter the response (Y) data, shown previously in **Figure 16.5**.

🖰   Rather than re-enter the data into the generated design data table, open the sample data table Flrpaste.jmp.

**Note**: You see the values for Strength, shown to the right. These values correspond to the order of the data in the Flrpaste.jmp sample data table, which is different than the table you generated with the Screening Design facility.

| Strength |
|----------|
| 7.8 |
| 17.4 |
| 8.2 |
| 3.7 |
| 6.5 |
| 7.1 |
| 4.3 |

| |
|---|
| 6.1 |
| 4.8 |
| 3.5 |
| 4.3 |
| 2.9 |
| 3.7 |
| 9 |
| 1.8 |
| 4.1 |

You are now ready to analyze the data.

### Examine the Response Data

As usual, a good place to start is by examining the distribution of the response, Y, which is the peel strength.

🖰   Choose **Analyze > Distribution** for the variable Strength, the Y variable.

🖰   When the histogram appears, select **Normal Quantile Plot** from the popup menu on the title bar of the histogram.

You should now see the plots shown to the right.

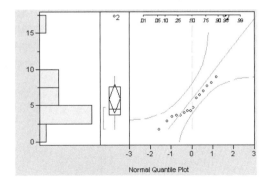

- Drag downward on the upper part of the histogram axis to increase the maximum so that the extreme points show clearly.

- Click on the highest point to identify it as row number 2.

The box plot and the Normal quantile plot are useful for identifying runs that have extreme values. In this case, run 2 has an unusually high peeling strength.

## Analyze a Screening Model

Of the nine factors in the flour paste experiment, there may be only a few that stand out in comparison with the others. The goal of the experiment is to find the factor combinations that optimize the predicted response (peeling strength), it is not to show statistical significance of model effects. This kind of experimental situation lends itself to an effect screening analysis.

- Choose **Analyze > Fit Model**.

- When the Fit Model dialog appears, select Strength as the response (Y) variable.

- Shift-click all the columns from Flour to Coat in the variable selection list and click **Add** to make nine effects in the model.

**Figure 16.7** shows the completed dialog.

- Make sure that **Effect Screening** is selected on the Emphasis popup menu, and then click **Run Model**.

**Figure 16.7**  Fit Model Dialog for Screening Analysis of Flour Paste Main Effects

The Analysis of Variance table to the right shows that the model as a whole is not significant ($p = 0.1314$). The most significant factors are **Flour** and **Type** (of liquid). Note in the Summary of Fit table that the standard deviation of the error (Root Mean Square Error) is estimated as 2.6, a high value relative to the scale of the response. The $R^2$ of 0.79 is not particularly high.

**Note:** You can double-click columns in any report to specify the number of decimals to display.

Following the statistical report, you see a Prediction Profiler, which shows the predicted Y response for each combination of factor settings. **Figure 16.8** shows three manipulations of the Prediction Profiler for the flour experiment. The settings of each factor are connected by a line, called the *prediction trace* or *effect trace*. You can grab and move each vertical dotted line in the Prediction Profile plots to change the factor

### Response Strength

#### Summary of Fit

| | |
|---|---|
| RSquare | 0.794129 |
| RSquare Adj | 0.485323 |
| Root Mean Square Error | 2.641654 |
| Mean of Response | 5.95 |
| Observations (or Sum Wgts) | 16 |

#### Analysis of Variance

| Source | DF | Sum of Squares | Mean Square | F Ratio |
|---|---|---|---|---|
| Model | 9 | 161.51000 | 17.9456 | 2.5716 |
| Error | 6 | 41.87000 | 6.9783 | Prob > F |
| C. Total | 15 | 203.38000 | | 0.1314 |

#### Parameter Estimates

| Term | Estimate | Std Error | t Ratio | Prob>|t| |
|---|---|---|---|---|
| Intercept | 5.95 | 0.660413 | 9.01 | 0.0001 |
| Flour[White] | 1.6875 | 0.660413 | 2.56 | 0.0432 |
| Sifted[No] | -0.625 | 0.660413 | -0.95 | 0.3805 |
| Type[Milk] | -1.525 | 0.660413 | -2.31 | 0.0603 |
| Salt[No] | 0.775 | 0.660413 | 1.17 | 0.2851 |
| Liquid[4] | 1.1375 | 0.660413 | 1.72 | 0.1358 |
| Clamp[Loose] | -0.775 | 0.660413 | -1.17 | 0.2851 |
| Temp[Cool] | -0.6125 | 0.660413 | -0.93 | 0.3895 |
| Sugar[No] | 1 | 0.660413 | 1.51 | 0.1807 |
| Coat[Thick] | 0.8125 | 0.660413 | 1.23 | 0.2646 |

settings. The predicted response automatically recomputes and shows on the vertical axis, and the prediction traces are redrawn.

**Figure 16.8**   Screening Model Prediction Profiler for Flour Paste Experiment

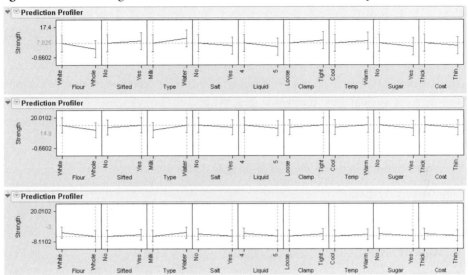

The Prediction Profiler lets you look at the effect on the predicted response of changing one factor setting while holding the other factor settings constant. It can be useful for judging the importance of the factors.

The effect traces in the plot at the top of **Figure 16.8** show a larger response for "White" than for "Whole wheat" and for "Water" than for "Milk", which indicates that changing them to their higher positions increases peel strength.

The second plot in **Figure 16.8** shows what happens if you click and move the effect trace to the high response of each factor; the predicted response changes from 7.825 to 14.9. This occurs with sifted white flour, 4 teaspoons warm water, no salt, no sugar, pasted applied thickly, and clamped tightly while drying.

If you had the opposite settings for these factors (bottom plot in **Figure 16.8**), then the linear model would predict a surface strength of –3 (a clearly impossible response). The prediction is off the original scale, though you can double-click in the *y*-axis area to change the *y* scale of the plot.

Commands in the popup menu on the Response Strength title bar give you other analyses.

Select **Normal Plot** from the **Effect Screening** submenu on the analysis title bar to display the Normal Plot of the parameter estimates shown in **Figure 16.9**.

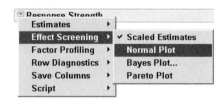

The *Normal Plot* is a Normal quantile plot (Daniel 1959), which shows the parameter estimates on the vertical axis and the Normal quantiles on the horizontal axis. In a screening experiment, you expect most of the effects to be inactive, to have little or no effect on the response. If that is true, then the estimates for those effects are be a realization of random noise centered at zero. What you want is a sense of the magnitude of an effect you should expect when it is truly active instead of just noise. On a Normal Plot, the active effects appear as outliers that lie away from the line that represents Normal noise.

Looking at responses in this way is a valid thing to do for two-level balanced designs, because the estimates are uncorrelated and all have the same variance. The Normal Plot is a useful way (and about the only way) to evaluate the results of a saturated design with no degrees of freedom for estimating the error.

The Normal Plot (**Figure 16.9**) also shows the straight line with slope equal to the *Lenth's PSE* (pseudo standard error) estimate (Lenth 1989). This estimate is formed by taking 1.5 times the median absolute value of the estimates after removing all the estimates greater than 2.75 times the median absolute estimate in the complete set of estimates. Lenth's PSE is computed using the Normalized estimates and disregards the intercept. Effects that deviate substantially from this Normal line are automatically labeled on the plot.

**Figure 16.9** Normal Plot Shows Most Influential Effect

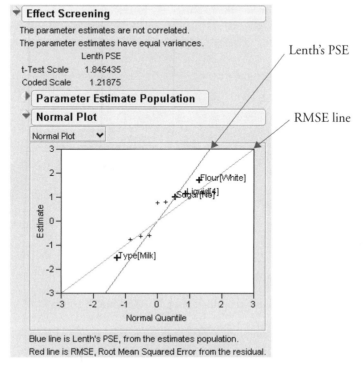

Usually, most effects in a screening analysis have small values; a few have larger values. In the flour paste example, there appears to be only noise, because none of the effects separate from the Normal lines more than would be expected from a Normal distribution of the estimates. This experiment actually has the opposite of the usual distribution. There is a vacant space near the middle of the distribution—none of the estimates are very near zero. (Note in your analysis tables that the Whole-Model $F$-test shows that the model as a whole is not significant.)

So far, the screening experiment has not been conclusive for most of the factors, although white flour does appear to give better results than whole wheat flour. The experiment might bear repeating with better control of variability and a better experimental procedure.

# Screening for Interactions: The Reactor Data

Box, Hunter, and Hunter (1978) discuss a study of nuclear reactors that has five two-level factors, **Feed Rate**, **Catalyst**, **Stir Rate**, **Temperature**, and **Concentration**. The purpose of the study is to find the best combination of settings for optimal reactor output. It is also known that there may be interactions among the factors.

A full factorial for five factors requires $2^5 = 32$ runs. You can generate the design table in JMP, as described in the previous example, by choosing the 32-run design using the Screening Design dialog. Or, you can use the **Full Factorial** selection on the **DOE** main menu as your design choice and define the five factors, as shown to the right.

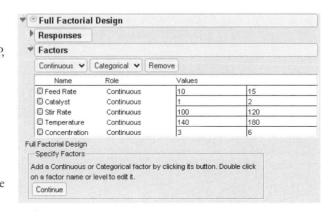

The design generated by JMP DOE and the results of this experiment are in the sample JMP data table called Reactor 32 Runs.jmp.

🖐  To analyze the reactor data, open the table called Reactor 32 Runs.JMP.

**Figure 16.10** shows a partial listing of the data.

**Figure 16.10**   Design Table and Data for Reactor Example

| | Pattern | Feed Rate | Catalyst | Stir Rate | Temperature | Concentration | Y |
|---|---|---|---|---|---|---|---|
| 1 | ----- | 10 | 1 | 100 | 140 | 3 | 61 |
| 2 | ----+ | 10 | 1 | 100 | 140 | 6 | 56 |
| 3 | ---+- | 10 | 1 | 100 | 180 | 3 | 69 |
| 4 | ---++ | 10 | 1 | 100 | 180 | 6 | 44 |
| 5 | --+-- | 10 | 1 | 120 | 140 | 3 | 53 |
| 6 | --+-+ | 10 | 1 | 120 | 140 | 6 | 59 |
| 7 | --++- | 10 | 1 | 120 | 180 | 3 | 66 |
| 8 | --+++ | 10 | 1 | 120 | 180 | 6 | 49 |
| 9 | -+--- | 10 | 2 | 100 | 140 | 3 | 63 |
| 10 | -+--+ | 10 | 2 | 100 | 140 | 6 | 70 |
| 11 | -+-+- | 10 | 2 | 100 | 180 | 3 | 94 |
| 12 | -+-++ | 10 | 2 | 100 | 180 | 6 | 78 |

It is useful to begin the analysis with a quick look at the response data.

🖐  Choose **Analyze > Distribution** to look at the distribution of the response variable Y.

🖐  Select the **Normal Quantile Plot** option from the popup menu on the histogram title bar to see the Normal Plot shown below.

For each run, the **Pattern** variable in the
design table shows the factors' values as a
string of minus signs and plus signs. When
this column is used as a label variable, you can
see the pattern of factor values by clicking on
the point, or by highlighting rows and
selecting **Label/Unlabel** in the **Rows** menu.
The figure to the right shows labels for the
runs with extreme high or low value.

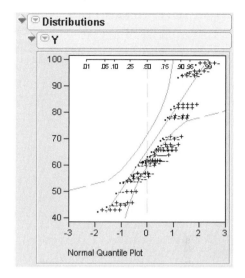

When you use JMP DOE to create a design
table, the table includes a Table Property
called **Model**, which is a JSL script that
generates the correct Fit Model dialog for the
design. The script is accessed by the popup
icon next to the Table Property name, as shown here.

🖑   When you select **Run Script** from this menu, the completed Fit Model dialog in
     **Figure 16.11** appears.

The Model Effects include all Main Effects and all two-factor interactions.

**Figure 16.11**   Fit Model Dialog for Full Factorial Analysis

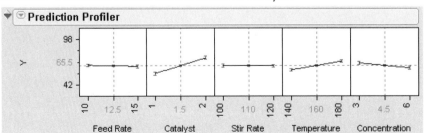

🖰   Click **Run Model** to continue with the analysis.

When you look at the resulting Prediction Profile plot in **Figure 16.12**, you can see that Feed Rate and Stir Rate appear flat. Concentration and Temperature are somewhat weak. Catalyst seems to have the strongest effect over its range.

However, if there are interactions, then this one-at-a-time profile plot may be misleading. The traces can shift their slope and curvature as you change current values. That is what interaction is all about.

**Figure 16.12**   Prediction Profiler for the Reaction Analysis

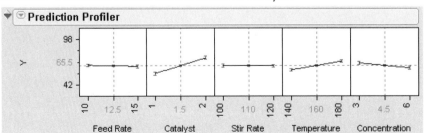

You can find out more about possible interactions with effect screening options.

Choose **Effect Screening > Normal Plot** from the popup menu on the analysis title bar to see what the Normal plot of the parameter estimates has to say (**Figure 16.13**).

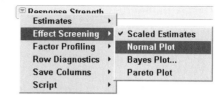

**Figure 16.13** Normal Plot of the Five-Factor Reactor Effects

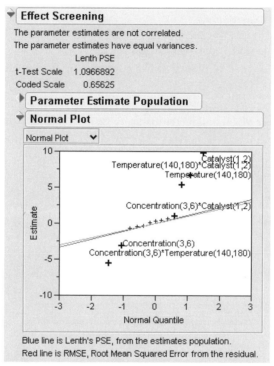

If all effects are due to random noise, they follow a straight line with slope σ, the standard error. The line with slope equal to the Lenth's PSE estimate shows in blue. If there are degrees of freedom for error, a line with slope equal to the root mean squared error (RMSE) shows in red.

The Normal Plot for the Reactor experiment helps you pick out effects that deviate from the Normal lines. In this example, the **Stir Rate** factor does not appear important. However, not only are the three factors **Temperature**, **Catalyst**, and **Concentration** active, but **Temperature** also appears to interact with both **Catalyst** and **Concentration**.

Let's look closer at the interaction of **Concentration** and **Temperature**.

🖰    Scroll again see the Profile Plots for the report.

🖰    In the profile plot, click the levels of **Temperature** repeatedly to alternate its setting from 140 to 180.

Now watch the slope on the profile for **Concentration**. The slopes change dramatically as temperature is changed, which indicates an interaction. When there is no interaction, only the heights of the profile should change, not the slopes. Watch the slope for **Catalyst**, too, because it also interacts with **Temperature**.

The Prediction Profile plots at the top of **Figure 16.14** show responses when temperature is at its low setting. The lower set of plots show what happens when temperature is at its higher setting.

**Figure 16.14**   Effect of Changing Temperature Levels

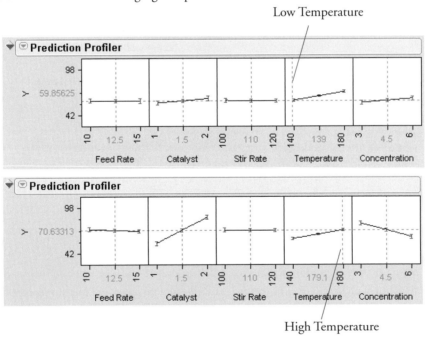

This slope change that is caused by interaction can be seen for all interactions in one picture as follows:

🖰    Use **Factor Profiling > Interaction Plot** from the popup menu on the analysis title bar to **Figure 16.15**.

These are profile plots that show the interactions between all pairs of variables in the reactor experiment.

**Figure 16.15**   Interaction Plots for Five-Factor Reactor Experiment

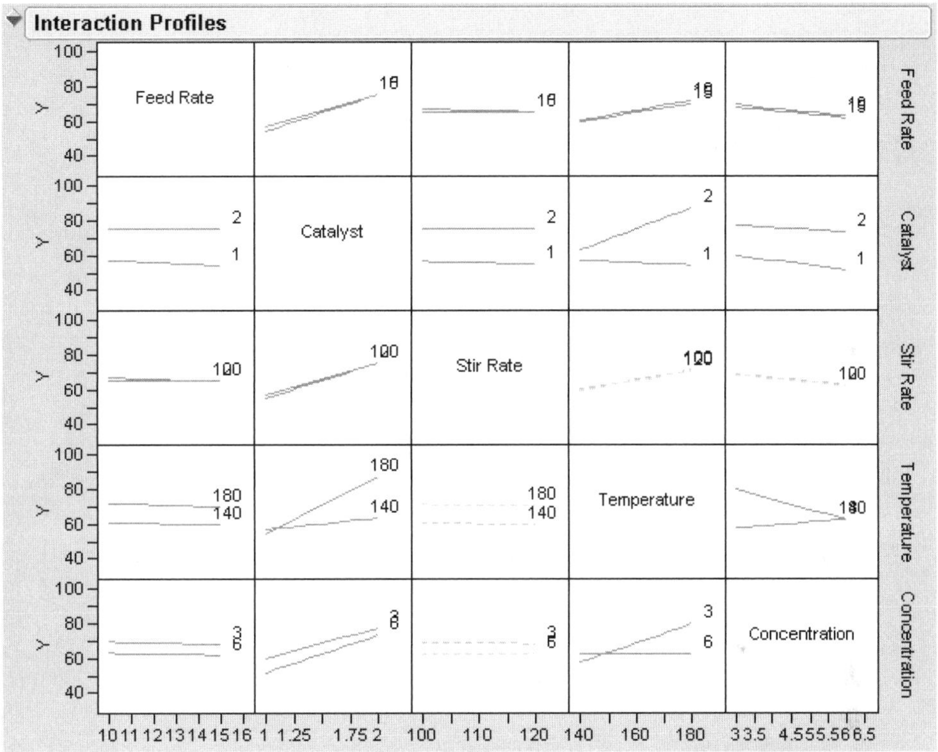

In an interaction plot, the *y*-axes are the response. Each small plot shows the effect of two factors on the response. One factor (associated with the column of the matrix of plots) is on the *x*-axis. This factor's effect shows as the slope of the lines in the plot. The other factor becomes multiple prediction profiles (lines) as it varies from low to high. This factor shows its effect on the response as the vertical separation of the profile lines. If there is an interaction, then the slopes are different for different profile lines, like those in the Temperature by Catalyst plot.

**Note:** The lines of a cell in the interaction plot are dotted when there is no corresponding interaction term in the model.

The Prediction Profile plots in **Figure 16.14** indicated that Temperature is active and Stir Rate is inactive. In the matrix of Interaction Profile plots, look at the Stir Rate column for row Temperature; Temperature causes the lines to separate, but doesn't determine the slope of the lines. Look at the plot when the factors are reversed (row Stir Rate and column Temperature); the lines are sloped, but they don't separate.

Recall that Temperature interacted with Catalyst and Concentration. This is evident by the differing slopes showing in the Temperature by Catalyst and the Temperature by Concentration Interaction Profile plots.

# Response Surface Designs

Response surface designs are useful for modeling a curved (quadratic) surface to continuous factors. If a minimum or maximum response exists inside the factor region, a response surface model can pinpoint it. Three distinct values for each factor are necessary to fit a quadratic function, so the standard two-level designs cannot fit curved surfaces.

The most popular response surface design is the *central composite design*, illustrated by the diagram to the left. It combines a two-level fractional factorial and two other kinds of points defined as follows:

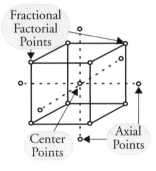

- *Center points*, for which all the factor values are at the midrange value.

- *Axial* (or *star*) points, for which all but one factor are set at midrange and one factor is set at outer (axial) values.

The Box-Behnken design is an alternative to central composite designs. The illustration to the left is of a Box-Behnken design. It combines a fractional factorial with incomplete block designs in such a way as to avoid the extreme vertices and to present an approximately rotatable design with only three levels per factor.

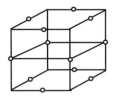

Another important difference between the two design types is that the Box-Behnken design has no points at the vertices of the cube defined by the ranges of the factors. This is sometimes useful when it is desirable to avoid extreme points due to engineering considerations. The price of this characteristic is the higher uncertainty of prediction near the vertices compared to the Central Composite design.

## Response Surface Designs in JMP

To generate a response surface design,

🖰  Choose **DOE > Response Surface Design**, or select Response Surface Design from the DOE tab on the JMP Starter window.

🖰  In the resulting dialog, enter one factor (giving a total of three factors, shown to the left in **Figure 16.16**) and click **Continue**.

This shows the design selection list shown to the right in **Figure 16.16**.

When you generate a response surface design list, you can choose from a Box-Behnken design and two types of central composite designs, called *uniform precision* and *orthogonal*. These properties relate to the number of center points in the design and to the axial values.

> **Note**: You can also generate all types of designs (including Response Surface designs) with the Custom Designer. The Custom Designer is introduced at the end of this chapter. Once you understand the fundamentals of experimental designs, the Custom Designer is the recommended approach for designing.

• Uniform precision means that the number of center points is chosen so that the prediction variance at the center is approximately the same as at the design vertices.

• For orthogonal designs, the number of center points is chosen so that the second order parameter estimates are minimally correlated with the other parameter estimates.

🖰  Select the 20-run central composite design with a block size of 6 (**CCD-Orthogonal Blocks**).

**Figure 16.16**  Design Dialog to Specify Factors and Choose Design Type

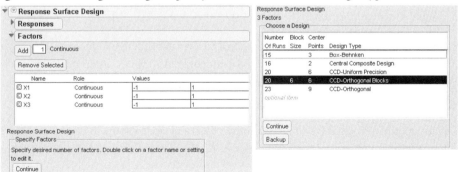

🖰  After selecting a design, again click **Continue**, which appends the Display and Modify Design panel to the Response Surface Design dialog, as shown in **Figure 16.17**.

🖰 Click the **Rotatable** radio button, which generates a design in which the variance of prediction depend only on the scaled distance from the center of the design.

Because the selected design includes a blocking factor, the default run order is **Randomize within Blocks**.

**Figure 16.17**   Design Dialog to Modify Order of Runs

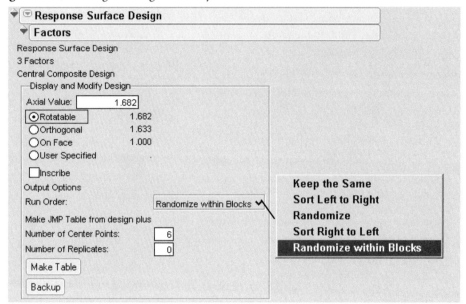

🖰 Click **Make Table** to see a 20-run central composite JMP design table similar to the one shown in **Figure 16.18**.

**Note**: Because you changed from an orthogonal to a rotatable design, a message advises you that the resulting design might no longer be orthogonal. Also, your design might appear in a different order because of the randomization within blocks. However, it will have the same number of runs.

This design table has eight factorial runs in two blocks of size four, one block of size six with additional center points, and has a resolution of 6. This design is rotatable as specified, but might not be orthogonal.

**Figure 16.18**   Design Table for a Central Composite Response Surface Design

| # | Pattern | Block | X1 | X2 | X3 | Y |
|---|---------|-------|-----|-----|-----|---|
| 1 | -++ | 1 | -1 | 1 | 1 | • |
| 2 | 000 | 1 | 0 | 0 | 0 | • |
| 3 | --- | 1 | -1 | -1 | -1 | • |
| 4 | +-+ | 1 | 1 | -1 | 1 | • |
| 5 | 000 | 1 | 0 | 0 | 0 | • |
| 6 | ++- | 1 | 1 | 1 | -1 | • |
| 7 | +-- | 2 | 1 | -1 | -1 | • |
| 8 | -+- | 2 | -1 | 1 | -1 | • |
| 9 | +++ | 2 | 1 | 1 | 1 | • |
| 10 | 000 | 2 | 0 | 0 | 0 | • |
| 11 | 000 | 2 | 0 | 0 | 0 | • |
| 12 | --+ | 2 | -1 | -1 | 1 | • |
| 13 | 00a | 3 | 0 | 0 | -1.6817928 | • |
| 14 | 000 | 3 | 0 | 0 | 0 | • |
| 15 | 0a0 | 3 | 0 | -1.6817928 | 0 | • |
| 16 | a00 | 3 | -1.6817928 | 0 | 0 | • |
| 17 | A00 | 3 | 1.68179283 | 0 | 0 | • |
| 18 | 0A0 | 3 | 0 | 1.68179283 | 0 | • |
| 19 | 000 | 3 | 0 | 0 | 0 | • |
| 20 | 00A | 3 | 0 | 0 | 1.68179283 | • |

(Side panels: Central Compos… / Design Central Co / Model; Columns (6/0): Pattern, Block, X1, X2, X3, Y; Rows: All Rows 20, Selected 0, Excluded 0, Hidden 0, Labelled 0)

Like all JMP tables generated by the DOE facility, the Table Property called **Model** contains the JSL script to generate the completed Fit Model dialog to analyze the design after data are collected.

 Select **Run Script** from the **Model** Table Property to see the completed dialog in **Figure 16.19**.

This completed Fit Model dialog includes all the effects for a response surface model. Each main effect has &RS appended to its name, and there is a crossed effect for each combination of main effects, which tells JMP to analyze the data as a response surface model.

The script completes the dialog automatically, but you can generate a response surface analysis for any effects by selecting them in the Select Columns list and choosing **Response Surface** from the effect **Macros**, as illustrated in **Figure 16.19**.

**Figure 16.19**  Fit Model Dialog for Response Surface Design

① Select variables here

**Model Specification**

Select Columns
- Pattern
- Block
- X1
- X2
- X3
- Y

Pick Role Variables

Y    | Y
     | optional

Weight | optional Numeric
Freq   | optional Numeric
By     | optional

Personality: Standard Least Squares
Emphasis: Effect Screening

Help    Run Model
Remove

② Select **Response Surface** macro

- Full Factorial
- Factorial to degree
- Factorial sorted
- **Response Surface**
- Mixture Response Surface
- Polynomial to Degree

Construct Model Effects

Add
Cross
Nest
Macros ✓
Degree 2
Attributes
☐ No Intercept

X1& RS
X2& RS
X3& RS
X1*X2
X1*X3
X2*X3
X1*X1
X2*X2
X3*X3
Block

The completed design is generated

# A Box-Behnken Design Example

Suppose the objective of an industrial experiment is to minimize the unpleasant odor of a chemical. It is known that the odor varies with temperature (temp), gas-liquid ratio (gl ratio), and packing height (ht). The experimenter wants to collect data over a wide range of values for these variables to see if a response surface can identify values that give a minimum odor (John, 1971).

First, generate a response surface experimental design:

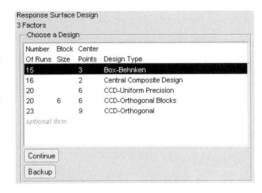

🖱 Choose **DOE > Response Surface Design** and add one factor for a total of three factors (two factors are show by default), then click **Continue**.

🖱 Open the Response panel and edit the name of the response. Change it from Y to Odor.

Response Surface Design
3 Factors
Choose a Design

| Number Of Runs | Block Size | Center Points | Design Type |
|---|---|---|---|
| 15 | | 3 | Box-Behnken |
| 16 | | 2 | Central Composite Design |
| 20 | | 6 | CCD-Uniform Precision |
| 20 | 6 | 6 | CCD-Orthogonal Blocks |
| 23 | | 9 | CCD-Orthogonal |
| optional item | | | |

Continue
Backup

Select the 15-run Box-Behnken design, then again click **Continue**.

The Box-Behnken design selected for three effects generates the design table of 15 runs shown in **Figure 16.20**. When the experiment is conducted, the responses are entered into the JMP table.

Suppose the factor names have been edited, the experiment has been conducted, and the results entered into the Odor column, as shown in **Figure 16.20**.

Open Odor.jmp to see the finished data table ready for analysis.

**Figure 16.20** A Three-Factor Box-Behnken Design Table with Data for Analysis

Odor
Design Box-Behnken
Model

Columns (5/0)
Pattern
temp
gl ratio
ht
odor

Rows
All Rows 15
Selected 0
Excluded 0
Hidden 0
Labelled 0

| | Pattern | temp | gl ratio | ht | odor |
|---|---|---|---|---|---|
| 1 | --0 | -1 | -1 | 0 | 66 |
| 2 | -+0 | -1 | 1 | 0 | 58 |
| 3 | +-0 | 1 | -1 | 0 | 65 |
| 4 | ++0 | 1 | 1 | 0 | -31 |
| 5 | 0-- | 0 | -1 | -1 | 39 |
| 6 | 0-+ | 0 | -1 | 1 | 17 |
| 7 | 0+- | 0 | 1 | -1 | 7 |
| 8 | 0++ | 0 | 1 | 1 | -35 |
| 9 | -0- | -1 | 0 | -1 | 43 |
| 10 | +0- | 1 | 0 | -1 | -5 |
| 11 | -0+ | -1 | 0 | 1 | 43 |
| 12 | +0+ | 1 | 0 | 1 | -26 |
| 13 | 000 | 0 | 0 | 0 | 49 |
| 14 | 000 | 0 | 0 | 0 | -40 |
| 15 | 000 | 0 | 0 | 0 | -22 |

Before you analyze the data, look at the design structure with the Spinning Plot platform:

With Odor.jmp as the current data table, choose **Graph > Spinning Plot** for the three factors temp, gl ratio, and ht.

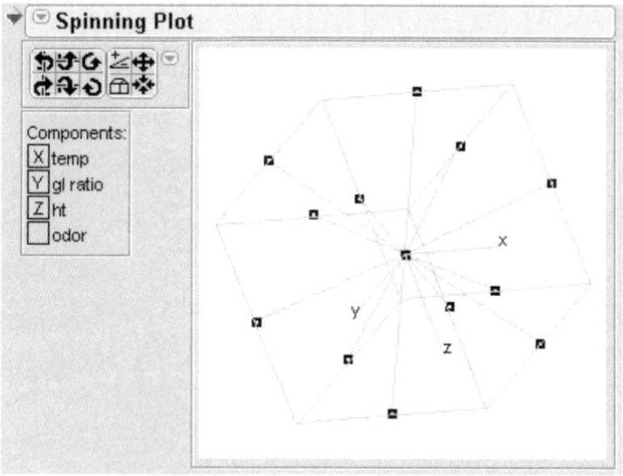


To see the Spinning Plot shown to the right:

Select the **Axis**, **Box** and **Rays** options found in the popup menu on the spin control panel.

🖰    Rotate the plot with the hand tool or the spin controls.

🖰    Highlight the points and change the markers with **Row > Markers.**

🖰    Use the zoom button on the spin control panel to move the points to the edges of the box.

 Zoom Button

As you can see from the spinning plot, this design has points that are extreme in two factors, but midrange in the third. For a three-factor design, this has the effect of locating points midway along the edges connecting the extreme vertices of the factor space.

To analyze the data, use the JSL script automatically generated by the DOE facility and saved with the JMP design table.

🖰    Select **Run Script** from the **Model** popup menu.

The effects appear in the Model Effects list as shown in **Figure 16.21**, with the &RS notation on the main effects (temp, gl ratio, and ht). This notation indicates that these terms are to be subjected to a curvature analysis.

🖰    Click **Run Model** on the Fit Model dialog to see the analysis.

The standard least squares analysis tables appear with an additional report outline level called Response Surface.

🖰    Open the Response Surface outline level to see the tables shown in **Figure 16.21**.

- The first table is a summary of the parameter estimates.

- The Solution table lists the critical values of the surface and tells the kind of solution (maximum, minimum, or saddle point). The critical values are where the surface has a slope of zero, which could be an optimum depending on the curvature.

- The Canonical Curvature table shows eigenvalues and eigenvectors of the effects. The eigenvectors are the directions of the principal curvatures. The eigenvalue associated with each direction tells whether it is decreasing slope, like a maximum (negative eigenvalue), or increasing slope, like a minimum (positive eigenvalue).

The Solution table in this example shows the solution to be a saddle point and also warns that the critical values given by the solution are outside the range of data values.

**Figure 16.21**  Response Surface Model and Analysis Results

Construct Model Effects

| | |
|---|---|
| Add | temp& RS |
| | gl ratio& RS |
| Cross | ht& RS |
| Nest | temp*gl ratio |
| | temp*ht |
| Macros ▼ | gl ratio*ht |
| Degree 2 | temp*temp |
| Attributes ▽ | gl ratio*gl ratio |
| ☐ No Intercept | ht*ht |

▼ ⊟ **Response odor**

▼ **Response Surface**

Coef

| | temp | gl ratio | ht | odor |
|---|---|---|---|---|
| temp | 25.291667 | -22 | -5.25 | -25.875 |
| gl ratio | . | 18.541667 | -5 | -23.5 |
| ht | . | . | -7.208333 | -10.625 |

▼ **Solution**

| Variable | Critical Value |
|---|---|
| temp | 0.769112 |
| gl ratio | 0.910289 |
| ht | -1.332783 |

Solution is a SaddlePoint

Critical values outside data range

Predicted Value at Solution  -17.89921

▼ **Canonical Curvature**

Eigenvalues and Eigenvectors

| Eigenvalue | 33.4325 | 11.1061 | -7.9136 |
|---|---|---|---|
| temp | 0.80549 | 0.57914 | 0.12561 |
| gl ratio | -0.59241 | 0.79249 | 0.14497 |
| ht | -0.01559 | -0.19119 | 0.98143 |

# Plotting Surface Effects

If there are more than two factors, you can see a contour plot of any two factors at intervals of a third factor by using the Contour Profiler.

🖰  Choose **Factor Profiling > Contour Profiler** found in the popup menu on the report title, as shown here.

The **Contour Profiler** displays a panel that lets you use interactive sliders to vary one factor's values and observe the effect on the other two factors. You can also vary one factor

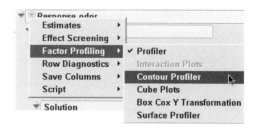

and see the effect on a mesh plot of the other two factors. **Figure 16.22** shows contours of ht as a function of temp and gl ratio. The mesh plots on the right in **Figure 16.22** show the response surface in the combinations of the three factors taken two at a time.

🖰  Enter zero as the Hi Limit specification.

Entering a value defines and shades the region of acceptable values for the three variables.

🖱 Optionally, use the **Contour Grid** option in the Contour Profiler popup to add grid
lines with specified values to the contour plot.

**Figure 16.22**   Contour Profiler with Contours and Mesh Plot

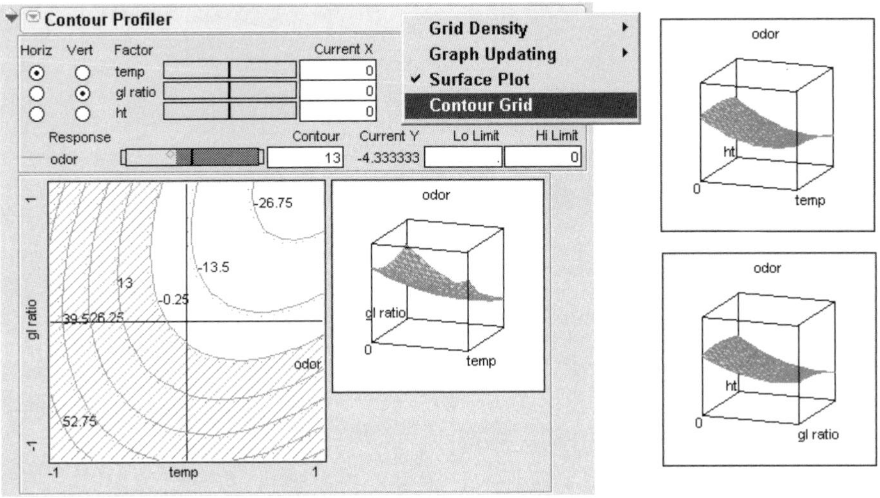

# Design Issues

So far, the discussion has been on the particulars of how to get certain designs. But why are
these designs good? Here is some general advice, though all these points have limitations or
drawbacks as well.

*Keep the design balanced.* Allocate an equal number of runs to each factor value. There are two
reasons to do this:

- A balanced design achieves the most power.

- A balanced design keeps the estimates uncorrelated so that tests on one parameter are
  independent of tests on another.

*Take values at the extremes of the range.* This is done for two reasons:

- It maximizes the power. By separating the values, you are increasing the parameter for
  the difference, making it easy to distinguish it from zero.

- It keeps the applicable range of the experiment wide. When you use the prediction
  formula outside the range of the data, its variance is high, and it is unreliable for other
  reasons.

*Put runs at all combinations of extreme levels.* If you can't afford this, try to come close to this by designing so that many subsets of variables cover all combinations of levels.

*Put a few runs in the center too,* if this makes sense.

- This allows for estimation of curvature.

- If you have several runs at the same point, you can estimate pure error and do a lack-of-fit test.

*Randomize the assignment of runs.* This is an attempt to neutralize any inadvertent effects that may be present in a given sequence.

## Balancing

Let's illustrate the balancing issue: In the simplest case of two groups, the standard error of the difference between the two means is proportional to the square root of the sum of the reciprocals of the sample size, as shown in the formula to the right.

$$\sqrt{\frac{1}{n1} + \frac{1}{n2}}$$

The plot shown to the right is a graph of this formula as a function of *n1*, where *n1* + *n2* is kept constant at 20. You can see that the smallest variance occurs when the design is balanced; where *n1* = *n2* = 10.

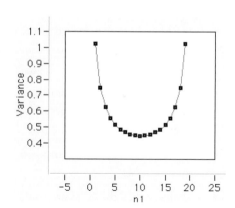

You can clearly see that if the design is balanced, you minimize the standard errors of differences, and thus maximize the power of the tests.

## Wide Range

Let's illustrate why taking a wide range is good.

The data table shown to the right is a simple arrangement of 12 points. There are 4 points at extreme values of X, 4 points at moderate range, and 4 points at the center. Examine subsets of the 12 points, to see which ones perform better:

| | X | Y |
|---|---|---|
| 1 | -4 | -1 |
| 2 | -4 | -3 |
| 3 | -2 | -2 |
| 4 | -2 | 0 |
| 5 | 0 | -1 |
| 6 | 0 | -1 |
| 7 | 0 | 1 |
| 8 | 0 | 1 |
| 9 | 2 | 0 |
| 10 | 2 | 2 |
| 11 | 4 | 1 |
| 12 | 4 | 3 |

- Exclude the extremes of ±4.

- Exclude the intermediate points of ±2.

- Exclude the center points.

To exclude points,

⌐ Highlight the rows to exclude in the data table and select **Rows > Exclude**.

**Figure 16.23** compares the confidence interval of the regression line in the three subsets. Each situation has the same number of observations, the same error variance, and the same residuals. The only difference is in the spacing of the X values, but the subsamples show the following differences:

- The fit excluding the points at the extremes of ±4 has the widely flared confidence curves. The confidence curves cross the horizontal line at the mean, indicating that it is significant at 0.05.

- The fit excluding the intermediate range points of ±2 has much less flared confidence curves. The confidence curves do not cross the horizontal line at the mean, indicating that it is not significant at 0.05.

- The fit excluding the center points was the best, though not much different than the previous case.

- All the fits had the same size confidence limits in the middle.

For given sample size, the better design is the one that spaces out the points farther.

**Figure 16.23**  Confidence Interval of the Regression Line

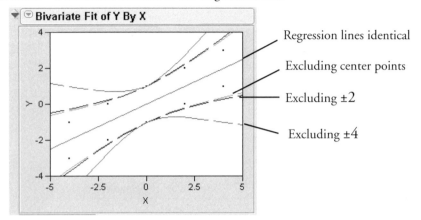

Regression lines identical

Excluding center points

Excluding ±2

Excluding ±4

## Center Points

Using the same data as before, in the same three scenarios, look at a quadratic curve instead of a line. Note the confidence limits for the curve that excluded the center points have a much wider interval at the middle. Curvature is permitted by the model but is not well supported by the design.

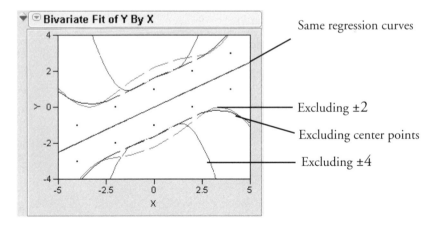

Same regression curves

Excluding ±2

Excluding center points

Excluding ±4

Center points are also usually replicated points that allow for an independent estimate of pure error, which can be used in a lack-of-fit test. For the same number of points, the design with center points can detect lack-of-fit and nonlinearities.

# The JMP Custom Designer

The JMP Custom Designer builds a design for your specific problem that is consistent with your resource budget. You can use the Custom Designer for routine factor screening, response optimization, and mixture problems. Also, the Custom Designer can find designs for special conditions not covered in the lists of predefined designs.

**DOE**

- 🗗 Custom Design ————— Create a design to solve a problem
- 🗗 Screening Design ——
- ✛ Response Surface Design
- 🗗 Full Factorial Design ——— Catalog of classical designs
- ⬡ Taguchi Arrays
- △ Mixture Design ———
- 🗗 Augment Design ——————— Change an existing design
- 🗗 Sample Size and Power

This section guides you through the interface of the JMP Custom Designer. You interact with this facility to describe your experimental situation, and JMP creates a design that fits your requirements. It shows how to use the Custom Design interface to build a design using this easy, step-by-step approach.

Using the Custom Designer involves these key steps:

1. Enter and name one or more responses, if needed. The DOE dialog always begins with a single response, called Y, and the Response panel is closed by default.

2. Use the Factors panel to name and describe the types of factors you have.

3. Enter factor constraints, if there are any.

4. Choose a model.

5. Modify the sample size alternatives.

6. Choose the run order.

7. Optionally, add center points and replicates.

   You can use the Custom Design dialog to enter main effects, then add interactions, and specify center points and replicates.

### Define Factors in the Factors Panel

When you select **Custom Design** from the DOE menu, or from the DOE tab on the JMP Starter, the dialog in **Figure 16.24** appears. One way to enter factors is to click the **Add N Factors** text edit box and enter the number of continuous factors you want. If you want other kinds of factors, click **Add Factor** and select a factor type: **Continuous, Categorical, Blocking, Covariate, Mixture,** or **Constant.**

**Figure 16.24** Select Custom Design and Enter Factors

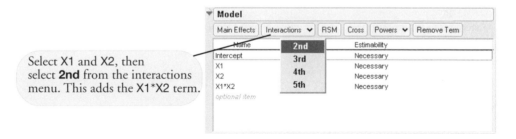

When you finish defining factors, Click **Continue** in the Factors panel to proceed to the next step.

### Describe the Model in the Model Panel

When you click **Continue**, the Model panel initially appears with only the main effects corresponding to the factors you entered. Next, you might want to enter additional effects to estimate. That is, if you do not want to limit your model to main effects, you can add factor interactions or powers of continuous factors to the model.

The example shown above has two continuous factors, X1 and X2. If you click **Continue**, the current Model panel appears with only those factors. The Model panel has buttons for you to add specific factor types to the model. For example, when you select **2nd** from the **Interactions** popup menu, the X1*X2 interaction term is added to the list of model effects.

### The Design Generation Panel

As you add effects to the model, the Design Generation panel shows the minimum number of runs needed to perform the experiment. It also shows alternate numbers of runs, or lets you choose your own number of runs. Balancing the cost of each run with the information gained by extra runs you add is a judgment call that you control.

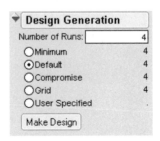

The Design Generation panel has the following radio buttons:

- **Minimum** is the number of terms in the design model. The resulting design is saturated (no degrees of freedom for error). This is the most risky choice. Use it only when the cost of extra runs is prohibitive.

- **Default** is a custom design suggestion for the number of runs. This value is based on heuristics for creating balanced designs with the least number of additional runs above the minimum.

- **Compromise** is a second suggestion that is more conservative than the **Default**. Its value is generally between **Default** and **Grid**.

- **Grid**, in most cases, shows the number of points in a full-factorial design. Exceptions are for mixture and blocking designs. Generally, **Grid** is unnecessarily large and is included as an option for reference and comparison.

- **User Specified** highlights the **Number of Runs** text box. You key in a number of runs that is at least the minimum.

After specifying the number of runs, click **Make Design** to see the factor design layout and the Design panel appended to the Model panel in the DOE dialog.

# Modify a Design Interactively

There is a **Back** button at several stages in the design dialog that allows you to go back to a previous step and modify the design. For example, you can modify a design by adding quadratic terms to the model, removing the center points, and removing the replicate. **Figure 16.25** shows the steps to modify a design interactively.

When you click **Continue**, the Design panel shows with 8 runs as default. If you choose the **Grid** option, the design that results has 9 runs.

**Figure 16.25** Back up to Interactively Modify a Design

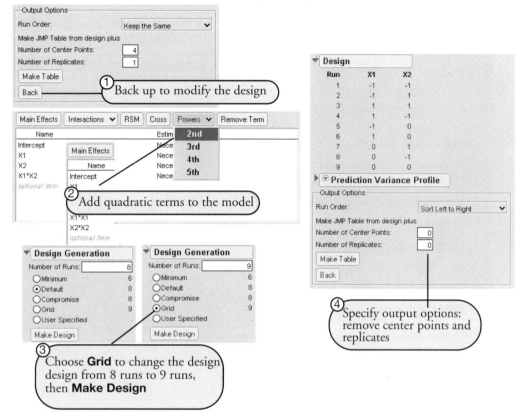

# The Prediction Variance Profiler

Although most design types have at least two factors, the following examples have a single continuous factor and compare designs for quadratic and cubic models. The purpose of these examples is to introduce the prediction variance profile plot.

### A Quadratic Model

Follow these steps to create a simple quadratic model with a single continuous factor.

- Select **DOE > Custom Design.** Add one continuous factor and click **Continue**.

- Select **2nd** from the Powers popup menu in the Model panel to create a quadratic term.

- Use the default number of runs, 6, and click **Make Design**.

When the design appears,

🖰 Open the **Prediction Variance Profile**.

For continuous factors, the initial setting is at the mid-range of the factor values. For categorical factors, the initial setting is the first level. If the design model is quadratic, then the prediction variance function is quartic. The three design points are –1, 0, and 1. The prediction variance profile shows that the variance is a maximum at each of these points, on the interval –1 to 1. The $y$-axis is the relative variance of prediction of the expected value of the response.

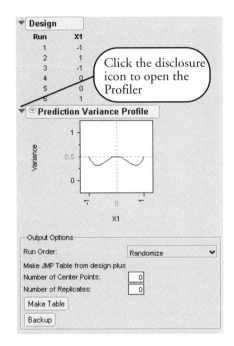

The prediction variance is relative to the error variance. When the prediction variance is 1, the absolute variance is equal to the error variance of the regression model.

When you choose a sample size, you are deciding how much variance in the expected response you are willing to tolerate. As the number of runs increases, the prediction curve (prediction variance) decreases.

🖰 To compare profile plots, **Back** and choose **Minimum** in the Design Generation panel, which gives a sample size of 3.

This produces a curve that has the same shape as the previous plot, but the maxima are at 1 instead of 0.5. **Figure 16.26** compares plots for sample size 6 and sample size 3 for this quadratic model example. You can see the prediction variance increase as the sample size decreases. These profiles are for middle variance and lowest variance, for sample sizes 6 (top charts) and 3 (bottom charts).

**Figure 16.26**   Comparison of Prediction Variance Profiles

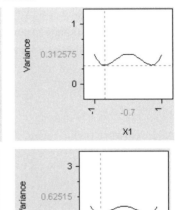

**Note:** You can double-click on the axis to set a factor level precisely.

🖱  For a final look at the Prediction Variance Profile for the quadratic model, **Back** and enter a sample size of 4 in the Design Generation panel and click **Make Design**.

The sample size of 4 adds a point at −1 (**Figure 16.27**). Therefore, the variance of prediction at −1 is lower (half the value) than the other sample points. The symmetry of the plot is related to the balance of the factor settings. When the design points are balanced, the plot is symmetric, like those in **Figure 16.26**. When the design is unbalanced, the prediction plot is not symmetric, as shown in **Figure 16.27**.

**Figure 16.27**   Sample Size of Four for the One-Factor Quadratic Model

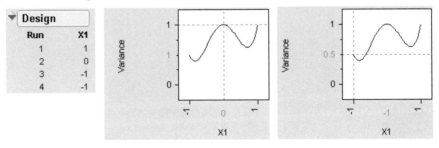

## A Cubic Model

The runs in the quadratic model are equally spaced. This is not true for the single-factor cubic model shown in this section. To create a one-factor cubic model, follow the same steps as in "A Quadratic Model" on page 437.

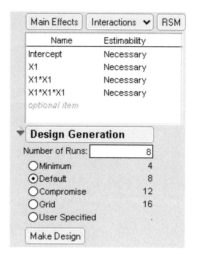

🖰   In addition, add a cubic term to the model with the **Powers** popup menu.

Use the Default number of runs in the Design Generation panel.

🖰   Click **Make Design** to continue.

🖰   Open the Prediction Variance Profile Plot to see the Prediction Variance Profile and its associated design shown in **Figure 16.28**.

The cubic model has a variance profile that is a 6th degree polynomial.

Note that the points are not equally-spaced in X. It is not obvious that this design has a better prediction variance profile than the equally-spaced design with the same number of runs.

**Figure 16.28** Comparison of Prediction Variance Profiles

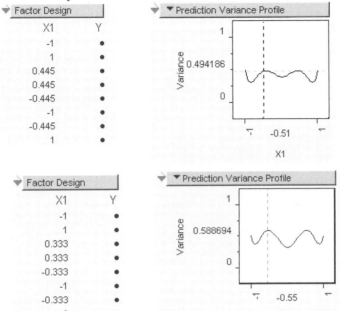

# Routine Screening Using Custom Designs

You can use the Screening designer to create screening designs, as illustrated at the beginning of this chapter. You can also generate screening designs with the Custom Designer. The Custom Designer is a general-purpose design environment. The first example below shows the steps to generate a main-effects-only screening design, an easy design to create and analyze.

### Main Effects Only

🖱 Select **DOE > Custom Design**.

🖱 Enter the number of factors you want (six for this example) into the Factors panel and click **Continue** as shown in **Figure 16.29**.

This example uses six factors. Because there are no complex terms in the model, no further action is needed in the Model panel. The default number of runs (8) is correct for the main-effects-only model.

The result is a resolution-3 screening design. All main effects are estimable but are confounded with two-factor interactions.

🖱 Click **Make Design** to see the Factor Design table in **Figure 16.29.**

**Figure 16.29**  A Main Effects Only Screening Design

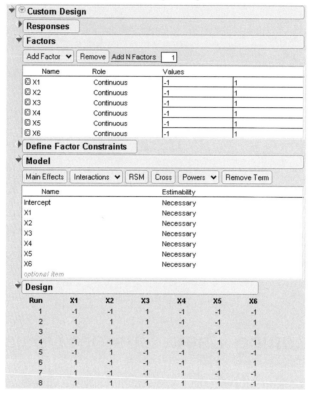

## All Two-Factor Interactions Involving Only One Factor

Sometimes there is reason to believe that some two-factor interactions may be important. The following example illustrates adding all the two-factor interactions involving a single factor. The example has five continuous factors.

This design is a resolution-4 design equivalent to folding over on the factor for which all two-factor interactions are estimable.

To get a specific set of crossed factors (rather than all interactions or response surface terms),

🖱 Select the factor to cross (**X1**, for example) in the Factors table.

🖱 Select the other factors in the Model Table and click **Cross** to see the interactions in the model table, as shown in **Figure 16.30**.

The default sample size for designs with only two-level factors is the smallest power of two that is larger than the number of terms in the design model. For example, in **Figure 16.30**, there are 9 terms in the model, so $2^4=16$ is the smallest power of two that is greater than 9.

**Figure 16.30** Two-Factor Interactions That Involve Only One of the Factors

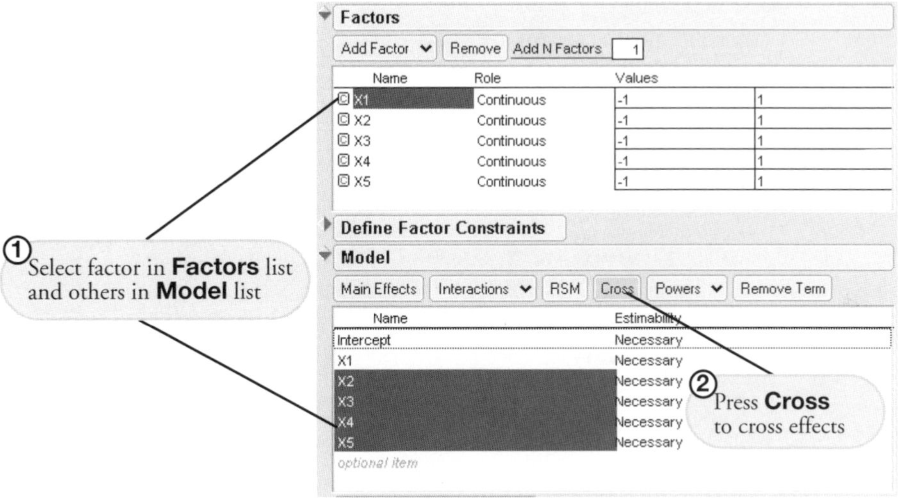


Design

| Run | X1 | X2 | X3 | X4 | X5 |
|---|---|---|---|---|---|
| 1 | -1 | -1 | -1 | 1 | -1 |
| 2 | 1 | -1 | 1 | -1 | -1 |
| 3 | 1 | 1 | -1 | -1 | -1 |
| 4 | 1 | -1 | -1 | 1 | -1 |
| 5 | -1 | -1 | -1 | -1 | 1 |
| 6 | 1 | 1 | 1 | 1 | 1 |
| 7 | -1 | 1 | -1 | -1 | -1 |
| 8 | -1 | -1 | 1 | -1 | -1 |
| 9 | 1 | 1 | 1 | 1 | -1 |
| 10 | -1 | -1 | 1 | 1 | 1 |
| 11 | 1 | -1 | -1 | 1 | 1 |
| 12 | 1 | -1 | 1 | -1 | 1 |
| 13 | -1 | 1 | 1 | 1 | -1 |
| 14 | 1 | 1 | -1 | -1 | 1 |
| 15 | -1 | 1 | -1 | 1 | 1 |
| 16 | -1 | 1 | 1 | -1 | 1 |

## All Two-Factor Interactions

In situations where there are few factors and experimental runs are cheap, you can run screening experiments that allow for estimating all the two-factor interactions. The Custom Design interface makes this simple (see **Figure 16.31**).

- Enter the number of factors (5 in this example).
- Click **Continue** and choose **2nd** from the **Interactions** popup in the Model outline.

🖰 Click **Make Design**.

**Figure 16.31** shows a partial listing of the two-factor design with all interactions. The default design has the minimum-power-of-two sample size consistent with fitting the model.

**Figure 16.31** All Two-Factor Interactions

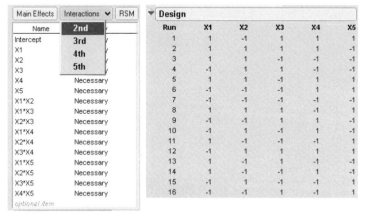

# Special Topic: How the Custom Designer Works

The Custom Designer starts with a random design with each point inside the range of each factor. The computational method is an iterative algorithm called *coordinate exchange*. Each iteration of the algorithm involves testing every value of each factor in the design to determine if replacing that value increases the optimality criterion. If so, the new value replaces the old. Iteration continues until no replacement occurs in an entire iterate.

To avoid converging to a local optimum, the whole process is repeated several times using a different random start. The designer displays the best of these designs.

Sometimes a design problem can have several equivalent solutions. Equivalent solutions are designs having equal precision for estimating the model coefficients as a group. When this is true, the design algorithm will generate different (but equivalent) designs if you press the **Back** and **Make Design** buttons repeatedly.

# Exploratory Modeling 17

## Overview

*Exploratory modeling* (sometimes known as *data mining*) is the process of exploring large amounts of data, usually using an automated method, to find patterns and discoveries. JMP has two platforms especially designed for exploratory modeling: The Partition platform and the Neural Net platform.

The Partition platforms recursively partitions data, automatically splitting the data at optimum points. The result is a decision tree that classifies each observation into a group. The classic example is turning a table of symptoms and diagnoses of a certain illness into a hierarchy of assessments to be evaluated on new patients.

The Neural Net platform implements a standard type of neural network. Neural nets are used to predict one or more response variables from a flexible network of functions of input variables. They can be very good predictors, and are useful when the underlying functional form of the response surface is not important.

# The Partition Platform

The Partition platform is used to recursively partition a data set in ways similar to CART[TM], CHAID[TM], and C4.5. The technique is often taught as a data mining technique, because

- it is good for exploring relationships without having a good prior model

- it handles large problems easily, and

- the results are very interpretable.

The factor columns ($X$'s) can be either continuous or categorical. If an $X$ is continuous, then the splits (partitions) are created by a *cutting value*. The sample is divided into values below and above this cutting value. If the $X$ is categorical, then the sample is divided into two groups of levels.

The response column ($Y$) can also be either continuous or categorical. If $Y$ is continuous, then the platform fits means, and creates splits which most significantly separate the means by examining the sums of squares due to the means differences. If $Y$ is categorical, then the response rates (the estimated probability for each response level) become the fitted value, and the most significant split is determined by the largest likelihood-ratio chi-square statistic. In either case, the split is chosen to maximize the difference in the responses between the two branches of the split.

The Partition platform displays slightly different outputs, depending on whether the $Y$-variables in the model are continuous or categorical. In **Figure 17.1**, each point represents a response from the category and partition it is in. The $y$-position is random within the $y$ partition, and the $x$-position is a random permutation so the points are in the same rectangle but at different positions in successive analyses. **Figure 17.2** shows the corresponding case for a continuous response.

**Figure 17.1**  Output with Categorical Response

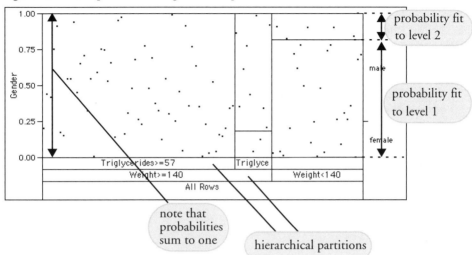

**Figure 17.2**  Output for Continuous Responses

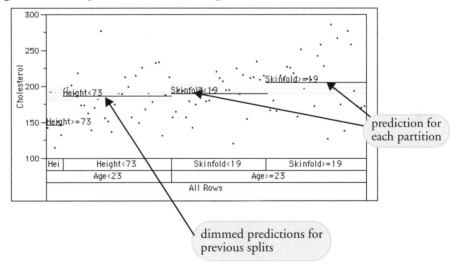

# Modeling with Recursive Trees

As an example of a typical analysis, open the Lipid.jmp data table. This data contains results from blood tests, physical measurements, and medical history for 95 subjects.

Cholesterol tests are invasive (requiring the extraction of blood) and require laboratory procedures to obtain their results. Suppose these researchers are interested in using noninvasive, external measurements and information from questionnaires to determine which patients are likely to have high cholesterol levels. Specifically, they want to predict the values stored in the Cholesterol column with information found in the Gender, Age, weight, skinfold, systolic BP, and diastolic BP columns.

To begin the analysis,

🖰   Select **Analyze > Modeling > Partition**.

🖰   Assign the variables as shown in **Figure 17.3**.

🖰   Click **OK**.

**Figure 17.3**  Partition Dialog

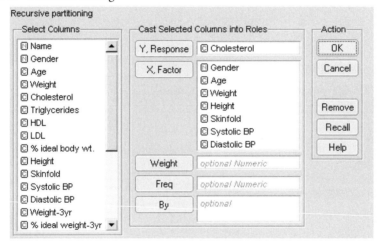

The initial Partition report appears, as in **Figure 17.4**. By default, the **Candidates** node of the report is closed, but is opened here for illustration. Note that no partitioning has happened yet—all of the data are placed in a single group whose estimate is the mean cholesterol value. In order to begin the partitioning process, you must interactively request splits.

**Figure 17.4**   Initial Lipid Partition Report

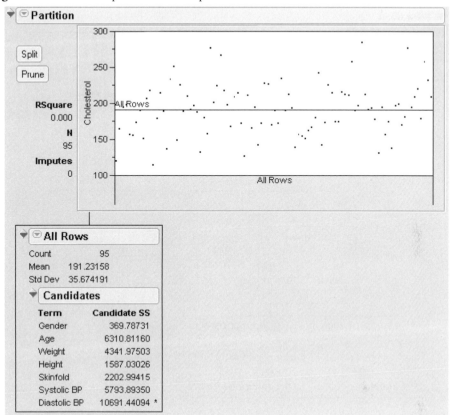

To determine the optimum split, each *x*-value is considered. The one that results in the highest reduction in total sum of squares is the optimum split, and is used to create a new branch of the tree. In this example, a split using Diastolic BP results in a reduction of 10691.44094 in the total SS, so it is used as the splitting variable.

🖱  Click the **Split** button to cause the split to take place.

The resulting report is shown in **Figure 17.5**. As expected, the Diastolic BP variable is involved, splitting at the value 80. People with a diastolic blood pressure less than 80 tend to have lower cholesterol (in fact, a mean of 183.3) than those with blood pressure above 80 (whose mean is 205.4).

**Figure 17.5**   First Split of Lipid Data

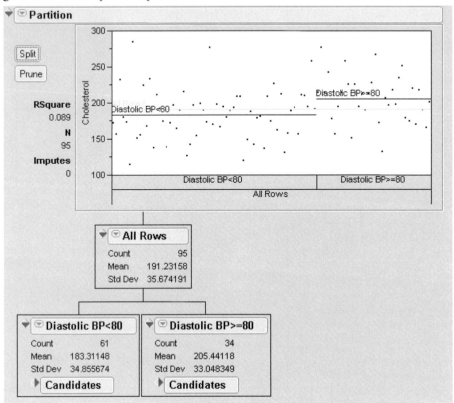

An examination of the candidates report (**Figure 17.6**) shows the possibilities for the second split. Under the **Diastolic BP<80** leaf, a split in the weight variable would produce a 10839.83-unit reduction in the sum of squares. Under the **Diastolic BP>80** leaf, a split in the Age variable would produce a 6138.13-unit reduction. Therefore, pressing the **Split** button splits under the **Diastolic BP<80** leaf.

**Figure 17.6**   Candidates for Second Split

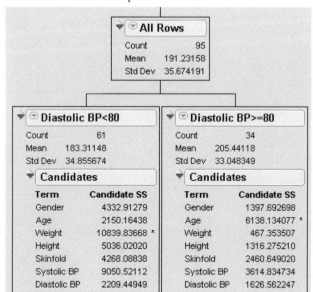

🖰   Press the **Split** button to conduct the second split.

The resulting report is shown in **Figure 17.7**, with its corresponding tree shown in **Figure 17.8**.

**Figure 17.7**   Plot After Second Split

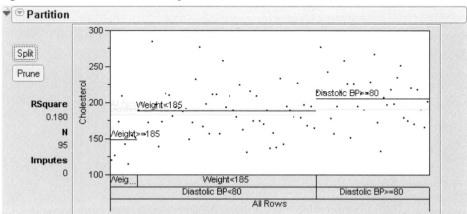

**Figure 17.8**  Tree After Second Split

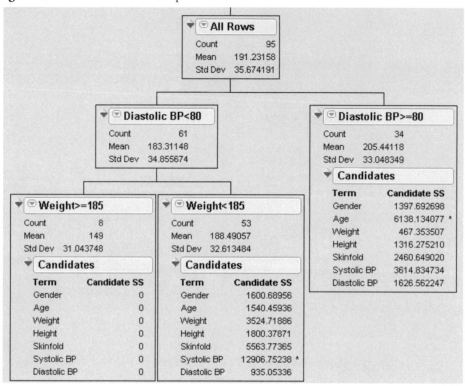

This second split shows that of the people with diastolic blood pressure less than 80, their weight is the best predictor of cholesterol. The model predicts that those who weigh more than 185 pounds have a cholesterol of 149—less than those that weigh more than 185 pounds, whose average cholesterol is predicted as 188.5.

Splitting can continue in this manner until you are satisfied with the predictive power of the model. As opposed to software that continues splitting until a criterion is met, JMP allows you to be the judge of the effectiveness of the model.

🖰  Press the **Split** button two more times, to produce a total of four splits.

## Viewing Large Trees

With several levels of partitioning, tree reports can become quite large. JMP has several ways to ease the viewing of these large trees.

- Use the scroll tool on the tool bar ( ✥ ) or **Tools** menu to easily scroll around the report.

- Close the upper nodes of the tree. When closed, the split condition still shows in the title bar of each leaf. As an example, **Figure 17.9** shows the current lipid data set after four splits.

**Figure 17.9**   Lipid Data After Four Splits

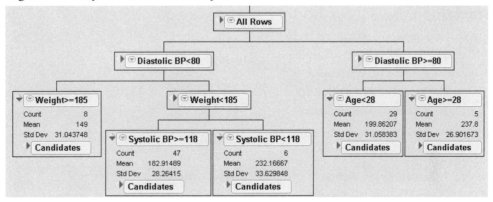

- Select **Small Tree View** from the menu on the title bar of the partition report. This option toggles a compact view of the tree, appended to the right of the main partition graph. **Figure 17.10** shows the Small Tree View corresponding to **Figure 17.9**.

**Figure 17.10**   Small Tree View

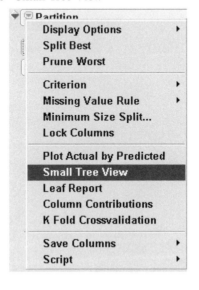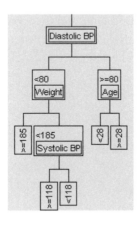

**Note:** The font used for the small tree view is controlled by the **Small** font, found on the Fonts tab of the JMP preferences. **Figure 17.10** uses an 8-point Arial font.

## Saving Results

The **Save Columns** submenu shows the options for saving results. All commands create a new column in the data table for storing their values. The commands that contain the word Formula (**Save Prediction Formula**, for example) store a formula in the column, where the other commands save values only. As an example,

> 🖰 Select **Save Prediction Formula** from the **Save Columns** submenu.

This adds a column to the report named Cholesterol Predictor. It

| Display Options ▶ |
| Split Best |
| Prune Worst |
| Criterion ▶ |
| Missing Value Rule ▶ |
| Minimum Size Split... |
| Lock Columns |
| Plot Actual by Predicted |
| Small Tree View |
| Leaf Report |
| Column Contributions |
| K Fold Crossvalidation |
| Save Columns ▶ |
| Script ▶ |

| Save Residuals |
| Save Predicteds |
| Save Leaf Numbers |
| Save Leaf Labels |
| Save Prediction Formula |
| Save Leaf Number Formula |
| Save Leaf Label Formula |

contains a formula that duplicates the partitions of the tree. To see the formula,

> 🖰 Right-click (Control-click on the Macintosh) in the title area of the Cholesterol Predictor column and select **Formula** from the menu that appears.

The formula for this example is shown in **Figure 17.11**. A series of If statements duplicates the model.

**Figure 17.11**    Prediction Formula

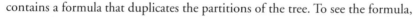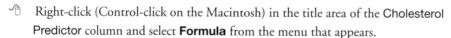

$$\text{If}\left[ \begin{array}{l} Diastolic\ BP<80 \Rightarrow \text{If}\left[ \begin{array}{l} Weight>=185 \Rightarrow 149 \\ else \qquad \Rightarrow \text{If}\left[ \begin{array}{l} Systolic\ BP>=118 \Rightarrow 182.914893617021 \\ else \qquad\qquad \Rightarrow 232.166666666667 \end{array} \right] \end{array} \right] \\ else \qquad\qquad \Rightarrow \text{If}\left[ \begin{array}{l} Age<28 \Rightarrow 199.862068965517 \\ else \quad \Rightarrow 237.8 \end{array} \right] \end{array} \right]$$

Other Save commands are as follows.

• **Save Leaf Numbers** saves the leaf numbers of the tree to a column in the data table.

- **Save Leaf Labels** saves leaf labels of the tree to the data table. The labels document each branch that the row would trace along the tree, with each branch separated by &.

- **Save Prediction Formula** saves the prediction formula to a column in the data table. The formula is made up of nested conditional clauses.

- **Save Leaf Number Formula** saves a formula that computes the leaf number to a column in the data table.

- **Save Leaf Label Formula** saves a formula that computes the leaf label to a column in the data table.

# Other Platform Commands

The Partition platform menu also contains the following commands.

**Display Options** gives a submenu consisting of three items that toggle report elements on and off. **Show Points** affects the points in the plot. **Show Tree** affects the large tree of partitions. **Show Graph** affects the partition graph.

**Split Best** splits the tree at the optimal split point. This is the same action as the **Split** button.

**Prune Worst** removes the lowermost split that has the least discrimination ability.

**Criterion** offers two choices for splitting. **Maximize Split Statistic** splits based on the raw value of sum of squares (continuous cases) or $G^2$ (categorical cases). **Maximize Significance** calculates significance values for each split candidate, and uses these rather than the raw values to determine the best split.

When **Maximize Significance** is chosen, the candidate report includes a new column titled **LogWorth**. This is the negative log of the adjusted $p$-value. The adjusted log-$p$-value is calculated in a complex manner that takes into account the number of different ways splits can occur. It is reported on the logarithmic scale because many significance values could easily underflow the numbers representable in machine floating-point form. It is reported as a negative log so that large values are associated with significant terms.

The essential difference between the two methods is that **Maximize Significance** tends to choose splits involving fewer levels, compared to the **Maximize Split Statistic** method, which allows a significant result to happen by chance when there are many opportunities.

For continuous responses, **Maximize Significance** tends to choose groups with small within-sample variances, whereas **Maximize Split Statistic** only takes effect size into account, rather than residual variance.

**Missing Value Rule** is a submenu with two choices. The default, **Closest**, assigns a missing value to the group that the response fits best with. **Random** is an option used when there are many missing values, and a more realistic measure of the goodness of fit is needed. With the **Random** option, JMP assigns the missing values randomly with probability based on the non-missing sample sizes in the partition being split. The default is zero.

**Minimum Size Split** presents a dialog box where you enter a number or a fractional portion of the total sample size which becomes the minimum size split allowed.

**Lock Columns** brings up a check box table to allow you to interactively lock columns so that they are not considered for splitting.

**Plot Actual by Predicted** produces a plot of actual values by predicted values.

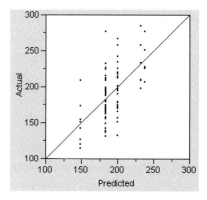

**Small Tree View** displays a smaller version of the partition tree to the right of the scatterplot.

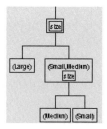

**Column Contributions** brings up a report showing how each input column contributed to the fit, including how many times it was split and the total G2 or Sum of Squares attributed to that column.

**K Fold Crossvalidation** randomly assigns all the (nonexcluded) rows to one of $k$ groups, then calculates the error for each point using means or rates estimated from all the groups except the group to which the point belongs. The result is a cross-validated $R^2$.

**Color Points** colors the points based on their response classification. This option is only available for categorical predictors. To make this helpful command easily accessible, a button for it is also provided. The button is removed once the command is performed.

**Script** is the typical JMP script submenu, used to repeat the analysis or save a script to various places.

At each split, the submenu has these commands.

- **Split Best** finds and executes the best split at or below this node.

- **Split Here** splits at the selected node on the best column to split by.

- **Split Specific X** allows specification of which $X$ column to split by. You can also specify the cut values or group levels to split by.

- **Prune Below** eliminates the splits below the selected node.

- **Prune Worst** finds and removes the worst split below the selected node.

- **Select Rows** selects the data table rows corresponding to this node.

- **Close All Below** closes all the displayed branches below the selected node.

- **Open All Below** opens all the displayed branches below the selected node.

# Neural Networks

The **Neural Net** platform implements a standard type of neural network. Neural nets are used to predict one or more response variables from a flexible network of functions of S-shaped input variables. Neural networks can be very good predictors when it is not necessary to know the functional form of the response surface. Technical details of the particular functions used in the implementation are found in Chapter 23, "Neural Nets", of the *JMP Statistics and Graphics Guide, version 5*.

This section uses the Peanuts.jmp sample data file, from an experiment concerning a device for automatically shelling peanuts. A reciprocating grid is used to automatically shell the peanuts. The length and frequency of this stroke, as well as the spacing of the peanuts, are factors in the experiment. Kernel damage, shelling time, and the number of unshelled peanuts need to be predicted. We illustrate the procedure with unshelled peanuts, leaving the other two responses as exercises.

🖑 Open the Peanuts.jmp sample data file.

🖑   Select **Analyze > Modeling > Neural Net**.

🖑   Assign Unshelled to Y and Length, Freq, and Space to X.

🖑   Click **OK**.

When the Neural Net control panel appears,

🖑   Select **Diagram** from the platform popup menu, located on the title bar of the panel.

The Neural Net diagram appears, as shown in **Figure 17.12**. The diagram illustrates that the factor columns are "squished" through three hidden nodes, whose outputs are used to form the predicted values.

**Figure 17.12**    Neural Net Diagram

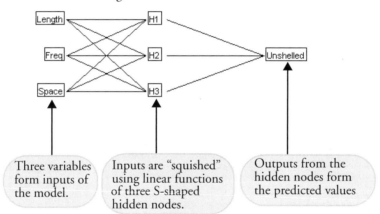

The Neural Net control panel has the following options that allow you to quickly explore several models.

• **Hidden Nodes** is the most important number to specify. A value too low underfits the model, while a number too high overfits. There are no hard-and-fast rules on how many hidden nodes to specify; experience and exploration usually reveal the number to use.

• **Overfit Penalty** helps prevent the model from overfitting. When a neural net is overfit, the parameters are too big, so this criterion helps keep the parameters (weights) small. The penalty is often called *lambda* or *weight decay*. A zero value causes overfitting and causes difficulty in convergence. With lots of data, the overfit penalty has less of an effect. With very little data, the overfit penalty will damp down the estimates a lot. We recommended that you try several values between 0.0001 and 0.01.

- **Number of Tours** sets the number of tours. Neural nets tend to have a lot of local minima, so that in order to more likely find global minima, the model is fit many times (tours) at different random starting values. Twenty tours is recommended. If this takes up too much time, then specify fewer. If you don't trust that you have a global optimum, then specify more.

- **Max Iterations** is the number of iterations JMP takes on each tour before reporting nonconvergence.

- **Converge Criterion** is the relative change in the objective function that an iteration must meet to decide it has converged.

- **Log the tours** shows the best objective function value at each tour.

- **Log the iteration** shows the objective function for each iteration.

- **Log the estimates** shows the estimates at each iteration.

- **Save iterations in table** saves the estimates at each iteration in a data table.

## Modeling with Neural Networks

🖱 Click **Go** to begin the fitting process.

When the iterations finish, results of the fit are shown as in **Figure 17.13**.

**Note:** These results are based on random starting points, so your results may vary from those shown here.

**Figure 17.13**  Fitting Results

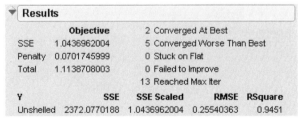

In this example, two of the twenty tours converged at a maximum. Five others converged, but at a non-maximal point. Thirteen iterations did not converge after reaching the maximum number of iterations.

Following the results summary, parameter estimates and diagnostic plots are included (closed by default). The parameter estimates are rarely interesting (JMP doesn't even bother to calculate standard errors for them).

The Actual by Predicted plot and the Residual plot (**Figure 17.14**) are used similarly to their counterparts in linear regression. Use them to judge the predictive power of the model. In this example, the model fits fairly well, with no glaring problems in the residual plot.

**Figure 17.14**   Neural Net Plots

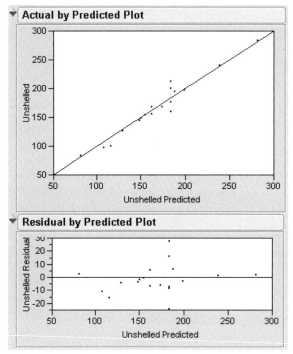

## Profiles in Neural Nets

The slices through the response surface are informative.

  ✎  From the platform drop-down menu, select **Profiler**.

The Prediction Profiler (**Figure 17.15**) clearly shows the nonlinear nature of the model. Running the model with more hidden nodes increases the flexibility of these curves; running with fewer stiffens them.

**Figure 17.15** Prediction Profiler

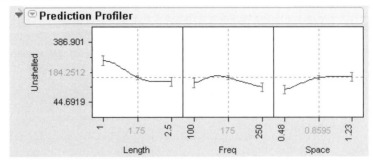

The Profiler retains all of the features used in analyzing linear models and response surfaces (discussed in "Analyze a Screening Model" on page 411 of this book and in the *JMP Statistics and Graphics Guide*). Since we are interested in minimizing the number of unshelled peanuts, we utilize the Profiler's **Desirability Functions**.

🖰 Select **Desirability Functions** from the drop-down menu on the Prediction Profiler title bar.

This appends the desirability functions to the profiler, as shown in **Figure 17.16**.

**Figure 17.16** Desirability Functions

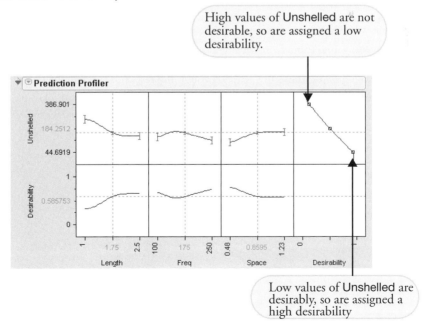

To have JMP automatically compute the optimum value,

🖱  Select **Maximize Desirability** from the platform drop-down menu.

JMP computes the maximum desirability, which in this example is a low value of Unshelled. Results are shown in **Figure 17.17**.

**Figure 17.17**   Maximized Desirability

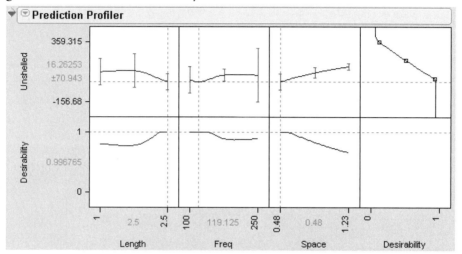

Optimal settings for the factors are shown (on screen in red) below the plots. In this case, optimal Unshelled values came from setting Length = 25, Freq = 119.12, and Space = 0.48.

In addition to seeing two-dimensional slices through the response surface, the Contour Profiler can be used to visualize contours (sometimes called level curves) and mesh plots of the response surface.

🖱  Select **Contour Profiler** from the platform drop-down menu.

The contours and mesh plot for this example are shown in **Figure 17.18**.

**Figure 17.18**   Contour Profiler

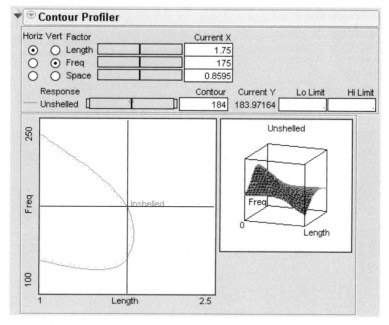

You can also get a good view of the response surface using the Surface Profiler.

🖰   Select **Surface Profiler** from the platform drop-down menu.

**Figure 17.19**  Surface Profiler

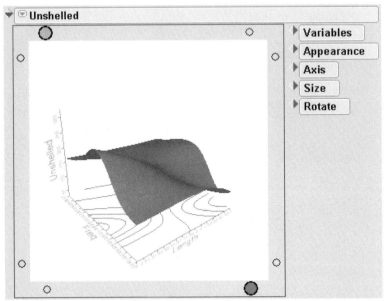

## Saving Columns

Results from Neural Net analyses can be saved to columns in the data table. The following save options are available.

**Save Hidden and Scaled Cols** makes new columns containing the scaled data used to make the estimates, the values of the hidden columns, and of the predicted values. No formulas are saved.

**Save Predicted and Limits** creates three new columns in the data table for each response variable. One column holds the predicted values, while the other two hold upper and lower 95% confidence intervals.

**Save Formulas** creates a new column in the data table for each response variable. This column holds the prediction formula for each response, so predicted values are calculated for each row. This option is useful if rows are added to the data table, since predicted values are automatically calculated. Use **Save Predicted and Limits** if formulas are not desired. In this example, this command produces columns for each hidden layer, plus the following prediction formula.

$$\left[\begin{array}{l}0.39510127474469 \\ +\text{-}2.6342652689551*H1\ Formula \\ +\text{-}3.6970949824481*H2\ Formula \\ +2.89460096379126*H3\ Formula\end{array}\right]*47.6735334276125+168.15$$

**Save Profile Formulas** saves formulas almost equivalent to the formulas made with the **Save Formulas** command, but the hidden layer calculations are embedded into the final predictor formulas rather than made through an intermediate hidden layer. In this example, a column with the following formula is added.

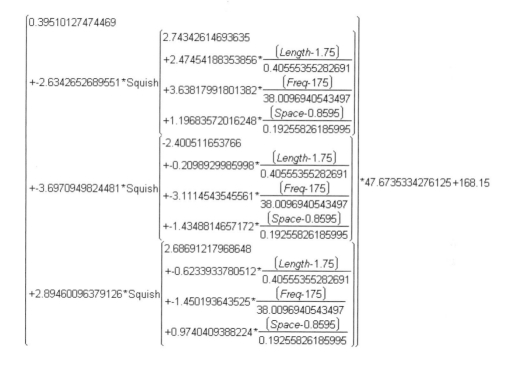

# Exercises

1. As shown in this chapter, the Lipids.jmp sample data set contains blood measurements, physical measurements, and questionnaire data from subjects in a California hospital. Repeat the Partition analysis of this chapter to explore models for

    (a)   HDL cholesterol

    (b)   LDL cholesterol

(c)  Triglyceride levels

2. Use the Peanuts.jmp sample data set and the Neural Net platform to complete this exercise. The factors of Freq, Space, and Length are described earlier in this chapter. Come up with a model using these factors for

(a)  Time, the time to complete the shelling process. Use the Profiler and Desirability Functions to find the optimum settings for the factors that minimize the time of shelling.

(b)  Damaged, the number of damaged peanuts after shelling is complete. Find values of the factors that minimize the number of damaged peanuts.

(c)  Compare the values found in the text of the chapter (**Figure 17.17**) and the values you found in parts (a) and (b) of this question. What settings would you recommend to the manufacturer?

# Discriminant and Cluster Analysis

## Overview

This chapter illustrates two multivariate techniques known as *discriminant analysis* and *cluster analysis*. Although JMP has the capability to do several kinds of cluster analysis (including *k*-means clustering, Normal mixtures clustering, and self-organizing maps), this chapter only covers the simplest type of cluster analysis known as hierarchical clustering. Throughout the chapter, the term cluster analysis is used to mean hierarchical clustering. Details on more advanced clustering methods are found in the online help or in the *JMP Statistics and Graphics Guide*.

Both discriminant analysis and cluster analysis classify observations into groups. The fundamental difference between the two analyses is that discriminant analysis has actual pre-defined groups to predict, whereas cluster analysis forms groups of points that are close together.

# Discriminant Analysis

Discriminant analysis is appropriate for situations where you want to classify a categorical variable based on values of continuous variables. For example, you may be interested in the voting preferences (Democrat, Republican, or Other) of people of various ages and income levels. Or, you may want to classify animals into different species based on physical measurements of the animal.

There is a strong similarity between discriminant analysis and logistic regression. In logistic regression, the classification variable is random and predicted by the continuous variables, whereas in discriminant analysis the classifications are fixed, and the continuous factors are random variables. However, in both cases, a categorical value is predicted by continuous variables.

The discrimination is most effective when there are large differences among the mean values of the different groups. Larger separations of the means make it easier to determine the classifications.

The classification of values is completed using a *discriminant function*. This function is quite similar to a regression equation—it uses linear combinations of the continuous values to assign each observation into a categorical group.

The example in this section deals with a trace chemical analysis of cherts. Cherts are rocks formed mainly of silicon, and are useful to archaeologists in determining the history of a region. By determining the original location of cherts, inferences can be drawn about the peoples that used them in tool making. Klawiter (2000) was interested in finding a model that predicted the location of a chert sample based on a trace element analysis. A subset of his data is found in the data table Cherts.jmp.

- 🖰 Open the Cherts.jmp data table.

- 🖰 Select **Analyze > Multivariate Methods > Discriminant**.

- 🖰 Assign all the chemical names as Y and location name as X.

- 🖰 Click **OK**.

The discriminant analysis report consists of two basic parts, the canonical plot and scores output.

## Canonical Plot

The Canonical Plot shows the points and multivariate means in the two dimensions that best separate the groups. The canonical plot for this example is shown in **Figure 18.1**. Note that the biplot rays, which show the directions of the original variables in the canonical space, has been moved to better show the canonical graph. Click in the center of the biplot rays and drag them to move them around the report.

**Figure 18.1** Canonical Plot of the Cherts Data

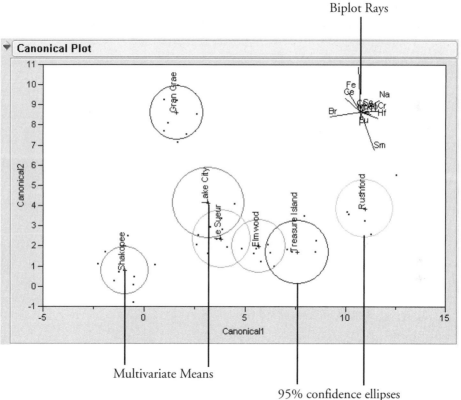

Each multivariate mean is surrounded by a 95% confidence ellipse, which appears circular in canonical space. In this example, the multivariate means for Shakopee, Gran Grae, and Rushford are more separated from the cluster of locations near the center of the graph.

## Discriminant Scores

The scores report shows how well each point is classified. The first five columns of the report represent the actual (observed) data values, showing row numbers, the actual classification, the

distance to the mean of that classification, and the associated probability. JMP graphs
−log(prob) in a histogram to show the loss in log-likelihood when a point is predicted poorly.
When the histogram bar is large, the point is being poorly predicted. A portion of the
discriminant scores for this example are shown in **Figure 18.2**.

**Figure 18.2**    Portion of Discriminant Scores Report

**Discriminant Scores**

| | | | | | | | |
|---|---|---|---|---|---|---|---|
| Number Misclassified | 2 | | | | | | |
| Percent Misclassified | 4.545 | | | | | | |
| -2LogLikelihood | 3.875 | | | | | | |

| Row Actual | Dist(Actual) | Prob(Actual) | -Log(Prob) | | Predicted | Prob(Pred) Others |
|---|---|---|---|---|---|---|
| 1 Gran Grae | 1.9398 | 0.9998 | 0.000 | | Gran Grae | 0.9998 |
| 2 Gran Grae | -3.7355 | 1.0000 | 0.000 | | Gran Grae | 1.0000 |
| 3 Gran Grae | 3.1250 | 0.9998 | 0.000 | | Gran Grae | 0.9998 |
| 4 Gran Grae | 2.9987 | 1.0000 | 0.000 | | Gran Grae | 1.0000 |
| 5 Gran Grae | 17.9437 | 1.0000 | 0.000 | | Gran Grae | 1.0000 |
| 6 Gran Grae | -1.5159 | 1.0000 | 0.000 | | Gran Grae | 1.0000 |
| 7 Gran Grae | -6.0120 | 1.0000 | 0.000 | | Gran Grae | 1.0000 |
| 8 Lake City | 6.5249 | 0.2069 | 1.576 | | * Gran Grae | 0.7931 |
| 9 Lake City | -0.2071 | 0.6913 | 0.369 | | Lake City | 0.6913 Le Sueur 0.30 |
| 10 Lake City | -1.4026 | 0.9998 | 0.000 | | Lake City | 0.9998 |
| 11 Lake City | 3.4283 | 0.9938 | 0.006 | | Lake City | 0.9938 |
| 12 Le Sueur | -5.2518 | 0.9601 | 0.041 | | Le Sueur | 0.9601 |
| 13 Le Sueur | 3.8099 | 0.9997 | 0.000 | | Le Sueur | 0.9997 |
| 14 Le Sueur | 2.9406 | 0.3018 | 1.198 | | * Elmwood | 0.6975 |
| 15 Le Sueur | 15.9743 | 1.0000 | 0.000 | | Le Sueur | 1.0000 |
| 16 Le Sueur | 0.2139 | 1.0000 | 0.000 | | Le Sueur | 1.0000 |
| 17 Le Sueur | -5.2969 | 0.8840 | 0.123 | | Le Sueur | 0.8840 Elmwood 0.11 |

The prediction for rows 8 and 14 are incorrect, notated by an asterisk to the right of the plot.
Why were these rows misclassified? Examining them in the canonical plot gives some insight.

🖰   From the platform menu on the title bar of the Discriminant report, select **Score
Options > Select Misclassified Rows**.

🖰   Select **Rows > Markers** and select the square marker for these rows.

The result of this selection is shown in **Figure 18.3**.

**Figure 18.3**   Misclassified Rows

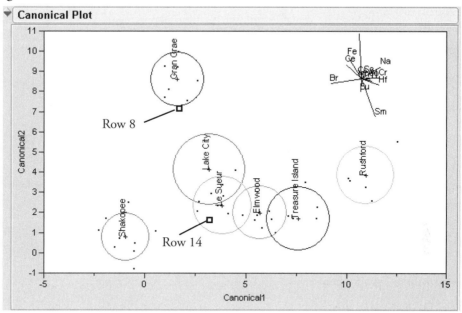

Row 8, although actually from Lake City, is very close to Gran Grae in canonical space. This closeness is the likely reason it was misclassified. Row 14, on the other hand, is close to Le Sueur, its actual value. It was misclassified because of the lack of separation of the means in this area of the canonical space.

Another quick way to examine misclassifications is to look at the Counts report (**Figure 18.4**) found below the discrimination scores. Zeros on the non-diagonal entries indicate perfect classification. The misclassified rows 8 and 14 are represented by the 1's in the non-diagonal entries.

**Figure 18.4**   Counts report

Counts: Actual Rows by Predicted Columns

| | Elmwood | Gran Grae | Lake City | Le Sueur | Rushford | Shakopee | Treasure Island |
|---|---|---|---|---|---|---|---|
| Elmwood | 7 | 0 | 0 | 0 | 0 | 0 | 0 |
| Gran Grae | 0 | 7 | 0 | 0 | 0 | 0 | 0 |
| Lake City | 0 | 1 | 3 | 0 | 0 | 0 | 0 |
| Le Sueur | 1 | 0 | 0 | 5 | 0 | 0 | 0 |
| Rushford | 0 | 0 | 0 | 0 | 6 | 0 | 0 |
| Shakopee | 0 | 0 | 0 | 0 | 0 | 9 | 0 |
| Treasure Island | 0 | 0 | 0 | 0 | 0 | 0 | 5 |

Misclassified Rows

# Other Discriminant Options

The following commands are available from the platform popup menu.

**Stepwise Variable Selection** returns to the stepwise control panel.

**Score Data** shows or hides the Discriminant Scores portion of the report.

**Canonical Plot** shows or hides the Canonical Plot.

**Specify Priors** brings up a dialog that allows the specification of prior probabilities for each level of the X variable.

**Consider New Levels** is used when you have some points that may not fit any known group, but instead may be from an unscored, new group.

**Save Discrim Matrices** creates a global list (`DiscrimResults`) for use in JMP's scripting language.

**Show Within Covariances** shows or hides the Covariance Matrix report.

**Show Group Means** shows or hides a table with the means of each variable. Means are shown for each level and for all levels of the $x$-variable.

## Score Options

These options deal with the scoring of the observations.

**Show Interesting Rows Only** shows rows that are misclassified and those where $p > 0.05$ and $p < 0.95$) for any $p$, the attributed probability.

**Show Classification Counts** shows a matrix of actual by predicted counts for each category. When the data are perfectly predicted, the off-diagonal elements are zero.

**Select Misclassified Rows** selects the misclassified rows in the original data table.

**Select Uncertain Rows** selects the rows which have uncertain classifications in the data table. When this option is selected, a dialog box appears so you can specify the difference (0.1 is the default) to be marked as uncertain.

**Save Formulas** saves formulas to the data table. The distance formulas are Dist[0], needed in the Mahalanobis distance calculations, and a Dist[ ] column for each $x$-level's Mahalanobis distance. Probability formulas are Prob[0], the sum of the exponentials of $-0.5$ times the Mahalanobis distances, and a Prob[ ] column for each $x$-level's posterior probability of being in that category. The Pred column holds the most likely level for each row.

### Canonical Options

These options all affect the Canonical Plot.

**Show Points** shows or hides the points in the plot.

**Show Means CL Ellipses** shows or hides the 95% confidence ellipse of each mean. The ellipses appear as a circle because of scaling. Categories with more observations have smaller circles.

**Show Normal 50% Contours** shows or hides the Normal contours which estimate the region where 50% of the level's points lie.

**Show Biplot Rays** shows or hides the biplot rays. These rays indicate the directions of the variables in the canonical space.

**Color Points** colors the points based on levels of the *x*-variable. This statement is equivalent to selecting **Rows > Color or Mark by Column** and selecting the *x*-variable.

**Show Canonical Details** shows or hides the Canonical details.

**Save Canonical Scores** creates columns in the data table holding the canonical score for each observation. The new columns are named Canon[ ].

# Cluster Analysis

Cluster analysis is the process of dividing a set of observations into a number of groups. JMP provides several methods of clustering, including *k*-means clustering, Normal mixtures clustering, and self-organizing maps. Some advanced clustering methods are beyond the scope of this introductory text. The interested reader is referred to Chapter 20 of the *JMP Statistics and Graphics Guide*, which gives complete details on all clustering methods.

In this book, we examine methods of hierarchical clustering. These methods group observations into clusters based on some measure of distance. JMP measures distance in the simple Euclidean way. There are dozens of measures of the proximity of observations, but the essential point is that observations that are "close" to each other are joined together in groups. Each of them has the objective of minimizing within-cluster variation and maximizing between-cluster variation.

The clustering process is actually quite simple.

- Find the two points that are closest together in multivariate space.

- Replace these two points with a single point at their multivariate mean.
- Repeat this process with the next two closest points.
- Continue the process until all points are subsumed in one group.

This process is illustrated in **Figure 18.5**.

**Figure 18.5**   Illustration of Clustering Process

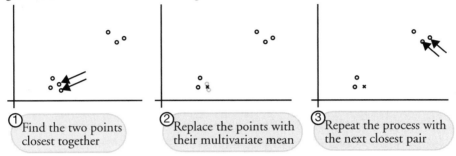

As a simple example, examine the SimulatedClusters.jmp data table.

- Open the SimulatedClusters.jmp data table.

- Select **Analyze > Fit Y By X**.

- Assign Y to **Y** and X to **X**, then click **OK**.

The results are shown in **Figure 18.6**.

**Figure 18.6**   Scatterplot of Simulated Data

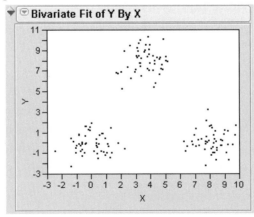

Obviously, the data clump together into three clusters. To analyze them with the clustering platform,

- ✲ Select **Analyze > Multivariate Methods > Cluster**.

- ✲ Assign X and Y to **Y, Columns**.

- ✲ Click **OK**.

The report appears as in **Figure 18.7**.

**Figure 18.7**   Dendrogram and Scree Plot

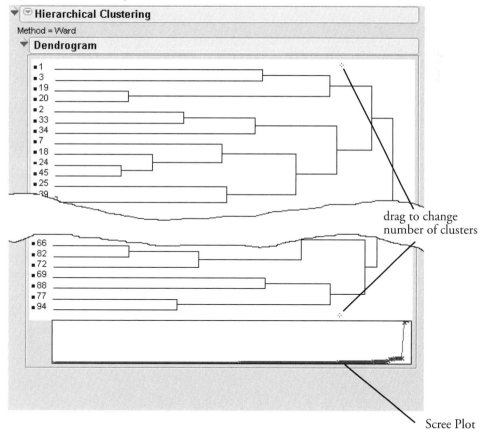

The top portion of the report shows a dendrogram, a visual tree-like representation of the clustering process. Branches that merge on the left were joined together earlier in the iterative algorithm.

Note the small diamonds at the bottom and the top of the dendrogram. This draggable diamond adjusts the number of clusters in the model.

Although there is no standard criterion for the best number of clusters to include, the Scree Plot (shown below the dendrogram) is shown to offer some guidance. Scree is a term for the rubble that accumulates at the bottom of steep cliffs, which this plot resembles. The place where the Scree Plot changes from a sharp downward slope to a more level slope (not always as obvious as in this example) is an indication of the number of clusters to include.

This example uses artificial data that was manufactured to have three clusters. Not surprisingly, the scree plot is very steep up to the point of three clusters, where it levels off.

  ✍ From the platform drop-down menu, select both **Color Clusters** and **Mark Clusters**.

This assigns a special color and marker to each observation which changes dynamically as you change the number of clusters in the model. To see this,

  ✍ Drag the windows so that you can see both the Fit Y By X scatterplot and the dendrogram at the same time

  ✍ Drag the number of cluster diamond to the right, observing the changes in colors and markers in the scatterplot.

  ✍ Move the marker to the point where there are three clusters. This should correspond to the level-off point of the Scree Plot.

The scatterplot should now look similar to the one in **Figure 18.8**.

**Figure 18.8**   Three Clusters

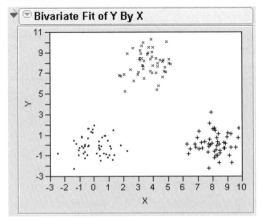

Once you decide that you have an appropriate number of clusters, you can save a column in the data table that holds the cluster value for each point.

🖰   From the platform pop-up menu, select **Save Clusters**.

The cluster values are often useful in subsequent analyses and graphs. For example, you can draw density ellipses around each group.

🖰   From the Fit Y By X report's drop-down menu, select **Group By**.

🖰   In the dialog that appears, select the Cluster column and click **OK**.

🖰   From the Fit Y By X drop-down menu select **Density Ellipse > .95**.

The display should now appear as in **Figure 18.9**.

**Figure 18.9**   Clusters with Density Ellipses

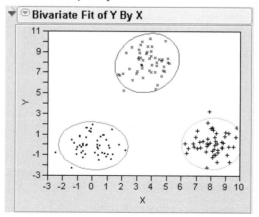

## A Real-World Example

The data set Teeth.jmp contains measurements of the number of certain teeth for a variety of mammals. A cluster analysis can show which of these mammals have similar dental profiles.

🖰   Open the data set Teeth.jmp.

🖰   Select **Analyze > Multivariate Methods > Cluster**.

🖰   Assign all the dental variables to **Y, Columns**.

🖰   Click **OK**.

When the dendrogram appears,

🖰   Select **Mark Clusters** from the platform drop-down menu.

The Scree Plot (bottom of **Figure 18.10**) does not show a clear number of clusters for the model. However, there seems to be some leveling-off at a point around six or seven clusters.

🖰   Drag the number of clusters diamond to the point corresponding to 6 clusters.

The report should appear as in **Figure 18.10**.

**Figure 18.10**   Dendrogram of Teeth data

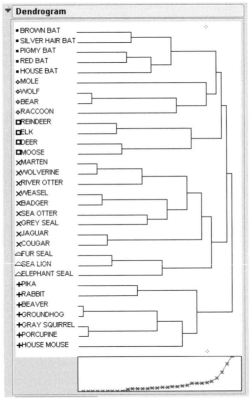

Examine the mammals classified in each cluster. Some conclusions based on these clusters are as follows.

• The five varieties of bats are in a cluster by themselves at the top of the dendrogram.

• Small mammals (like the rabbit, beaver, and squirrel) form a cluster at the bottom of the dendrogram.

- Seals and sea lions form a cluster, although not the same one as the sea otter and grey seal.

- Large mammals like the reindeer and moose form a cluster.

## Cluster Platform Options

The following options are available in the hierarchical clustering platform.

**Color Clusters** and **Mark Clusters** automatically assign colors and markers to the clusters identified by the position of the sliding diamond in the dendrogram.

**Save Clusters** saves the cluster number of each row in a new data table column for the number of clusters identified by the slider in the dendrogram.

**Save Display Order** creates a new column in the data table that lists the order of each observation the cluster table.

**Save Cluster Hierarchy** saves information needed if you are going to do a custom dendrogram with scripting. For each clustering it outputs three rows: the joiner, the leader, and the result, with the cluster centers, size, and other information.

**Number of Clusters** prompts you to enter a number of clusters and positions the dendrogram slider to the that number.

**Color Map** is an option to add a color map showing the values of all the data colored across it's value range. There are several color theme choices in a submenu. Another term for this feature is *heat map*.

**Geometric X Scale** is an option useful when there are many clusters and you want the clusters near the top of the tree to be more visible than those at the bottom. This option is the default now for more than 256 rows.

**Distance Graph** shows or hides the scree plot at the bottom of the histogram.

# Exercises

1. A classic example of discriminant analysis uses Fisher's Iris data, stored in the data table Iris.jmp. Three species of irises (setosa, virginica, and versicolor) were measured on four variables (sepal length, sepal width, petal length, and petal width). Use the discriminant analysis platform to make a model that classifies the flowers into their respective species using the four measurements.

2. The data set Birth Death.jmp contains mortality (*i.e.* birth and death) rates for several countries. Use the cluster analysis platform to determine which countries share similar mortality characteristics. What do you notice that is similar among the countries that group together?

# Statistical Quality Control

## Overview

Some statistics are for proving things. Some statistics are for discovering things. And some statistics are to keep an eye on things, watching to make sure something stays within specified limits.

The watching statistics are needed mostly in industry for systems of machines in production processes that sometimes stray from proper adjustment. These statistics monitor variation, and their job is to distinguish the usual random variation (called *common causes*) from abnormal change (called *special causes*).

These statistics are usually from a time series, and the patterns they exhibit over time are clues to what is happening to the production process. If they are to be useful, the data for these statistics need to be collected and analyzed promptly so that any problems they detect can be fixed.

This whole area of statistics is called *Statistical Process Control* (SPC) or *Statistical Quality Control* (SQC). The most basic tool is a graph called a *control chart* (or *Shewhart control chart*, named for the inventor, Walter Shewhart). In some industries, SQC techniques are taught to everyone—engineers, mechanics, shop floor operators, even managers.

The use of SQC techniques became especially popular in the 1980s as industry began to better-understand the issues of quality, after the pioneering effort of Japanese industry and under the leadership of Edward Deming and Joseph Juran.

# Control Charts and Shewhart Charts

Control charts are a graphical and analytical tool for deciding whether a process is in a state of statistical control. Control charts in JMP are automatically updated when rows are added to the current data table. In this way, control charts can be used to monitor an ongoing process.

**Figure 19.1** shows a control chart that illustrates characteristics of most control charts:

• Each point represents a summary statistic computed from a subgroup sample of measurements of a quality characteristic.

• The vertical axis of a control chart is scaled in the same units as the summary statistics specified by the type of control chart.

• The horizontal axis of a control chart identifies the subgroup samples.

• The center line on a Shewhart control chart indicates the average (expected) value of the summary statistic when the process is in statistical control.

• The upper and lower control limits, labeled UCL and LCL, give the range of variation to be expected in the summary statistic when the process is in statistical control.

• A point outside the control limits signals the presence of a special cause of variation.

**Figure 19.1**   Control Chart Example

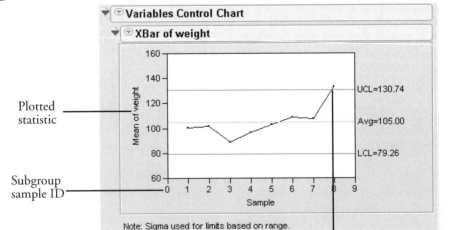

Control charts are broadly classified as *variables charts* and *attributes charts*.

## Variables Charts

Control Charts for variables (variables charts) are used when the quality characteristic to be monitored is measured on a continuous scale. There are different kinds of variables control charts based on the subgroup sample summary statistic plotted on the chart, which can be the mean, the range, or the standard deviation of a measurement, an individual measurement itself, or a moving range. For quality characteristics measured on a continuous scale, it is typical to analyze both the process mean and its variability by showing a Mean chart aligned above its corresponding $r$-(range) or $s$- (standard deviation) chart. If you are charting individual response measurements, the Individual Measurement chart is aligned above its corresponding Moving Range chart. JMP automatically arranges charts in this fashion.

## Attributes Charts

Control Charts for attributes (attributes charts) are used when the quality characteristic of a process is measured by counting the number or the proportion of nonconformities (defects) in an item, or by counting the number or proportion of nonconforming (defective) items in a subgroup sample.

# The Control Chart Launch Dialog

When you select **Control Charts** from the **Graph** menu, you see the Control Chart launch dialog in **Figure 19.2**. You use this dialog to specify the desired type of control chart. You can think of the control chart dialog as a composite of four panels in which you enter four kinds of information:

- Process information
- Chart type information
- Test requests
- Limits specification

**Figure 19.2**   Control Chart Dialog

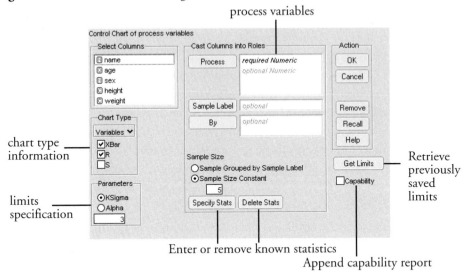

Specific information shown in the dialog varies according to the kind of chart you request. Through interaction with this dialog, you specify exactly how you want the charts created. The next sections discuss the kinds of information needed to complete the Control Chart dialog and talk about types of control charts.

# Process Information

The Process Information Panel displays a list of all columns in the current data table and has buttons to specify the variables to be analyzed, the subgroup sample size, and (optionally), the subgroup sample ID.

### Process

identifies variables for charting.

- For variables charts, specify the measurements.

- For attributes charts, specify the defect count or defect proportion.

### Sample Label

lets you specify a variable whose values label the horizontal axis and can also identify unequal subgroup sizes. If no sample ID variable is specified, the samples are identified by their sequence number.

- If the subsamples are the same size, check the Sample Size Constant radio button and enter the size into the text box. If you entered a Sample ID variable, its values are used to label the horizontal axis.

- If the subsamples have an unequal number of rows or have missing values, check the **Sample Grouped by Sample Label** radio button and remove sample size information from the Constant Sample Size text box.

For attributes charts (*p*, *np*, *c*, and *u*), this variable is the subgroup sample size. In Variable charts, it identifies the sample. When the chart type is IR, a Range Span text entry box appears. The range span specifies the number of consecutive measurements from which the moving ranges are computed. These chart types are described in more detail later.

## Chart Type Information

The Chart Type panel displays the popup menu shown here for chart type specification. Options in the Chart Type panel vary according to chart type. Shewhart controls charts broadly classify as variables charts and attributes charts. Moving average charts and Cusum charts can be thought of as special kinds of variables chart.

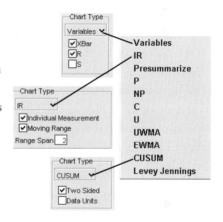

- The Shewhart Variable charts menu selection gives **XBar**, **R**, and **S** check boxes. The IR menu selection has check-box options for the Individual Measurement and Moving Range charts.

- Attributes chart selections are the **P**, **NP**, **C**, and **U** charts. There are no additional specifications for attributes charts.

- The uniformly weighted moving average (**UWMA**) and exponentially weighted moving average (**EWMA**) selections are special charts for means.

Descriptions and examples of specific kinds of charts are given later in this chapter.

## Limits Specification Panel

The Limits Specification Panel allows you to specify control limits computations by entering a value for *k* (**K Sigma**) or by entering a probability $\alpha$ (**Alpha**), or by retrieving a limits value from a previously created Limits Table (discussed later).

There must be a specification of either **K Sigma** or **Alpha**.

### K Sigma

allows specification of control limits in terms of a multiple of the sample standard error. K Sigma specifies control limits at $k$ sample standard errors above and below the expected value, which shows as the center line. To specify $k$, the number of sigmas, click **K Sigma** and enter a positive $k$ value into the box. The usual choice for $k$ is 3.

### Alpha

specifies control limits (also called probability limits) in terms of the probability that a single subgroup statistic exceeds its control limits, assuming that the process is in control. To specify $\alpha$, click the **Alpha** radio button and enter the probability you want. Reasonable choices for alpha are 0.01 or 0.001.

## Using Known Statistics

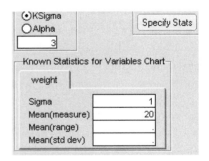

If you click the **Specify Stats** button on the Control Charts Launch dialog, a tab with editable fields is appended to the bottom of the launch dialog. This lets you enter known statistics for the process variable. The Control Chart platform uses those entries to construct control charts. The example to the right shows 1 as the standard deviation of the process variable and 20 as the mean measurement.

## Types of Control Charts for Variables

Control charts for variables are classified according to the subgroup summary statistic plotted on the chart:

- **XBar** charts display subgroup means (averages).
- **R** charts display subgroup ranges (maximum −minimum).
- **S** charts display subgroup standard deviations.

The **IR** selection gives two additional chart types:

- **Individual Measurement** chart type displays individual measurements.
- **Moving Range** chart type displays moving ranges of two or more successive measurements. Moving ranges are computed for the number of consecutive measurements you enter in the Range Span box. The default range span is 2. Because moving ranges are correlated, these charts should be interpreted with care.

## Mean, R, and S Charts

For quality characteristics measured on a continuous scale, a typical analysis shows both the process mean and its variability with a $\bar{X}$-chart aligned above its corresponding $r$- or $s$-chart.

The example in **Figure 19.3** uses the Coating.jmp data (taken from the *ASTM Manual on Presentation of Data and Control Charts*). The quality characteristic of interest is the Weight column. A subgroup sample of four is chosen. The $\bar{X}$-chart and an $r$- chart for the process show that sample six is above the upper control limit (UCL), which indicates that the process is not in statistical control. To check the sample values, you can click the sample summary point on either control chart and the corresponding rows highlight in the data table.

**Figure 19.3**   Variables Charts for Coating Data

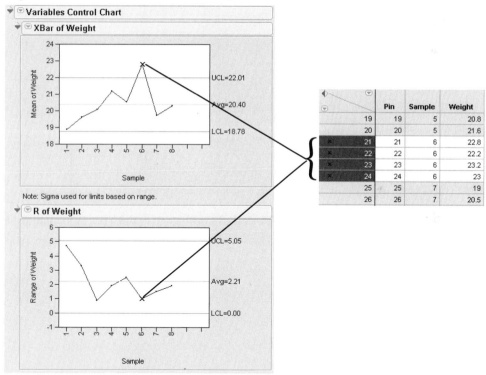

## Individual Measurement and Moving Range Charts

If you are charting individual measurements, the Individual Measurement chart shows above its corresponding Moving Range chart. Follow the example dialog in **Figure 19.4** to see these kinds of control charts.

  Open the Pickles.jmp data table.

The data show the acid content for vats of pickles. The pickles are produced in large vats, and high acidity can ruin an entire pickle vat. The acidity in four randomly selected vats was measured each day at 1:00, 2:00, and 3:00 PM. The data table records day (Date), time (Time), and acidity (Acid) measurements.

- Choose **Graph > Control Charts** and select IR from the Chart Type menu. Use Acid as the process variable, Date as the Sample Label, and enter a range span of 2 for the Moving Range chart (see **Figure 19.4**).

**Figure 19.4** Data and Control Chart Dialog for Individual Measurement Chart

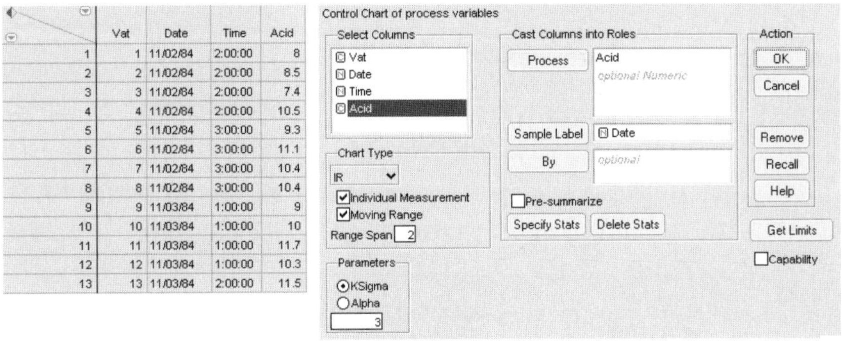

| | Vat | Date | Time | Acid |
|---|---|---|---|---|
| 1 | 1 | 11/02/84 | 2:00:00 | 8 |
| 2 | 2 | 11/02/84 | 2:00:00 | 8.5 |
| 3 | 3 | 11/02/84 | 2:00:00 | 7.4 |
| 4 | 4 | 11/02/84 | 2:00:00 | 10.5 |
| 5 | 5 | 11/02/84 | 3:00:00 | 9.3 |
| 6 | 6 | 11/02/84 | 3:00:00 | 11.1 |
| 7 | 7 | 11/02/84 | 3:00:00 | 10.4 |
| 8 | 8 | 11/02/84 | 3:00:00 | 10.4 |
| 9 | 9 | 11/03/84 | 1:00:00 | 9 |
| 10 | 10 | 11/03/84 | 1:00:00 | 10 |
| 11 | 11 | 11/03/84 | 1:00:00 | 11.7 |
| 12 | 12 | 11/03/84 | 1:00:00 | 10.3 |
| 13 | 13 | 11/03/84 | 2:00:00 | 11.5 |



- Click the **OK** button on the dialog to see the Individual Measurement and Moving Range charts shown in **Figure 19.5**.

In the pickle example, the Date variable labels the horizontal axis, which has been modified to better display the values. Tailoring axes is covered later in the chapter.

**Figure 19.5**  Individual Measurement Charts for Pickles Data

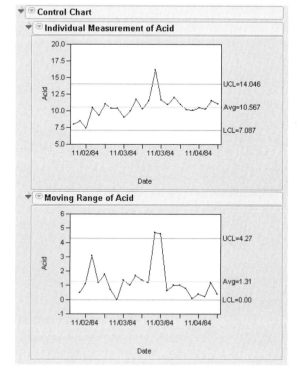

**Presummarize** charts summarize the process column before charting it. For an example, using the Coating.jmp data table,

- 🖱 Choose **Graph > Control Chart**.

- 🖱 Change the chart type to **Presummarize**.

- 🖱 Choose Weight as the **Process** variable and Sample as the **Sample Label**.

- 🖱 Check both **On Group Means** and **On Group Std Devs**.

The **Group by Sample Label** button is automatically selected when you choose a Sample Label variable before choosing **Presummarize**.

**Figure 19.6**  Example of Charting Pre-summarized Data

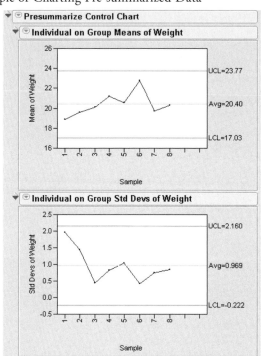

Although the points for $\bar{X}$-and $s$- charts are the same as the **Indiv of Mean** and **Indiv of Std Dev** charts, the limits are different because they are computed as Individual charts.

Here is another way to generate the pre-summarized charts:

- 🖱 Open Coating.jmp.

- 🖱 Choose **Tables > Summary**.

- 🖱 Assign Sample as the Group variable, then Mean(Weight) and Std Dev(Weight) as statistics, and click **OK.**

- 🖱 Select **Graph > Control Chart**.

- 🖱 Set the chart type again to **IR** and choose both Mean(Weight) and Std Dev(Weight) as process variables.

- 🖱 Click **OK**.

These new charts match the pre-summarized charts.

# Types of Control Charts for Attributes

Attributes charts, like variables charts, are classified according to the subgroup sample statistic plotted on the chart:

- *p-charts* display the proportion of nonconforming (defective) items in a subgroup sample.

- *np-charts* display the number of nonconforming (defective) items in a subgroup sample.

- *c-charts* display the number of nonconformities (defects) in a subgroup sample that usually consists of one inspection unit.

- *u-charts* display the number of nonconformities (defects) per unit in a subgroup sample with an arbitrary number of inspection units.

### *p*- and *np*-Charts

The Washers.jmp data table contains defect counts of 15 lots of 400 galvanized washers. The washers were inspected for finish defects such as rough galvanization and for exposed steel. The chart to the right illustrates an *np*-chart for the number of defects in the Washers data. A corresponding *p*-chart is identical except the vertical axis scale shows proportions instead of counts.

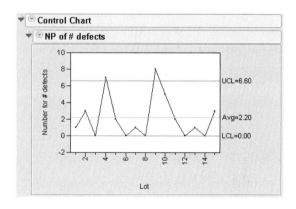

### *u*-Charts

The Braces.jmp data records the defect count in boxes of automobile support braces. A box of braces is one inspection unit. The number of defective braces found in a day is the process variable. The subgroup sample size is the number of boxes inspected in a day, which can vary.

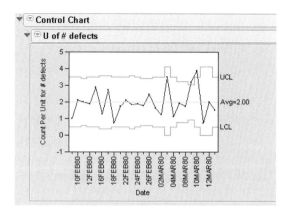

The *u*-chart shown to the right is monitoring the number of brace defects per unit. The upper and lower bounds vary according to the number of units (boxes of braces) inspected.

# Moving Average Charts

The control charts previously discussed plot each point based on information from a single subgroup sample. The Moving Average chart is different from other types because each point combines information from the current sample and from past samples. As a result, the Moving Average chart is more sensitive to small shifts in the process average. On the other hand, it is more difficult to interpret patterns of points on a Moving Average chart because consecutive moving averages can be highly correlated (Nelson 1983).

In a Moving Average chart, the quantities that are averaged can be individual observations instead of subgroup means. However, a Moving Average chart for individual measurements is not the same as a control (Shewhart) chart for individual measurements or moving ranges with individual measurements plotted.

## Uniformly Weighted Moving Average (UWMA) Charts

Each point on a Uniformly Weighted Moving Average (UWMA) chart is the average of the $w$ most recent subgroup means, including the present subgroup mean. When you obtain a new subgroup sample, the next moving average is computed by dropping the oldest of the previous $w$ subgroup means and including the newest subgroup mean. The constant $w$ is called the *span* of the moving average. There is an inverse relationship between $w$ and the magnitude of the shift that can be detected. Thus, larger values of $w$ allow the detection of smaller shifts. Complete the following steps to see an example.

   ⏺ Open the data table called Clips1.jmp.

A partial listing of the data is shown in **Figure 19.7**.

The measure of interest is the gap between the ends of manufactured metal clips. To monitor the process for a change in average gap, subgroup samples of five clips are selected daily, and a UWMA chart with a moving average span of three samples is examined. To see the UWMA chart, complete the Control Chart dialog.

   ⏺ Choose **Graph > Control Charts** and select **UWMA** from the Chart Type drop down menu.

   ⏺ Use Gap as the **Process** variable and Sample as the **Sample Label**.

   ⏺ Enter 3 as the moving average span and 5 as the Constant sample size.

The completed Control Chart dialog should look like the one shown in **Figure 19.7**.

**Figure 19.7**   Specification for UWMA Charts of CLIPS Data

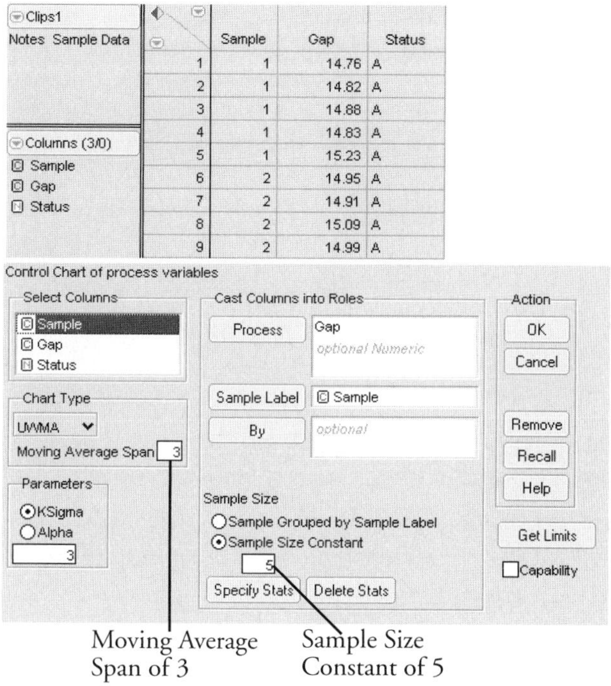

Moving Average
Span of 3

Sample Size
Constant of 5

🖱 Click the **OK** button to see the top chart in **Figure 19.8**.

**Figure 19.8**  UWMA Charts for the Clips Data

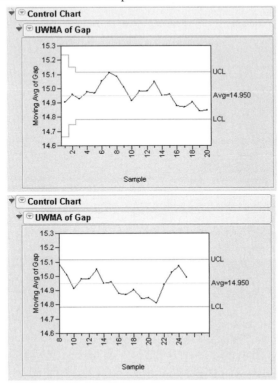

The point for the first day is the mean of the first subsample only, which consists of the five sample values taken on the first day. The plotted point for the second day is the average of subsample means for the first and second day. The points for the remaining days are the average of subsample means for each day and the two previous days.

Like all control charts, the UWMA chart updates dynamically when you add rows to the current data table.

Add rows to the Clips1 data table as follows:

- ⌐ Open the clipsadd.jmp data table.

- ⌐ Click at the top of both columns, Sample and Gap, to highlight them.

- ⌐ Choose **Edit > Copy** to copy the two columns to the clipboard.

- ⌐ Click on the Clips1 data table to make it the active table.

- ⌐ Click at the top of the Sample and Gap columns to highlight them in the Clips1 table.

🖱 Click in the cell immediately below the Sample measurement in the last row (in the first column that will become row 101).

🖱 Choose **Edit > Paste** to append the contents of the clipboard to the Clips1 table.

When you paste the new data into the table, the chart immediately updates, as shown in the bottom table in **Figure 19.8**.

### Exponentially Weighted Moving Average (EWMA) Chart

Each point on an Exponentially Weighted Moving Average (EWMA) chart, also referred to as a Geometric Moving Average (GMA) chart, is the weighted average of all the previous subgroup means, including the mean of the present subgroup sample.

The weights decrease exponentially going backward in time. The weight $(0 < r < 1)$ assigned to the present subgroup sample mean is a parameter of the EWMA chart. Small values of $r$ are used to guard against small shifts. If $r = 1$, the EWMA chart reduces to a Mean control (Shewhart) chart, previously discussed. The figure shown here is an EWMA chart for the same data used for **Figure 19.8**.

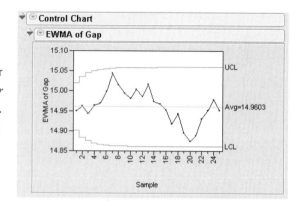

## Tailoring the Horizontal Axis

When you double-click the x-axis, a dialog appears that allows you to specify the number of ticks to be labeled.

For example, the Pickles.jmp example, seen previously, lists eight measures a day for three days. **Figure 19.9** shows Individual Measurement charts for the Pickles data. If there is no ID variable, the x-axis is labeled at every tick. Sometimes this gives illegible labels, as shown to the left in **Figure 19.9**. If you specify a label for every eighth tick mark, the x-axis is labeled once for each day, as shown in the plot on the right.

**Figure 19.9**   Example of Modified *x*-Axis Tick Marks

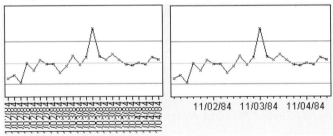

## Tests for Special Causes

You can select one or more tests for special causes (often called the *Western Electric rules*) in the Control Chart dialog or with the options popup menu beneath each plot. Nelson (1984, 1985) developed the numbering notation to identify special tests on control charts.

Table 19.1 lists and interprets the eight tests, and **Figure 19.10** illustrates the tests.

The following rules apply to each test:

- The area between the upper and lower limits is divided into six zones, each with a width of one standard deviation.

- The zones are labeled A, B, C, C, B, A with Zone C nearest the center line.

- A point lies in Zone B or beyond if it lies beyond the line separating zones C and B.

- Any point lying on a line separating two zones is considered to belong to the outermost zone.

Tests 1 through 8 apply to Mean and Individual Measurement charts. Tests 1 through 4 can also apply to *p-*, *np-*, *c-*, and *u-* charts.

**Table 19.1. Description of Special Causes Tests** (from Nelson, 1984,1985)

| | | |
|---|---|---|
| Test 1 | One point beyond Zone A | Detects a shift in the mean, an increase in the standard deviation, or a single aberration in the process. For interpreting Test 1, the $R$ chart can be used to rule out increases in variation. |
| Test 2 | Nine points in a row in Zone C or beyond | Detects a shift in the process mean. |
| Test 3 | Six points in a row steadily increasing or decreasing | Detects a trend or drift in the process mean. Small trends are signaled by this test before Test 1. |
| Test 4 | Fourteen points in a row alternating up and down | Detects systematic effects such as two alternately used machines, vendors, or operators. |
| Test 5 | Two out of three points in a row in Zone A or beyond | Detects a shift in the process average or increase in the standard deviation. Any two out of three points provide a positive test. |
| Test 6 | Four out of five points in Zone B or beyond | Detects a shift in the process mean. Any four out of five points provide a positive test. |
| Test 7 | Fifteen points in a row in Zone C, above and below the center line | Detects stratification of subgroups when the observations in a single subgroup come from various sources with different means. |
| Test 8 | Eight points in a row on both sides of the center line with none in Zone C | Detects stratification of subgroups when the observations in one subgroup come from a single source, but subgroups come from different sources with different means. |

Tests 1, 2, 5, and 6 apply to the upper and lower halves of the chart separately. Tests 3, 4, 7, and 8 apply to the whole chart.

**Figure 19.10**  Illustration of Special Causes Tests, from Nelson (1984, 1985)

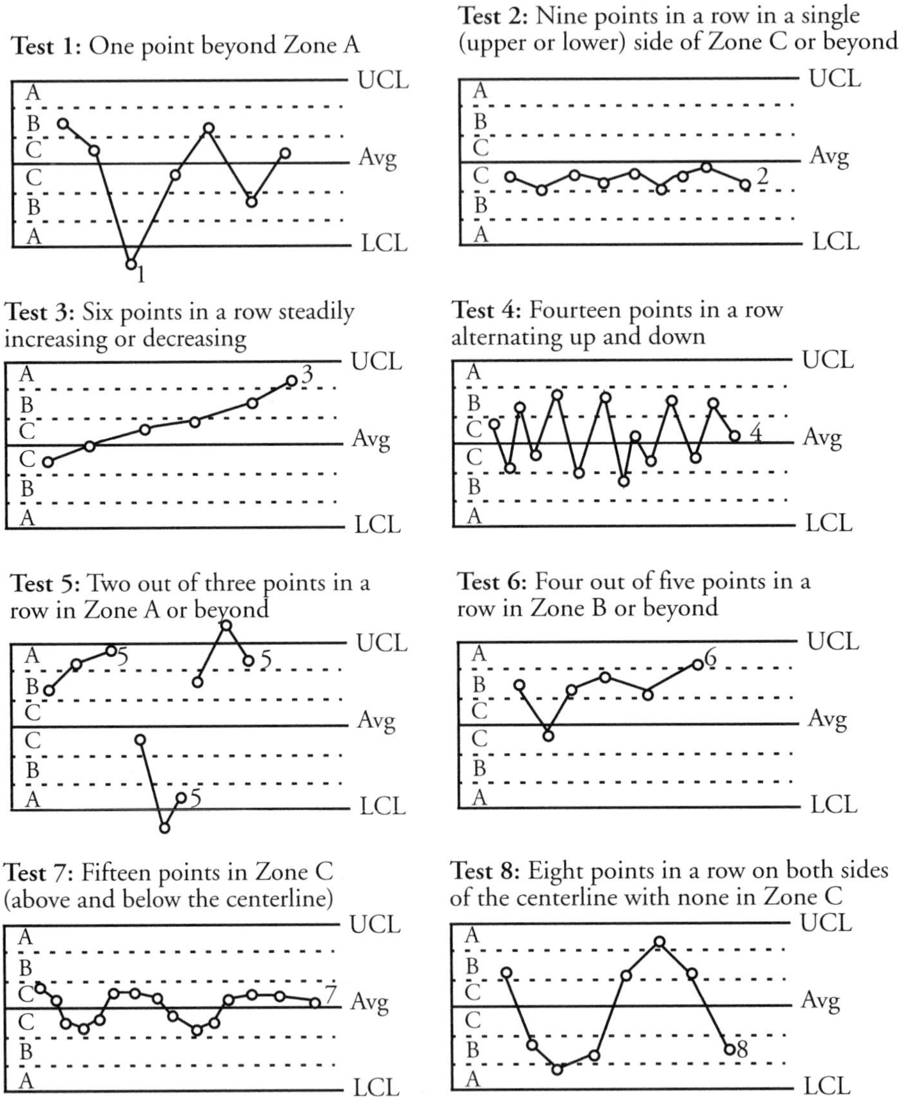

**Test 1:** One point beyond Zone A

**Test 2:** Nine points in a row in a single (upper or lower) side of Zone C or beyond

**Test 3:** Six points in a row steadily increasing or decreasing

**Test 4:** Fourteen points in a row alternating up and down

**Test 5:** Two out of three points in a row in Zone A or beyond

**Test 6:** Four out of five points in a row in Zone B or beyond

**Test 7:** Fifteen points in Zone C (above and below the centerline)

**Test 8:** Eight points in a row on both sides of the centerline with none in Zone C

# Westgard Rules

Westgard rules are implemented under the **Westgard Rules** submenu of the Control Chart platform. The different tests are abbreviated with the decision rule for the particular test. For example, **1 2s** refers to a test that one point is two standard deviations away from the mean.

Because Westgard rules are based on sigma and not the zones, they can be computed without regard to constant sample size.

**Table 19.2.** Westgard Rules

**Rule 1 2s** is commonly used with Levey-Jennings plots, where control limits are set 2 standard deviations away from the mean. The rule is triggered when any one point goes beyond these limits.

**Rule 1 3s** refers to a rule common to Levey-Jennings plots where the control limits are set 3 standard deviations away from the mean. The rule is triggered when any one point goes beyond these limits.

**Rule 2 2s** is triggered when two consecutive control measurements are farther than two standard deviations from the mean.

**Rule 4s** is triggered when one measurement in a group is two standard deviations above the mean and the next is two standard deviations below.

**Rule 4 1s** is triggered when four consecutive measurements are more than one standard deviation from the mean.

**Rule 10 X** is triggered when ten consecutive points are on one side of the mean.

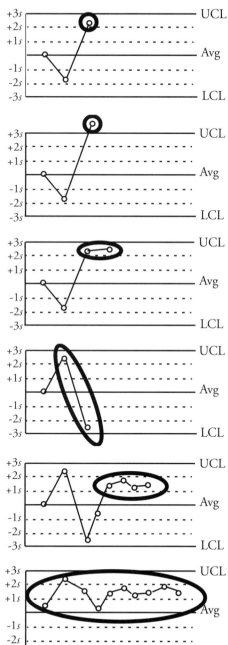

# Time Series

## Overview

When you have time series data, the best future prediction is usually some form of extrapolation of the historical behavior. For short-term forecasts, the behavior of interest is the short-term pattern of fluctuations. These fluctuations can be efficiently modeled by one of two methods:

- regression on recently-past values

- regression on past random errors.

This chapter focuses mainly on these ARIMA (also known as Box-Jenkins) models, those that model a series based on lagged values of itself. Other possibilities for modeling this type of data include frequency analysis and extensions of standard regression techniques.

# Introduction

When you take data from a process in the real world, the values can be a product of everything else that happened until the time of measurement. Time series data usually is non-experimental: it is the work of the world rather than the product of experimentally-controlled conditions. Furthermore, the data are not presented with all relevant covariates; they are often taken alone, without any other variables to co-analyze.

Time series methods have been developed to characterize how variables behave across time and how to forecast them into the future.

There are three general approaches to fitting a time series.

- Model the data as a function of time itself.

- Model the data as a function of its past (lagged) values.

- Model the data as a function of random noise.

The first approach emphasizes the structural part of the model, whereas the second and third approaches emphasizes the random part of the model. The first approach is generally modeled using regression, which is covered in other chapters of this book. The second and third methods are the focus of this chapter.

Techniques of time series analysis implemented in JMP assume that the series is made up of points that are equally-spaced in time. This doesn't mean that some of the values cannot be missing, but only that the data is collected on an regular schedule.

Models in this chapter are most useful for short-term predictions. For example, suppose you need to predict whether it will be raining one minute from now. Faced with this question, which information is more useful, what time it is, or if it is currently raining? If rain were structurally predictable (like a television schedule) you would rather know what time it is. Weather, however, does not behave in this fashion, so to predict if it will rain in a minute, it is more helpful to know if it is currently raining.

# Lagged Values

This idea of looking into the past to forecast the future is the idea behind a lagged variable. Put simply, a lagged variable is formed by looking at values of the variable that occurred in the past.

As an example, examine a series of chemical concentration readings taken every two hours.

🖑   Open the data file Seriesa.jmp[1].

The data set consists of only one variable, Cn. A lagged variable of Cn takes these values and shifts them forward or backward. Although it's not necessary for most time series analyses, we actually construct a variable that is lagged by one period for illustration.

🖑   Add a new column to the data table and call it Lag Cn.

🖑   Right-click (Control-click on the Macintosh) and select **Formula** from the menu that appears.

🖑   Select **Row > Lag** from the list of functions

🖑   Click on the Cn column in the **Table Columns** list to specify that we are lagging the Cn variable.

The Formula Editor should now appear as in **Figure 20.1**.

**Figure 20.1**   Formula Editor for Lagged Variable

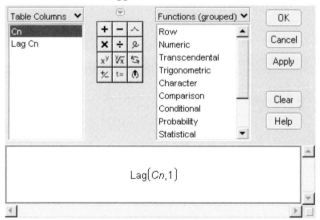

🖑   Click **OK**.

**Figure 20.2** shows the results of the formula. Lag Cn is merely Cn shifted by one time period. We could have specified another lag period by adjusting the second argument of the Lag() function in the formula.

---

1. The file is named SeriesA because it is the first data set presented in the classic time series book by Box and Jenkins. The book was so influential that time series modeling is often called Box-Jenkins analysis in their honor.

**Figure 20.2** Illustration of a Lagged Variable

| | Cn | Lag Cn |
|---|---|---|
| 1 | 17 | • |
| 2 | 16.6 | 17 |
| 3 | 16.3 | 16.6 |
| 4 | 16.1 | 16.3 |
| 5 | 17.1 | 16.1 |
| 6 | 16.9 | 17.1 |
| 7 | 16.8 | 16.9 |
| 8 | 17.4 | 16.8 |
| 9 | 17.1 | 17.4 |
| 10 | 17 | 17.1 |

We can now examine whether Cn is correlated with its lagged values. A variable that is correlated with its own lagged values is said to be *autocorrelated*.

- Select **Analyze > Fit Y by X**.
- Assign Cn to **Y** and Lag Cn to **X**.
- Click **OK**.
- When the bivariate plot appears, select **Density Ellipse > 0.95**.
- Click the blue disclosure icon to open the correlation report.

The correlation coefficient is shown to be 0.5712. If there were no autocorrelation present, we would expect this value to be near zero. The presence of autocorrelation is an indicator that time series methods are necessary.

**Figure 20.3**   Bivariate report showing autocorrelation

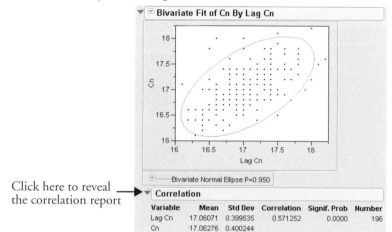

Click here to reveal
the correlation report

We are also interested in autocorrelations of lags greater than one. Rather than computing multiple lag columns, JMP provides a report that shows the correlation of many lagged values. To see this report,

- Select **Analyze > Modeling > Time Series**.

- Assign Cn to the **Y, Time Series** role.

- Click **OK**.

The lag-1 correlation (shown in **Figure 20.4**) is close to the one found in the bivariate platform[1]. Values for higher lags are also shown. Note that the lag-0 value (each value of the variable correlated with itself) is exactly one. This plot is the fundamental tool used to determine the appropriate model for the data.

Notice the blue lines on the autocorrelation report. These are confidence intervals on the null hypothesis that the autocorrelation at each lag is zero. Bars that extend beyond these lines are evidence that the autocorrelation at that lag is not zero.

---

1. The autocorrelation in the Time Series platform is slightly different from the one reported in the Bivariate platform because the Time Series tradition uses $n$ instead of $n-k$ as a divisor in calculating autocorrelations, where $n$ is the length of the series, and $k$ is the lag.

**Figure 20.4**  Initial time series report

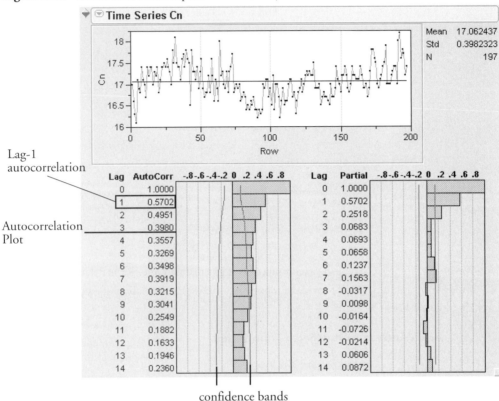

Lag-1 autocorrelation

Autocorrelation Plot

confidence bands

## Testing for Autocorrelation

The recommended method for observing autocorrelation is to examine the autocorrelation plots from the Time Series platform. However, there is a statistical test for lag-1 autocorrelation known as the *Durbin-Watson* test. It is available in the Fit Model platform.

The Durbin-Watson test examines the difference between consecutive errors compared to the values of the errors themselves. Specifically, the test statistic is

$$d = \frac{\sum_t (\hat{e}_t - \hat{e}_{t-1})^2}{\sum_t \hat{e}_t^2}$$

If you expand the numerator, this statistic becomes

$$d = \frac{\sum_t (\hat{e}_t^{\,2} - 2\hat{e}_t\hat{e}_{t-1} + \hat{e}_{t-1}^{\,2})}{\sum_t \hat{e}_t^{\,2}}$$

which allows you to see its value when there is no autocorrelation. With no autocorrelation, the middle term $2\hat{e}_t\hat{e}_{t-1}$ becomes zero, and the terms $\hat{e}_t^{\,2}$ and $\hat{e}_{t-1}^{\,2}$ combine to form $2\hat{e}_t^{\,2}$, so $d = 2$. If there is significant positive autocorrelation, $d$ becomes quite small. Similarly, significant negative autocorrelation causes $d$ to approach the value 4. With the null hypothesis that there is no correlation, then, we have a situation similar to the following.

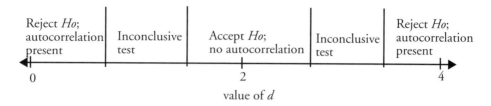

The particular boundaries for each region depend on the values of the variables under consideration. Luckily, JMP can calculate the value and significance of the test automatically. Use the SeriesA.jmp data table as an example.

- If the SeriesA data table is not front-most, bring it to the front using the **Window** menu.

- Select **Analyze > Fit Model** and assign Cn to the **Y** role.

- Click **Run Model**.

When the report appears,

- Select **Row Diagnostics > Durbin Watson** from the platform popup menu.

When the Durbin-Watson report appears,

- Select **Significance P Value** from the Durbin Watson popup menu.

- When JMP warns you that this may take a long time (it won't in this case), click **OK**.

You should now have the report in **Figure 20.5**.

**Figure 20.5**  Durbin-Watson Report

| Durbin-Watson | Number of Obs. | AutoCorrelation | Prob<DW |
|---|---|---|---|
| 0.8558983 | 197 | 0.5702 | 0.0000 |

The low *p*-value indicates that there is significant autocorrelation in this data. Note, however, that the Durbin-Watson test examines only first-order (lag-1) autocorrelation, so if there is autocorrelation beyond lag-1, this test provides no information.

# White Noise

The most fundamental of all time series is called *white noise* (so termed because values tend to enter at all frequencies, similar to white light). The white noise series is a series composed of values chosen from a random Normal distribution with mean zero and variance one. It should also have autocorrelation of (near) zero for all lags. To construct this series in JMP,

    Select **File >New > Data Table** (PC and Linux) or **File > New** (Macintosh).

In the data table that appears,

    Right-click (Control-click on the Macintosh) on the column header and select **Formula**.

In the Formula Editor,

    Select **Random > Random Normal**.

    Click **OK.**

    Select **Rows > Add Rows** and add 100 rows to the data table.

You should now have a single column that contains 100 random values.

    Select **Analyze > Modeling > Time Series.**

    Assign the column to the **Y, Time Series** role.

    Click **OK**.

None of the bars in the autocorrelation report extend beyond the blue confidence lines and the mean and standard deviation are as expected. None of the correlations appear to be different from zero. You should remember the look of the correlation plots for future reference, since reducing a time series to white noise is part of the modeling procedure.

**Figure 20.6**   White Noise Report

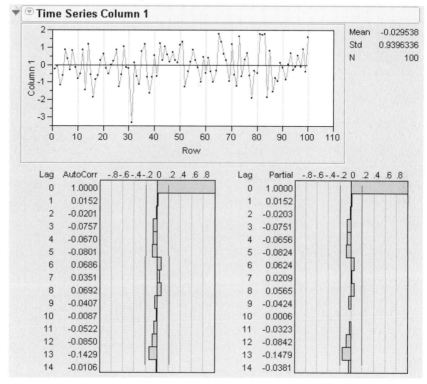

# Autoregressive Processes

Suppose you construct a series like the following.

$$x_1 = 1$$
$$x_n = x_{n-1} + \varepsilon$$

where $\varepsilon$ is random white noise. Such a series is called a *random walk*; a sample is stored in the data table Random Walk.jmp. Since each value is just a slight perturbation of its previous value, we expect to see significant lag-1 autocorrelation, slightly less lag-2 correlation, even less lag-3 correlation, and so on. **Figure 20.7** shows the time series report from this series.

**Figure 20.7**   Random Walk report

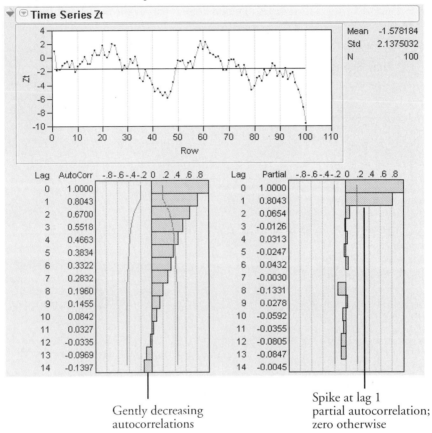

Gently decreasing
autocorrelations

Spike at lag 1
partial autocorrelation;
zero otherwise

The random walk is the simplest example of an *autoregressive process.* Each value is the sum of the previous value and random noise. Written mathematically, if $z_t$ represents the value of the series at time $t$,

$$z_t = \varphi z_{t-1} + \varepsilon$$

where $\varphi$ represents the influence of the lag value on the present value and $\varepsilon$ represents random (white noise) error. In other words, the series is composed of two parts—a fraction of its lag-1 values and noise. As shorthand, we refer to this as an AR(1) ("autoregressive lag-1") process. More generally, if we use $p$ to represent the number of lags, autoregressive processes of order $p$ are written AR($p$).

The plot to the right of the autocorrelation plot in **Figure 20.7** is called the *partial autocorrelation plot*. This plot shows the autocorrelations at several lag values *after all lower-valued lagged autocorrelations are taken into account*. In this example, there is a spike at lag 1, indicating that the model should contain a lag-1 term. All other partial autocorrelations aren't significantly different than zero, indicating that we don't need further autocorrelation terms. None of this is a surprise since we constructed this series out of lag-1 values.

## Correlation Plots of AR Series

The pattern in **Figure 20.7** is a typical example of an AR($p$) process. The partial autocorrelation plot has spikes at lags 1 to $p$ and is zero for lags greater than $p$. The autocorrelation plot shows a gently decreasing pattern.

# Estimating the Parameters of an Autoregressive Process

Our objective is to find the value of $\varphi$ that generated the model. To estimate this parameter in JMP,

🖱 Select **ARIMA** from the platform popup menu.

This produces the dialog shown in **Figure 20.8**.

**Figure 20.8**   Specify ARIMA Dialog

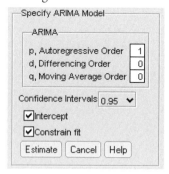

🖱 Enter a 1 beside **p, Autoregressive Order**.

🖱 Click **Estimate**.

This produces the report in **Figure 20.9**

**Figure 20.9**  AR(1) Report

| Parameter Estimates | | | | | | |
|---|---|---|---|---|---|---|
| Term | Lag | Estimate | Std Error | t Ratio | Prob>|t| | Constant Estimate |
| AR1 | 1 | 0.93332078 | 0.0455133 | 20.51 | <.0001 | -0.1451781 |
| Intercept | 0 | -2.1772613 | 1.4090472 | -1.55 | 0.1255 | |

The estimated lag-1 autoregressive coefficient is 0.93, very close to the actual value of 1. After accounting for this AR(1) term, all that should be left is white noise. To verify this,

🖑   Click on the disclosure icon beside **Residuals**.

This reveals a time series report of the residuals of the model. Since there are no significant autocorrelations or partial autocorrelations, we conclude that our residuals are indistinguishable from white noise, so the AR(1) model is adequate.

**Figure 20.10**  Residuals Report

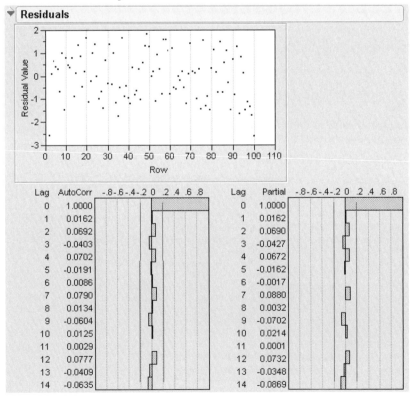

# Moving Average Processes

A *moving average* (MA) process models a time series on lagged values of a white noise series rather than lagged values of the series itself. In other words, MA series listen to white noise and derive their values from what they hear. Using $q$ to designate the number of lags involved in the moving average, we abbreviate "moving average of $q$ lags" as MA($q$).

To see a simple example of an MA series,

- Open Timema1.jmp

- Add 100 rows to the data table.

There are two columns in the data table. The Noise column holds simple white noise, generated with a formula. The MA1 column is computed using lagged values of the white noise. Examine a simple time series report of the value of MA1.

- Select **Analyze > Modeling > Time Series** and use MA1 as the **Y, Time Series** variable.

This produces the report in **Figure 20.11**.

**Figure 20.11** MA(1)Report

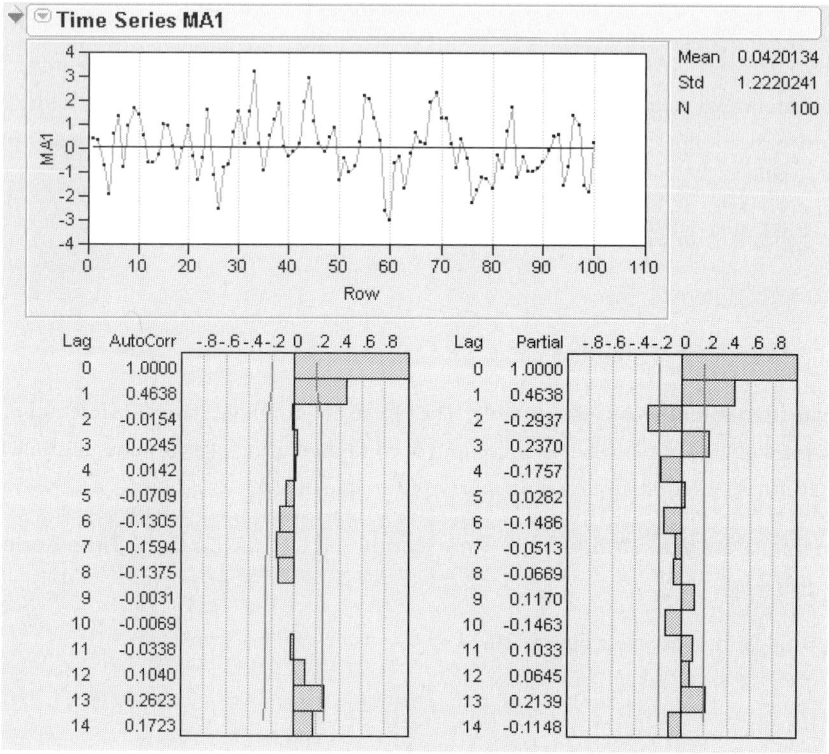


| Lag | AutoCorr | Lag | Partial |
|---|---|---|---|
| 0 | 1.0000 | 0 | 1.0000 |
| 1 | 0.4638 | 1 | 0.4638 |
| 2 | -0.0154 | 2 | -0.2937 |
| 3 | 0.0245 | 3 | 0.2370 |
| 4 | 0.0142 | 4 | -0.1757 |
| 5 | -0.0709 | 5 | 0.0282 |
| 6 | -0.1305 | 6 | -0.1486 |
| 7 | -0.1594 | 7 | -0.0513 |
| 8 | -0.1375 | 8 | -0.0669 |
| 9 | -0.0031 | 9 | 0.1170 |
| 10 | -0.0069 | 10 | -0.1463 |
| 11 | -0.0338 | 11 | 0.1033 |
| 12 | 0.1040 | 12 | 0.0645 |
| 13 | 0.2623 | 13 | 0.2139 |
| 14 | 0.1723 | 14 | -0.1148 |

## Correlation Plots of MA Series

Contrast the autocorrelation plot of this moving average process with the one from the autoregressive process of the last section. For an MA($q$) series, the autocorrelation plot goes to zero after $q$ lags, and the partial autocorrelation plot drops off slowly. In this case, "drops off slowly" means the magnitudes of the autocorrelations drop off slowly, even though they alternate in sign.

# Example of Diagnosing a Time Series

The data set SeriesD.jmp (Box, Jenkins, and Reinsel 1976) contains viscosity readings from a chemical process. Using the autocorrelation and partial autocorrelation functions, we try to determine the nature of the model.

Open the data file SeriesD.jmp.

🖰  Select **Analyze > Modeling > Time Series**.

🖰  Assign Viscosity to the **Y, Time Series** role.

🖰  Click **OK**.

Examining the plots, it is easy to see that the autocorrelation plot decreases slowly, while the partial autocorrelation plot has a spike at lag 1, but is (essentially) zero everywhere else. This is the condition we saw for an AR($p$) process ("Correlation Plots of AR Series" on page 511). Since the partial autocorrelation plot has a spike at lag 1, we guess that an AR(1) model is appropriate.

**Figure 20.12**   SeriesD Report

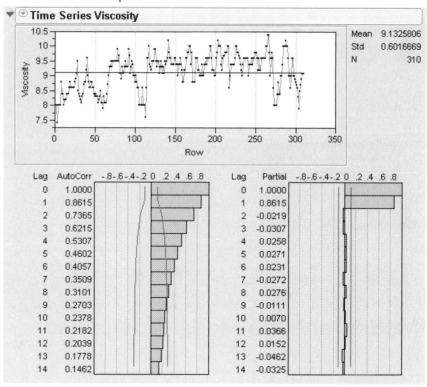

To estimate the parameters of the model,

🖰  Select **ARIMA** from the platform menu.

🖰  Enter a 1 in the **p, Autoregressive Order** box.

🖰   Click **Estimate**.

The resulting residuals (and their correlation plots) look like white noise, so we are satisfied with the model we have chosen.

**Figure 20.13**   SeriesD Residual Report

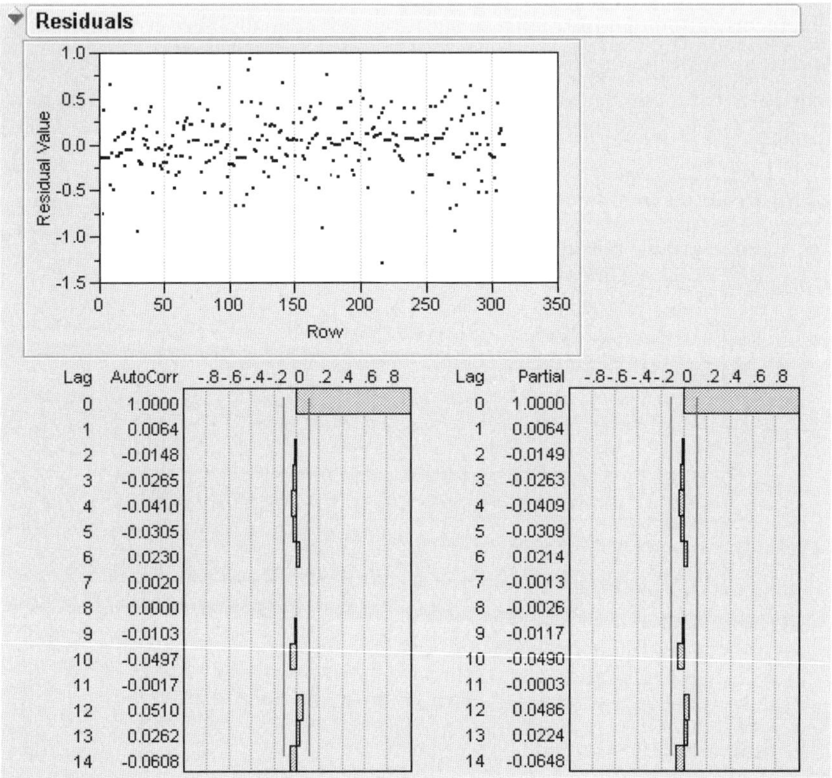

# ARMA Models and the Model Comparison Table

We cannot always model time series using AR or MA models alone. There may be some terms from both types of models. Notationally, we call these *ARMA* models and designate them as ARMA($p$, $q$) where $p$ represents the autoregressive order and $q$ represents the moving average order. When either of these numbers are zero, we drop their corresponding letters from the ARMA designation. So, for example, and ARMA(1,0) model is written merely as AR(1). As an example,

🖱  Open the data table TimeARMA.jmp.

🖱  Add 100 rows to the data table.

🖱  Produce a time series report of the ARMA variable.

**Figure 20.14**   TimeARMA Report

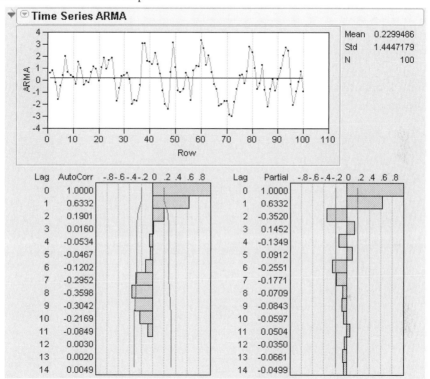

It is not immediately clear what order AR and MA coefficients to use, since neither autocorrelation plot has a familiar pattern. Both appear to drop to zero after a few lags, yet they both have spikes at later lags as well. If you ignore the later lags, the partial autocorrelation plot suggests an AR(2). Similarly, the autocorrelation plot suggests an MA(1). Since there is no clear-cut model, we fit several models and examine their residuals and fit statistics to find a suitable model.

🖱  Fit an MA(1), MA(2), AR(1), AR(2), and ARMA(1,1) model to the data.

At the top of the report, you find a model comparison table, as shown in **Figure 20.15**. As each model is fit, its corresponding statistics are appended to this table. These statistics give clues as to which model is best-fitting.

**Figure 20.15**  Model Comparison Table

| Model | DF | Variance | AIC | SBC | RSquare | -2LogLH |
|---|---|---|---|---|---|---|
| AR(2) | 97 | 1.1641027 | 18.14914 | 25.96465 | 0.459 | 12.662178 |
| AR(1) | 98 | 1.2698821 | 25.872133 | 31.082473 | 0.404 | 22.380767 |
| MA(1) | 98 | 1.0646884 | 8.2479493 | 13.45829 | 0.498 | 5.4450228 |
| MA(2) | 97 | 1.0586446 | 8.6530194 | 16.46853 | 0.507 | 3.679936 |
| ARMA(1,1) | 97 | 1.0168344 | 4.6235096 | 12.43902 | 0.526 | -0.157914 |

In general, higher values of $R^2$ are desirable, and lower values of AIC and SBC are desirable. So, for this example, the ARMA(1,1) model is the best choice among the candidates. This decision is supported by the residual plot for the ARMA(1,1) model, which looks fairly close to white noise.

**Figure 20.16**  TimeARMA Residuals

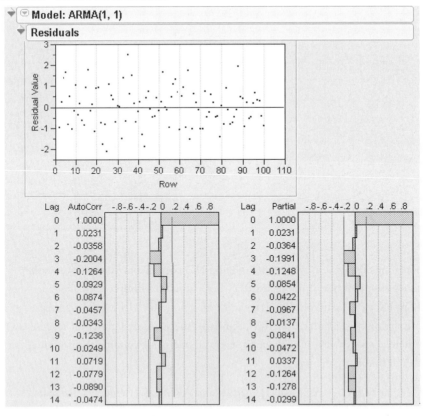

# Stationarity and Differencing

In all the examples so far, the series have had no noticeable linear trend. They seem to hover around a constant mean, and they do not show an overall increase or decrease as time progresses. The methods illustrated so far apply only to these *stationary* models. If a model exhibits a trend over time, it is termed *non-stationary*, and we must transform the data to attempt to remove the trend.

As an example,

🖰  Open the data table Seriesg.jmp.

🖰  Select **Analyze > Modeling > Time Series** and designate log passengers as the **Y, Time Series** variable.

The time series plot (**Figure 20.17**) shows a definite trend of increasing passengers over time.

**Figure 20.17**   SeriesG Plot

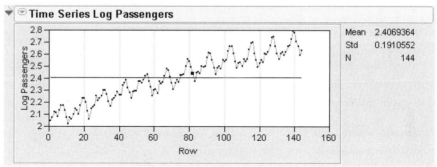

One approach to compensating for a trend is to *difference* the series. We compute a new series that is the difference between each value and its lag-1 value.

🖰  Create a new column in the data table called Difference.

🖰  Enter the following formula into the column.

$$\textit{Log Passengers} - \textit{Log Passengers}_{\text{Row}()-1}$$

Note: To enter a subscript, select **Subscript** from the **Row** group of formulas.

🖰  Produce a time series report of the new Difference variable.

As shown in **Figure 20.18**, the trend has disappeared.

**Figure 20.18**  Differenced Variable

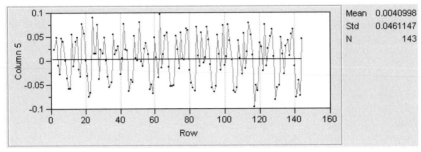

Differencing is a common task in time series analysis, so JMP contains an automatic **Difference** command in the report drop-down menu.

-  From the original **Log Passengers** report, select **Difference**.

-  In the dialog that appears, set the differencing order to 1.

-  Click **OK**.

You can see from the plot that this command performs the same function as our formula. As seen from the choices in the menu, differencing can be performed to several levels of lag, not just lag-1 as in this example.

When a model requires differencing, it is called an ARIMA model (where the I stands for *Integrated*). Symbolically, we write the model as ARIMA $(p, d, q)$, where $p$ and $q$ have the same functions as before, and $d$ represents the order of differencing.

With differencing, the process becomes stationary, and the other methods of this chapter can be used to determine the form of the model.

# Seasonal Models

Frequently, time series data are affected by a season—often, corresponding to seasons of the year or to business cycles. To model these seasonal factors, autoregressive and moving average factors of large lags are examined. For example, for yearly data, lags of length 12, 24, 36, and so on are often examined in addition to the nonseasonal terms.

As an example of a seasonal process, examine the file Steel Shipments.jmp. This file contains data on the monthly amount of steel shipped from U.S. steel mills from 1983 to 1994. From past experience, the researchers who collected this data expected a yearly cycle.

 🖱 Open the file Steel Shipments.jmp.

 🖱 Select **Analyze > Modeling > Time Series**.

 🖱 Assign Steel Shipments to the **Y, Time Series** role and Date as the **X, Time ID** role.

 🖱 Change the number of autocorrelation lags to 25.

 🖱 Click **OK**.

You should see the report in **Figure 20.19**.

**Figure 20.19**   Steel Shipments Initial Report

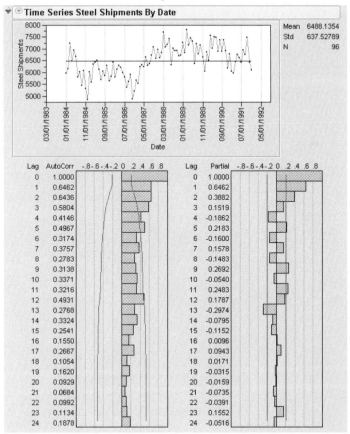

Note there is a spike in the autocorrelation plot near lag 12. This is a clue that there is a yearly seasonal element to the data. Since there is no corresponding spike at lag 24, the series may be seasonally stationary. The partial autocorrelation plot indicates that the nonseasonal component may be AR(2).

JMP designates seasonal components of an ARIMA model in a second set of parentheses listed after the nonseasonal components. Generally, the model is designated as ARIMA($p$, $d$, $q$)($P$, $D$, $Q$) where the capital letters indicate the order of the seasonal terms. In addition, the order of the seasonal terms (12 for a year-long monthly season, as in this example) is written after the model. So, for this example, we guess that an ARIMA(2, 0, 0)(1, 0, 0)12 model is appropriate in this case. To fit this model,

🖰    Select **Seasonal ARIMA** from the platform drop-down menu.

🖰    Fill out the resulting dialog so that it appears as in **Figure 20.20**.

**Figure 20.20**    Seasonal ARIMA Dialog

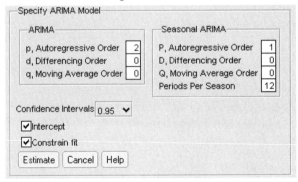

The resulting correlation plots (**Figure 20.21**) show that we do not have an adequate model, since the plots do not have the characteristics of white noise.

**Figure 20.21** Correlation Plots

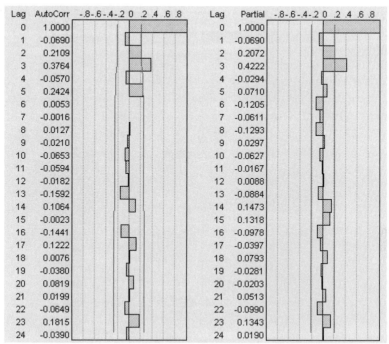

The autocorrelation plot of the seasonal model has a spike at lag 3, indicating a possible MA(3) portion of the model.

🖰 As another candidate, try an ARIMA(2, 0, 3)(1, 0, 0)12 model, resulting in the report shown in **Figure 20.22**.

**Figure 20.22**   ARIMA(2,0,3)(1,0,0)12 Model

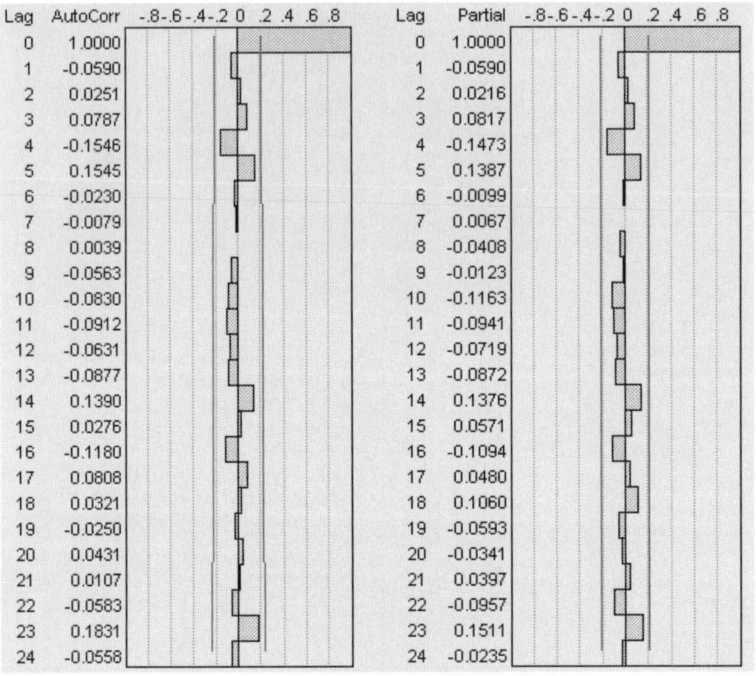

| Lag | AutoCorr | -.8 -.6 -.4 -.2 0 .2 .4 .6 .8 | Lag | Partial | -.8 -.6 -.4 -.2 0 .2 .4 .6 .8 |
|---|---|---|---|---|---|
| 0 | 1.0000 | | 0 | 1.0000 | |
| 1 | -0.0590 | | 1 | -0.0590 | |
| 2 | 0.0251 | | 2 | 0.0216 | |
| 3 | 0.0787 | | 3 | 0.0817 | |
| 4 | -0.1546 | | 4 | -0.1473 | |
| 5 | 0.1545 | | 5 | 0.1387 | |
| 6 | -0.0230 | | 6 | -0.0099 | |
| 7 | -0.0079 | | 7 | 0.0067 | |
| 8 | 0.0039 | | 8 | -0.0408 | |
| 9 | -0.0563 | | 9 | -0.0123 | |
| 10 | -0.0830 | | 10 | -0.1163 | |
| 11 | -0.0912 | | 11 | -0.0941 | |
| 12 | -0.0631 | | 12 | -0.0719 | |
| 13 | -0.0877 | | 13 | -0.0872 | |
| 14 | 0.1390 | | 14 | 0.1376 | |
| 15 | 0.0276 | | 15 | 0.0571 | |
| 16 | -0.1180 | | 16 | -0.1094 | |
| 17 | 0.0808 | | 17 | 0.0480 | |
| 18 | 0.0321 | | 18 | 0.1060 | |
| 19 | -0.0250 | | 19 | -0.0593 | |
| 20 | 0.0431 | | 20 | -0.0341 | |
| 21 | 0.0107 | | 21 | 0.0397 | |
| 22 | -0.0583 | | 22 | -0.0957 | |
| 23 | 0.1831 | | 23 | 0.1511 | |
| 24 | -0.0558 | | 24 | -0.0235 | |

These plots appear to be white noise, so we are satisfied with the fit. Notice how we tried a model, evaluated it, then tried another. This iterative method of refining the model is common; don't be surprised if several stages are required to find an adequate model.

# Spectral Density

The *spectral density plot* decomposes a series into frequencies, much like a prism does to a spectrum of white light. Spikes at certain frequencies of the spectral density plot give an indication of repeating cycles.

As an example,

- 🖱 Open Wolfer Sunspot.jmp.
- 🖱 Generate a time series report on the **wolfer** variable.
- 🖱 Select **Spectral Density** from the platform drop-down menu.

You should see plots as in **Figure 20.23**.

**Figure 20.23**   Spectral Density Plots

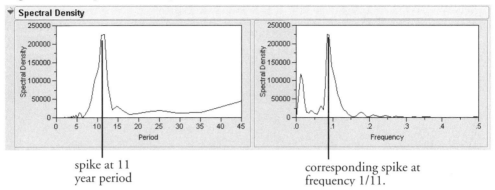

spike at 11
year period

corresponding spike at
frequency 1/11.

There is an obvious spike at a period of 11 years (corresponding to the now-known 11-year sunspot cycle). Since frequency is the reciprocal of period, there is a corresponding spike at frequency 1/11.

# Forecasting

Frequently, the purpose of fitting an ARIMA model to a time series is to forecast values into the future. In JMP, you must specify the number of periods into the future that you want forecasted. This number can be specified in the Time Series launch dialog (before the report is produced) or from the menu at the top level of the Time Series report (after the report is produced).

**Figure 20.24**   Changing the Number of Forecast Periods

Modeling a variable by its lagged values over time

Select Columns

Cn

Cast Selected Columns into Roles

Y, Time Series    *required Numeric*

X, Time ID    *optional Numeric*

By    *optional*

Data must be sorted by time, evenly spaced

Number of Autocorrelation Lags

Number of Forecast Periods    12

Action

OK

Cancel

Remove

Recall

Help

Time Series Cn

Graph                          ▶
✔ Autocorrelation
✔ Partial Autocorrelation
   Variogram
   AR Coefficients
   Spectral Density
   Save Spectral Density
   **Number of Forecast Periods**
   Difference
   ARIMA
   Seasonal ARIMA
   Smoothing Model          ▶
   Script                        ▶

Set the number of forecast periods in the launch dialog, or later using the menu.

For example, the Steel Shipments.jmp data with the ARIMA(2,0,3)(1,0,0)12 model has the following Forecast graph. Because the defaults were used when generating this report, forecasts extend 12 periods into the future. Confidence limits on the forecasts are also shown.

**Figure 20.25**   Forecast Graph

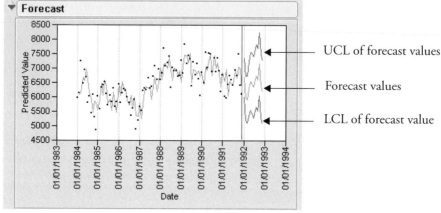

To add more periods to the forecast,

🖰   Select **Number of Forecast Periods** from the platform drop-down as shown in **Figure 20.24**.

🖰   Enter 25 in the dialog that appears.

The report now forecasts 25 periods into the future.

**Figure 20.26**   Forecast with 25 Periods

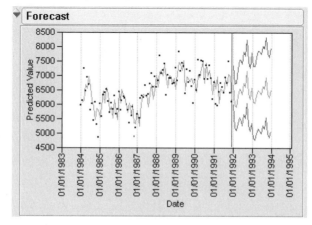

Aside from these graphical presentations of the forecasts, JMP can save the values to a new data table.

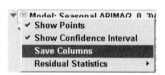

 🖑 Select **Save Columns** from the drop down menu on the model's outline bar.

This produces a new JMP table containing values for

* the time series itself
* the time variable, expressed in seconds since January 1, 1904
* predicted values for the time series and their standard errors
* residuals, representing the difference between the actual and predicted values, and
* upper and lower confidence intervals for each point.

This data table is often used for follow-up analyses on the series. For example, you can check the distribution of residuals (using the **Analyze > Distribution** command) or make overlay plots of the actual values with their confidence limits.

# Exercises

1. The Wolfer Sunspot.jmp data table was used as an example earlier in this chapter. Suppose two analysts conduct a time series analysis of this data, with one settling on an AR(2) model, while the other settles on an AR(3) model. Comment on their conclusions.

2. SeriesF.jmp is from the classic Box and Jenkins text. It can be sufficiently modeled with an ARMA model. Determine the AR and MA coefficients.

3. Simulated.jmp is a simulated time series. Find the ARIMA model used to generate it.

4. In order to test an automatic atomizer, Antuan Negiz (1994, "Statistical Monitoring and Control of Multivariate Continuous Processes") collected data from an aerosol mini-spray dryer device. The dryer injects a slurry at high speed. A by-product of his analysis was small dried particles left in his machine. The data from his study is stored in Particle Size.jmp. Fit an ARIMA model to the data.

5. The file RaleighTemps.jmp contains the average daily temperature for Raleigh, North Carolina, from January 1980 to December 1990. Temperature data have an obvious seasonal component to them. Determine this seasonal component and fit an ARIMA model to the data.

6. Nile.jmp consists of annual flow volumes of the Nile river at Aswan from 1871 to 1970.

    (a)   Does there appear to be a trend in this data?

(b)  Fit an appropriate ARIMA model.

7.  The Southern Oscillation Index is defined as the barometric pressure difference between Tahiti and the Darwin Islands at sea level. It is considered as an indicator of the El Niño effect, which affects fish populations and has been blamed for floods in the midwestern United States. The data are stored in Southern Oscillation.jmp. Use a Spectral Density plot to determine if there are cycles in the El Niño effect. (Scientists recognize one at around 12 months, as well as one around 50 months—slightly more than four years).

8.  This exercise examines moving average processes stored in Moving Average.jmp.

(a)  Examine the formulas in the columns Y1 and Y2 and determine the MA(1) coefficients.

(b)  Conduct a time series analysis on Y1 and Y2 and comment on what you see. Examine a time series text for a section on invertability to explain your findings.

# Machines of Fit

## Overview

This chapter is an essay on fitting for those of you who are mechanically inclined. If you have any talent for imagining how springs and tire pumps work, you can put it to work here in a fantasy in which all the statistical methods are visualized in simple mechanical terms.

The goal is to not only remember how statistics works, but also train your intuition so you will be prepared for new statistical issues.

Here is an illuminating trick that will help you understand and remember how statistical fits really work. It involves pretending that statistical fitting is performed by machines. If we can figure out the right machines and visualize how they behave, we can reconstruct all of statistics by putting together these simple machines into arrangements appropriate to the situation. We need only two machines of fit, the spring for fitting continuous Normal responses and the pressure cylinder for fitting categorical responses.

Readers interested in this approach should consult Farebrother (2002), who covers physical models of statistical concepts extensively.

# Springs for Continuous Responses

How does a spring behave? As you stretch the spring, the tension increases linearly with distance. The energy that you need to pull a spring a given distance is the integral of the force over the distance, which is proportional to the square of the distance.

Take $1/\sigma^2$ as the measure of how stiff a spring is. Then the graph and equations for the spring are as shown in **Figure 21.1**.

**Figure 21.1**   Behavior of Springs

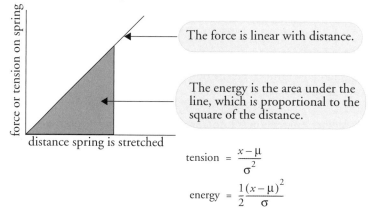

The force is linear with distance.

The energy is the area under the line, which is proportional to the square of the distance.

$$\text{tension} = \frac{x - \mu}{\sigma^2}$$

$$\text{energy} = \frac{1}{2}\frac{(x - \mu)^2}{\sigma}$$

In this way, springs will help us visualize least squares fits. They also help us do maximum likelihood fits when the response has a Normal distribution.

The formula for the log of the density of a Normal distribution is identical to the formula for energy of a spring centered at the mean, with a spring constant equal to the reciprocal of the variance. A spring stores and yields energy in exactly the way that Normal deviations get and give log-likelihood. So, maximum likelihood is equivalent to least squares, which is equivalent to minimizing energy in springs.

## Fitting a Mean

How do you fit a mean by least squares? Imagine stretching springs between the data points and the line of fit (see **Figure 21.2**) Then you move the line of fit around until the forces acting on it from the springs balance. That will be the point of minimum energy in the springs. For every minimization problem, there is an equivalent balancing (or orthogonality) problem, in which the forces (tensions, relative distances, residuals) add up to zero.

**Figure 21.2**   Fitting a Mean by Springs

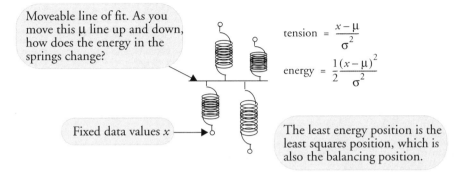

Moveable line of fit. As you move this $\mu$ line up and down, how does the energy in the springs change?

$$\text{tension} = \frac{x - \mu}{\sigma^2}$$

$$\text{energy} = \frac{1}{2} \frac{(x - \mu)^2}{\sigma^2}$$

Fixed data values $x$

The least energy position is the least squares position, which is also the balancing position.

## Testing a Hypothesis

If you want to test a hypothesis that the mean is some value, you force the line of fit to be that value and measure how much more energy you had to add to the springs (how much more the sum of squared residuals was) to constrain the line of fit. This is the sum of squares that is the main ingredient of the $F$-test. To test that the mean is (not) the same as a given value, find out how hard it is to move it there (see **Figure 21.3**).

**Figure 21.3**   Compare a Mean to a Given Value

sample mean

hypothesized mean

Here, the line of fit is the balance point of the springs; the energy in the springs is at a minimum when the system is balanced.

Here, the line of fit has been forced down to the hypothesized value for the mean, and it took a certain amount of energy to push it to this hypothesized value.

## One-Way Layout

If you want to fit several means, you can do so by balancing a line of fit with springs for each group. To test that the means are the same, you force the lines of fit to be the same, so that they balance as a single line, and measure how much energy you had to add to the springs to do this (how much greater the sum of squared residuals was). See **Figure 21.4.**

**Figure 21.4**  Means and the One-Way Analysis of Variance

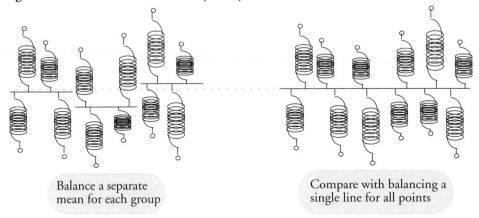

Balance a separate
mean for each group

Compare with balancing a
single line for all points

## Effect of Sample Size Significance

When you have a larger sample, there are more springs holding on to each mean estimate, and it is harder to pull them together. Larger samples lead to a greater energy expense (sum of squares) to test that the means are equal. The spring examples in **Figure 21.5** show how sample size affects the sensitivity of the hypothesis test.

Figure 21.5 A Larger Sample Helps Make Hypothesis Tests More Sensitive

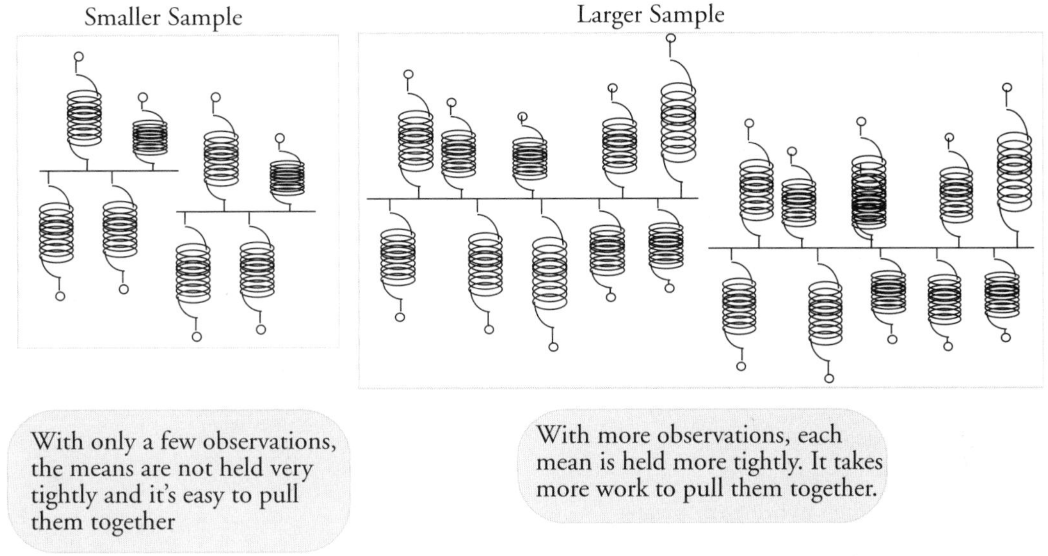

Smaller Sample

Larger Sample

With only a few observations, the means are not held very tightly and it's easy to pull them together

With more observations, each mean is held more tightly. It takes more work to pull them together.

## Effect of Error Variance on Significance

The spring constant is the reciprocal of the variance. Thus, if the residual error variance is small, the spring constant is bigger, the springs are stronger, it takes more energy to bring the means together, and the test is therefore more significant. The springs in **Figure 21.6** illustrate the effect of variance size.

**Figure 21.6**   Reduced Residual Error Variance Makes Hypothesis Tests More Sensitive

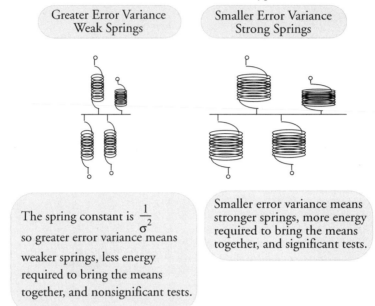

Greater Error Variance
Weak Springs

Smaller Error Variance
Strong Springs

The spring constant is $\frac{1}{\sigma^2}$ so greater error variance means weaker springs, less energy required to bring the means together, and nonsignificant tests.

Smaller error variance means stronger springs, more energy required to bring the means together, and significant tests.

## Experimental Design's Effect on Significance

If you have two groups, how do you arrange the points between the two groups to maximize the sensitivity of the test that the means are equal? Suppose that you have two sets of points loading two lines of fit, as in the one-way layout shown previously in **Figure 21.4**. The test that the true means are equal is done by measuring how much energy it takes to force the two lines together.

Suppose that one line of fit is suspended by a lot more points that the other. The line of fit that is suspended by few points will be easily movable and can be stretched to the other mean without much energy expenditure. The lines of fit would be more strongly separated if you had more points on this loosely sprung side, even at the expense of having fewer points on the more tightly sprung side. It turns out that to maximize the sensitivity of the test for a given number of observations, it is best to allocate points in equal numbers between the two groups. In this way both means are equally tight, and the effort to bring the two lines of fit together is maximized.

So the power of the test is maximized in a statistical sense by a balanced design, as illustrated in **Figure 21.7**.

**Figure 21.7**   Design of Experiments

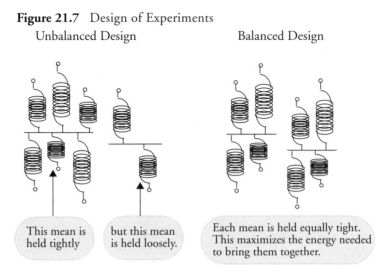

Unbalanced Design

This mean is held tightly

but this mean is held loosely.

Balanced Design

Each mean is held equally tight. This maximizes the energy needed to bring them together.

## Simple Regression

If you want to fit a regression line through a set of points, you fasten springs between the data points and the line of fit, such that the springs stay vertical. Then let the line be free so that the forces of the springs on the line balance, both vertically and rotationally (see **Figure 21.8**). This is the least-squares regression fit.

**Figure 21.8**   Fitting a Regression Line with Springs

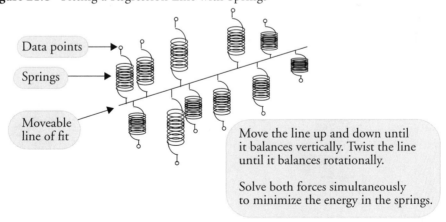

Data points

Springs

Moveable line of fit

Move the line up and down until it balances vertically. Twist the line until it balances rotationally.

Solve both forces simultaneously to minimize the energy in the springs.

If you want to test that the slope is zero, you force the line to be horizontal so that you're just fitting a mean and measure how much energy it took to constrain the line (the sum of squares due to regression). (See **Figure 21.9**)

**Figure 21.9**   Testing the Slope Parameter for the Regression Line

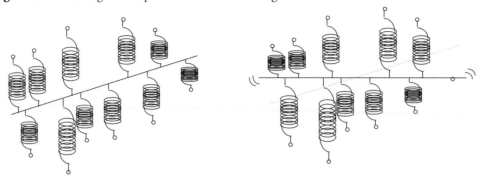

This line is where the forces governing the slope of the line balance. It is the minimum energy solution.

If you force the line to have a slope of zero, how much additional energy do you have to give the springs? How much work is it to move the line to be horizontal?

## Leverage

If most of the points that are suspending the line are near the middle, then the line can be rotated without much effort to change the slope within a given energy budget. If most of the points are near the end, the slope of the line of fit is pinned down with greatest resistance to force. That is the idea of leverage in a regression model. Imagine trying to twist the line to have a different slope. Look at **Figure 21.10** and decide which line would be easier to twist.

**Figure 21.10**   Leverage with Springs

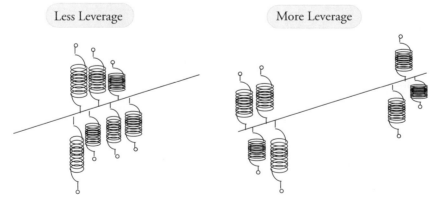

Less Leverage

More Leverage

## Multiple Regression

The same idea works for fitting a response to two regressors; the difference is that the springs are attached to a plane rather than a line. Estimation is done by adjusting the plane so that it balances in each way. Testing is done by constraining the plane.

**Figure 21.11**   Three-Dimensional Plot of Two Regressors and Fitted Plane

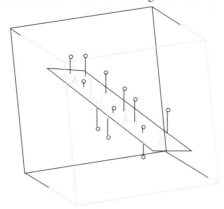

## Summary: Significance and Power

To get a stronger (more significant) fit, in which the line of fit is suspended more tightly, you must either have stiffer springs (have smaller variance in error), use more data (have more points to hang springs from), or move your points farther out on both ends of the *x*-axis (more leverage). The power of a test is how likely it is that you will be unable to move the line of fit given a certain energy budget (sum of squares) determined by the significance level.

# Machine of Fit for Categorical Responses

Just as springs are analogous to least-squares fits, gas pressure cylinders are analogous to maximum likelihood fits for categorical responses (see **Figure 21.12**).

## How Do Pressure Cylinders Behave?

Using Boyle's law of gases (pressure times volume is constant), the pressure in a gas cylinder is proportional to the reciprocal of the distance from the bottom of the cylinder to the piston. The energy is the force integrated over the distance (starting from a distance, $p$, of 1), which turns out to be $-\log(p)$.

**Figure 21.12**   Gas Pressure Cylinders to Equate –log(probability) to Energy

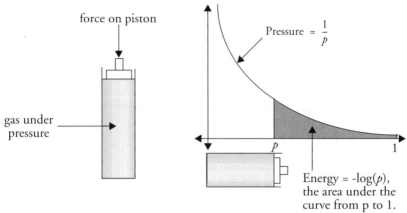

Now that you know how pressure cylinders work, start thinking of the distance from the bottom of the cylinder to the piston as the probability that some statistical model attributes to some response. The height of 1 will mean no stored energy, no surprise, a probability of 1. The height of zero will mean infinite stored energy, an impossibility, a probability of zero.

When stretching springs, we measured energy by how much work it took to pull a spring, which turned out to the be square of the distance. Now we measure energy by how much work it takes to push a piston from distance 1 to distance $p$, which turns out to be $-\log(p)$, the logarithm of the probability. We used the logarithm of the probability before in categorical problems when we were doing maximum likelihood. The maximum likelihood method estimates the response probabilities so as to minimize the sum of the negative logarithms of the probability attributed to the responses that actually occurred. This is the same as minimizing the energy in gas pressure cylinders, as illustrated in **Figure 21.13**.

**Figure 21.13**   Gas Pressure Cylinders Equate –log(probability) to energy

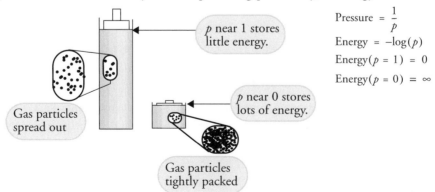

## Estimating Probabilities

Now we want to estimate the probabilities by minimizing the energy stored in pressure cylinders. First, we need to build a partitioned frame with a compartment for each response category and add the constraint that the sum of the heights of the partitions is 1. We will be moving the partitions around so that the compartments for each response category can get bigger or smaller (see **Figure 21.14**).

For each observation on the categorical response, put a pressure cylinder into the compartment for that response. After you have all the pressure cylinders in the compartments, start moving the partitions around until the forces acting on the partitions balance out. This will be the solution to minimize the energy stored in the cylinders. It turns out that the solution for the minimization is to make the partition sizes proportional to the number of pressure cylinders in each compartment.

For example, suppose you did a survey in which you asked 13 people what brand of car they preferred, and 5 chose Asian, 2 chose European, and 6 chose American brands. Then you would stuff the pressure cylinders into the frame as in **Figure 21.14**, and the partition sizes that would balance the forces would work out to 5/13, 2/13, and 6/13, which sum to 1.

To test that the true probabilities are some specific values, you move the partitions to those values and measure how much energy you had to add to the cylinders.

**Figure 21.14**  Gas Pressure Cylinders Estimate Probabilities for a Categorical Response

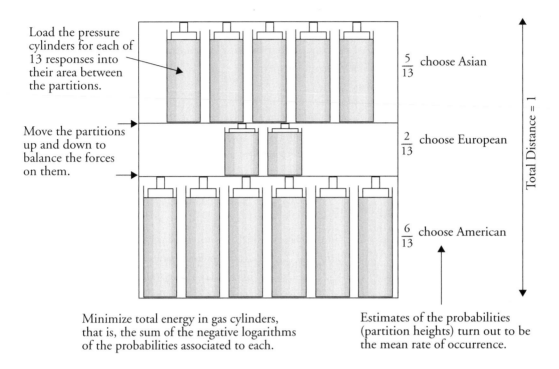

Load the pressure cylinders for each of 13 responses into their area between the partitions.

$\frac{5}{13}$  choose Asian

Move the partitions up and down to balance the forces on them.

$\frac{2}{13}$  choose European

$\frac{6}{13}$  choose American

Total Distance = 1

Minimize total energy in gas cylinders, that is, the sum of the negative logarithms of the probabilities associated to each.

Estimates of the probabilities (partition heights) turn out to be the mean rate of occurrence.

## One-Way Layout for Categorical Data

If you have different groups, you can fit a different response probability to each group. The forces acting on the partitions balance independently for each group. The plot shown in **Figure 21.15** (which should remind you of a mosaic plot) helps maintain the visualization of pressure compartments. As an alternative to pressure cylinders, you can visualize with free gas in each cell.

**Figure 21.15**   Gas Pressure Cylinder Estimate Probabilities for a Categorical Response

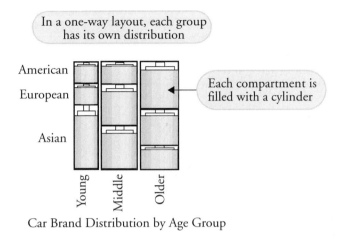

Car Brand Distribution by Age Group

How do you test the hypothesis that the true rates are the same in each group and that the observed differences can be attributed to random variation? You just move the partitions so that they are in the position corresponding to the ungrouped population and measure how much more energy you had to add to the gas-pressure system to force the partitions to be in the same positions.

**Figure 21.16** shows the three kinds of results you can encounter, corresponding to perfect fit, significant difference, and nonsignificant difference. To the observer, the issue is whether knowing which group you are in will tell you which response level you will have. When the fit is near perfect, you know with near certainty. When the fit is intermediate, you have more information if you know the group you are in. When the fit is inconsequential, knowing which group you are in doesn't matter. To a statistician, though, what is interesting is how firmly the partitions are held by the gases, how much energy it would take to move the partitions, and what consequences would result from removing boundaries between samples and treating it as one big sample.

**Figure 21.16**  Degrees of Fit

Three kinds of statistical results are possible.
There are three response levels and three factor categories.

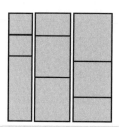

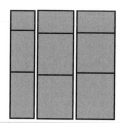

Almost perfect fit, when most probabilities of actual events attributed to be near 1, lots of space for the gas.

-log likelihood near zero. Huge differences with ungrouped case. LR test highly significant.

Intermediate relationship, when the probabilities of occurrence are not near 1, but they do differ among groups.

-log likelihood has intermediate value. Significant difference with ungrouped case by LR test.

Homogeneous case, when the probabilities of occurrence are almost the same in each group.

-log likelihood with large value. Not much difference with ungrouped cases. Non-significant LR test.

## Logistic Regression

Logistic regression is the fitting of probabilities over a categorical response to a continuous regressor. Logistic regression can also be visualized with pressure cylinders (see **Figure 21.17**). The difference with contingency tables is that the partitions change the probability as a continuous function of the *x*-axis. The distance between lines is the probability for one of the responses. The distances sum to a probability of 1. **Figure 21.18** shows what weak and strong relationships look like.

**Figure 21.17**   Logistic Regression as the Balance of Cylinder Forces

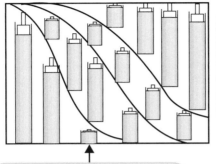

The pressure cylinders push on the logistic curves that partition the responses, moving them so that the total energy in the cylinders is minimized. The more cylinders in an area, the more probability room it pushes for.

This is an outlier, a response that is holding a lot of energy, having been squeezed down to a small probability

**Figure 21.18**   Strengths of Logistic Relationships

Near Perfect Fit

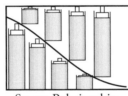

Strong Relationship

Weak Relationship

The probabilities are all near one. No cylinder is hot with energy.

Some cylinders must compete with nearby cylinders from a different response. More energy in some cylinders.

The probabilities are squeezed to near the horizontal case of homogeneity. All cylinders are warm with energy.

# References and Data Sources

Agresti, A. (1984), *Analysis of Ordinal Categorical Data*, New York: John Wiley and Sons, Inc.

Agresti, A. (1990), *Categorical Data Analysis*, New York: John Wiley and Sons, Inc.

Allison, T., and Cicchetti, D. V. (1976), "Sleep in Mammals: Ecological and Constitutional Correlates." Science, November 12, 194, 732-734.

American Society for Testing and Materials. *ASTM Manual on Presentation of Data and Control Chart Analysis*. STP No. 15-D. Philadelphia, 1976.

Anderson, T.W. (1971), *The Statistical Analysis of Time Series*, New York: John Wiley and Sons.

Anscombe, F.J. (1973), *American Statistician*, 27, 17-21.

*Aviation Consumer Home Page*. U.S. Government, Department of Transportation (http://www.dot.gov/airconsumer/).

Belsley, D.A., Kuh, E., and Welsch, R.E. (1980), *Regression Diagnostics*, New York: John Wiley and Sons.

Berger, R. L., and Hsu, J. C. (1996), "Bioequivalence Trails, Intersection-Union Tests, and Equivalence Confidence Sets," *Statistical Science*, 11, 283-319.

Box G.E.P. and Jenkins (1976), *Time Series Analysis: Forecasting and Control*, San Francisco: Holden-Day.

Box, G.E.P., Hunter, W.G., and Hunter, J.S. (1978), *Statistics for Experimenters*, New York: John Wiley and Sons, Inc.

Chase, M. A., and Dummer, G. M. (1992), "The Role of Sports as a Social Determinant for Children," *Research Quarterly for Exercise and Sport*, 63, 418-424.

Cochran, W.G. and Cox, G.M. (1957), *Experimental Designs, 2nd edition*, New York: John Wiley and Sons.

Creighton, W. P. (2000), "Starch content's dependence on several manufacturing factors." Unpublished data.

Cryer, J. and Wittmer, J. (1999) Notes on the Theory of Inference. NCSSM Summer Statistics Institute, available at http://courses.ncssm.edu/math/Stat_Inst/Notes.htm.

Daniel, C. (1959), "Use of Half-Normal Plots in Interpreting Factorial Two-level Experiments," *Technometrics*, 1, 311-314.

*Data and Story Library*, http://lib.stat.cmu.edu/DASL/.

Ehrstein, James and Croarkin, M. Carroll. *Statistical Reference Datasets*. US Government, National Institute of Standards and Technology (http://www.nist.gov/itl/div898/strd/).

Eppright, E.S., Fox, H.M., Fryer, B.A., Lamkin, G.H., Vivian, V.M., and Fuller, E.S. (1972), "*Nutrition of Infants and Preschool Children in the North Central Region of the United States of America*," World Review of Nutrition and Dietetics, 14.

Eubank, R.L. (1988), *Spline Smoothing and Nonparametric Regression*, New York: Marcel Dekker, Inc.

Farebrother, R. W. (2002), *Visualizing Statistical Models and Concepts*, New York: Marcel Dekker, Inc.

Fortune Magazine (1990), *The Fortune 500 List*, April 23, 1990.

Gabriel, K.R. (1982), "Biplot," *Encyclopedia of Statistical Sciences, Volume 1*, eds. N.L.Johnson and S. Kotz, New York: John Wiley and Sons, Inc., 263-271.

Gosset, W.S. (1908),"The Probable Error of a Mean," *Biometrika*, 6, pp 1-25.

Hajek, J. (1969), *A Course in Nonparametric Statistics*, San Francisco: Holden-Day.

Henderson, H. V. and Velleman, P. F. (1981), "Building Regression Models Interactively." *Biometrics*, 37, 391-411. Data originally collected from Consumer Reports.

Hosmer, D.W. and Lemeshow, S. (1989), *Applied Logistic Regression*, New York: John Wiley and Sons.

Iman, R.L. (1995), *A Data-Based Approach to Statistics*, Belmont, CA: Duxbury Press.

Iman, R.L. and Conover, W.J. (1979), "The Use of Rank Transform in Regression," *Technometrics*, 21, 499-509.

Isaac, R. (1995) The Pleasures of Probability. New York: Springer-Verlag.

John, P.M. (1971), *Statistical Design and Analysis of Experiments*, New York, Macmillan.

Kaiser, H.F. (1958), "The varimax criterion for analytic rotation in factor analysis" *Psychometrika*, 23, 187-200.

Kemp, A.W. and Kemp, C.D. (1991),"Weldon's dice data revisited," *The American Statistician*, 45 216-222.

Klawiter, B. (2000), *An Investigation into the Potential for Geochemical / Geoarchaeological Provenance of Prairie du Chien Cherts*. Master's Thesis: University of Minnesota.

Koehler, G. and Dunn, J.D. (1988), "The Relationship Between Chemical Structure and the Logarithm of the Partition," *Quantitative Structure Activity Relationships*, 7.

Koopmans, L. (1987),*Introduction to Contemporary Statistical Methods*, Duxbury Press, p 86.

Ladd, T. E.(1980 and 1984) and Carle, R. H. (1996), Clerks of the House of Representatives. *Statistics of the Presidential and Congressional Elections.* US Government. Available at http://clerkweb.house.gov/histrecs/history/elections/elections.htm.

Larner, M. (1996), Mass and its Relationship to Physical Measurements. MS305 Data Project, Department of Mathematics, University of Queensland.

Lenth, R.V. (1989), "Quick and Easy Analysis of Unreplicated Fractional Factorials," *Technometrics,* 31, 469-473.

Linnerud (see Rawlings (1988))

McCullagh, P. and Nelder, J. A. (1983), *Generalized Linear Models.* From *Monographs on Statistics and Applied Probability,* Cox, D. R. and Hinkley, D. V., eds. London: Chapman and Hall, Ltd.

Miller, A.J. (1990), *Subset Selection in Regression,* New York: Chapman and Hall.

Moore, D.S. and McCabe, G. P. (1989), *Introduction to the Practice of Statistics,* New York and London: W. H. Freeman and Company.

Myers and McCaulley pp 46-48 *Myers-Briggs test reference.*

Nelson, L. (1984), "The Shewhart Control Chart - Tests for Special Causes," *Journal of Quality Technology,* 15, 237-239.

Nelson, L. (1985), "Interpreting Shewhart X Control Charts," *Journal of Quality Technology,* 17, 114-116.

Perkiömäki, M. (1995) *Track and Field Statistics* (http://mikap.iki.fi/sport/index.html)

Smyth, G. (2000), "Selling Price of Antique Grandfather Clocks,"*OzDASL* web site (http://www.maths.uq.edu.au/~gks/data/general/antiques.html)

Rasmussen, M. (1998), Observations on Behavior of Dolphins. University of Denmark, Odense, via OzDASL (http://www.maths.uq.edu.au/~gks/data/index.html).

Rawlings, J.O. (1988), *Applied Regression Analysis: A Research Tool,* Pacific Grove CA: Wadsworth and Books/Cole.

Sall, J.P. (1990), "Leverage Plots for General Linear Hypotheses," *American Statistician,* 44, (4), 303-315.

SAS Institute (1986), *SAS/QC User's Guide, Version 5 Edition,* SAS Institute Inc., Cary, NC.

SAS Institute (1987), *SAS/STAT Guide for Personal Computers, Version 6 Edition,* Cary NC: SAS Institute Inc.

SAS Institute (1988), *SAS/ETS User's Guide, Version 6 Edition,* Cary NC: SAS Institute Inc.

SAS Institute (1989), *SAS/Technical Report P-188: SAS/QC Software Examples, Version 6 Edition.* Cary, NC: SAS Institute Inc.

SAS Institute Inc. (1989), *SAS/STAT User's Guide, Version 6, Fourth Edition, Volume 2,* Cary, NC: SAS Institute Inc., 1165-1168.

Schiffman, A. (1982), *Journal of Counseling and Clinical Psychology.*

Schuirmann, D.L. (1981), "On hypothesis testing to determine if the mean of a normal distribution is contained in a known interval," *Biometrics* 37, 617.

Snedecor, G.W. and Cochran, W.G. (1967), *Statistical Methods*, Ames Iowa: Iowa State University Press.

Simpson, E.H. (1951), The interpretation of interaction in contingency tables, *JRSS* B13: 238-241.

Stichler, R.D., Richey, G.G. and Mandel, J.(1953), "Measurement of Treadwear of Commercial Tires," *Rubber* Age, 73:2.

Stigler, S.M. (1986), *The History of Statistics*, Cambridge: Belknap Press of Harvard Press.

Stigler, S. M. (1977), Do Robust Estimators Work with Real Data? *The Annals of Statistics* 5:4, 1075.

Swift, Jonathan (1735), *Gulliver's Travels*. Quote is from p. 44 of the *Norton Critical Edition*, (1961) Robert A. Greenwood, ed. New York: W.W. Norton & Co.

Negiz, A. (1994) "Statistical Monitoring and Control of Multivariate Continuous Responses". NIST/SEMATECH e-Handbook of Statistical Methods (http://www.itl.nist.gov/div898/handbook/pmc/section6/pmc621.htm).

Neter, J. and Wasserman, W. (1974), *Applied Linear Statistical Models*, Homewood, IL: Richard D Irwin, Inc.

Theil and Fiebig, (1984), *Exploiting Continuity*, Cambridge Mass: Ballinger Publishing Co.

*Third International Mathematics and Science Study*, (1995). US Government: National Center for Educational Statistics and International Education Association.

Tukey, J. (1953), "A problem of multiple comparisons," Dittoed manuscript of 396 pages, Princeton University.

Tversky and Gilovich (1989), "The Cold Facts About the Hot Hand in Basketball," *CHANCE*, 2, 16-21.

Wardrop, Robert (1995), "Simpson's Paradox and the Hot Hand in Basketball", *American Statistician*, Feb 49:1, 24-28.

Westlake, W.J. (1981), "Response to R.B.L. Kirdwood: bioequivalence testing--a need to rethink", *Biometrics* 37, 589-594.

Yule, G.U. (1903), "Notes on the theory of association of attributes in statistics," *Biometrika* 2: 121-134.

# Technology License Notices

JMP software contains portions of the file translation library of MacLinkPlus, a product of DataViz Inc., 55 Corporate Drive, Trumbull, CT 06611, (203) 268-0030.

JMP for the Power Macintosh was compiled and built using the CodeWarrior C++ compiler from MetroWorks Inc.

SAS INSTITUTE INC.'S LICENSORS MAKE NO WARRANTIES, EXPRESS OR IMPLIED, INCLUDING WITHOUT LIMITATION THE IMPLIED WARRANTIES OF MERCHANTABILITY AND FITNESS FOR A PARTICULAR PURPOSE, REGARDING THE SOFTWARE. SAS INSTITUTE INC.'S LICENSORS DO NOT WARRANT, GUARANTEE OR MAKE ANY REPRESENTATIONS REGARDING THE USE OR THE RESULTS OF THE USE OF THE SOFTWARE IN TERMS OF ITS CORRECTNESS, ACCURACY, RELIABILITY, CURRENTNESS OR OTHERWISE. THE ENTIRE RISK AS TO THE RESULTS AND PERFORMANCE OF THE SOFTWARE IS ASSUMED BY YOU. THE EXCLUSION OF IMPLIED WARRANTIES IS NOT PERMITTED BY SOME STATES. THE ABOVE EXCLUSION MAY NOT APPLY TO YOU.

IN NO EVENT WILL SAS INSTITUTE INC.'S LICENSORS AND THEIR DIRECTORS, OFFICERS, EMPLOYEES OR AGENTS (COLLECTIVELY SAS INSTITUTE INC.'S LICENSOR) BE LIABLE TO YOU FOR ANY CONSEQUENTIAL, INCIDENTAL OR INDIRECT DAMAGES (INCLUDING DAMAGES FOR LOSS OF BUSINESS PROFITS, BUSINESS INTERRUPTION, LOSS OF BUSINESS INFORMATION, AND THE LIKE) ARISING OUT OF THE USE OR INABILITY TO USE THE SOFTWARE EVEN IF SAS INSTITUTE INC.'S LICENSOR'S HAS BEEN ADVISED OF THE POSSIBILITY OF SUCH DAMAGES. BECAUSE SOME STATES DO NOT ALLOW THE EXCLUSION OR LIMITATION OF LIABILITY FOR CONSEQUENTIAL OR INCIDENTAL DAMAGES, THE ABOVE LIMITATIONS MAY NOT APPLY TO YOU. SAS INSTITUTE INC.'S LICENSOR'S LIABILITY TO YOU FOR ACTUAL DAMAGES FOR ANY CAUSE WHATSOEVER, AND REGARDLESS OF THE FORM OF THE ACTION (WHETHER IN CONTRACT, TORT (INCLUDING NEGLIGENCE), PRODUCT LIABILITY OR OTHERWISE WILL BE LIMITED TO $50.00.

# Index